Anorganische und allgemeine Chemie
in Einzeldarstellungen
Herausgegeben von Margot Becke-Goehring

Band IX

Manfred W. Breiter

Electrochemical Processes in Fuel Cells

With 98 Figures

Springer-Verlag New York Inc. 1969

MANFRED W. BREITER

General Electric Research and Development Center,
Schenectady, N. Y./USA

ISBN-13: 978-3-642-46157-6 e-ISBN-13: 978-3-642-46155-2
DOI: 10.1007/978-3-642-46155-2

 Library of Congress Catalog Card Number 69-17789.

Softcover reprint of the hardcover 1st edition 1969

Title No. 4276

Preface

The necessity for a better understanding of the basic processes that determine the operation of fuel cells became evident during the development of practical units in the last three decades. The search for efficient electrocatalysts in low-temperature fuel cells intensified the general study of the nature and the role of the electrode material. Research on the complex mechanisms of the anodic oxidation of different fuels and of the reduction of molecular oxygen on solid electrodes was stimulated, and the strong influence of adsorbed species on the electrode reaction in question was investigated. Suitable electrolytes had to be found for the high-temperature fuel cells. The use of electrodes with large internal surface lead to the development of models of porous electrode structures and to the mathematical analysis of the operation of these models under certain conditions.

While the chapters I to III introduce the reader to the general field of fuel cells, the progress made in the understanding of the basic problems in the electrochemistry of fuel cells since the end of the second world war is reviewed in chapters IV to XVI of this monograph. In contrast, the technological aspects necessary for the development of practical units are not covered here.

The open literature published as books or as papers in scientific journals has been considered up to the time of the writing of the final draft of the specific chapter, at least till the end of 1967.

I want to express my gratefulness to some of my colleagues at the General Electric Research and Development Center in Schenectady, N. Y., for their helpful discussions and suggestions during the writing of the book: Dr. W. T. Grubb, Dr. L. W. Niedrach, Dr. D. A. Vermilyea, and Dr. F. G. Will. Dr. Niedrach kindly supplied me with the two micrographs of porous Teflon-bonded platinum black electrodes. Prof. G. J. Janz, Rensselaer Polytechnic Institut, Troy, N. Y., advised me on questions concerning chapter XIII. I am very grateful for the comments on the monograph from Dr. E. L. Simons, and on chapter XIV from Mr. D. W. White, both of the R and D Center.

It is a pleasure to acknowledge my indebtedness to the management at the Research and Development Center for their enlightened understanding and support of my work.

Schenectady, N. Y., May 1969 Manfred W. Breiter

Contents

Main Symbols

A	apparent surface area (cm^2)
A_B	BET surface area (m^2/g)
A_p	effective surface in the one-dimensional continuum model of porous structures (cm^{-1})
a_j	activity of species j (dimensionless)
C_D	double layer capacity ($\mu F/cm^2$)
C_S	capacitance of electrode impedance in a series analog circuit ($\mu F/cm^2$)
C_p	capacitance of electrode impendance in a parallel circuit ($\mu F/cm^2$)
c_j	concentration of species j (Mol/cm^3) or (Mol/l)
${}_bc_j$	concentration of species j in the bulk of the solution (Mol/cm^3) or (Mol/l)
${}_sc_j$	concentration of species j adjacent to the electrode surface (Mol/cm^3) or (Mol/l)
D_j	diffusion coefficient of species j (cm^2/sec)
E	electromotive force (V)
$\mathbf{E}$	electric field vector (V/cm)
e	electron charge (coul)
F	Faraday (coul/Mol)
f_j	activity coefficient of species j (cm^3/Mol) or (l/Mol)
ΔG	free energy change (kcal/Mol)
$\Delta G^{\ddagger}_{2\nu+1,\,ref}$	activation energy of anodic processes with respect to the reference electrode ref (kcal/Mol), $\nu = 1, 2, \ldots$
$\Delta G^{\ddagger}_{2\nu+2,\,ref}$	activation energy of cathodic processes with respect to the reference electrode ref (kcal/Mol), $\nu = 1, 2, \ldots$
ΔH	free enthalpy change (kcal/Mol)
$\Delta H^{\ddagger}_{2\nu+1}$	free enthalpy change of anodic processes (kcal/Mol)
$\Delta H^{\ddagger}_{2\nu+2}$	free enthalpy change of cathodic processes (kcal/Mol)
h	Planck's constant (erg sec)
I	current (A)
I_ν	partial current due to reaction ν (A)
I_o	exchange current (A)
I_l	limiting current (A)
I_d	limiting diffusion current (A)
i	current density I/A (A/cm^2)
i_ν	current density of reaction ν (A/cm^2)

i_o	exchange current density (A/cm^2)
i_l	limiting current density (A/cm^2)
i_d	limiting diffusion current density (A/cm^2)
$\mathbf{j}_d$	diffusional flux vector (Mol/sec cm^2)
K_t	tortuosity factor (dimensionless)
k	Boltzmann's constant (erg/°C)
k_{2v+1}	rate constant of anodic processes, given with respect to a reference electrode, in A if $I \sim a_j$ and in A cm^3/Mol if $I \sim c_j$
k_{2v+2}	rate constant of cathodic processes (A) or (A cm^3/Mol)
L	characteristic length (cm)
M	Mol/l
Pe	Peclet number (dimensionless)
Pr	Prandtl number (dimensionless)
P_t	total porosity (dimensionless)
P_v	specific porosity concerning pores of type v (dimensionless)
p_j	pressure of gaseous species j (atm)
$_bp_j$	pressure of species j in the bulk of the solution (atm)
$_sp_j$	pressure of species j adjacent to electrode surface (atm)
Q_H	charge due to the removal or the formation of adsorbed H atoms (mcoul/cm^2)
$_sQ_H$	charge corresponding to a monolayer of H atoms (mcoul/cm^2)
$_sQ'_H$	charge corresponding to a monolayer of H atoms when other species are adsorbed simultaneously (mcoul/cm^2)
$Q_{O,an}$	charge due to the formation of an oxygen layer (mcoul/cm^2)
$Q_{O,cath}$	charge due to the reduction of an oxygen layer (mcoul/cm^2)
Q_F	charge due to the anodic formation of an adsorbed layer of carbonaceous species (mcoul/cm^2)
$_sQ_F$	charge due to the anodic formation of the maximum coverage with carbonaceous species (mcoul/cm^2)
Q_{Ox}	charge due to the anodic removal of an adsorbed layer of carbonaceous species (mcoul/cm^2)
$_sQ_{Ox}$	charge due to the anodic removal of the maximum coverage with carbonaceous species (mcoul/cm^2)
R	gas constant (Joule/Mol °C) or (kcal/Mol °C)
R_1	solution resistance per unit pore length (Ω cm^{-1})
R_2	resistance of electrocatalyst layer per unit pore length (Ω cm^{-1})
R_{el}	electrolytic resistance (Ω)
R_S	ohmic component of electrode impendance in an analog series circuit (Ω cm^2)
R_p	ohmic component of electrode impedance in a parallel circuit (Ω cm^2)
ΔS	entropy change (kcal/Mol)

T	temperature (°K)
ϑ	temperature (°C)
t	time (sec) or (min) or (hr)
t_j	transference number of species j (dimensionless)
U_{ref}	electrode-solution potential difference of test electrode less that of a reference electrode, for instance ref = cal for calomel electrode, (V)
U	electrode-solution potential difference of test electrode less that of the hydrogen electrode in the same solution (V)
$U_{rev, ref}$	electrode-solution potential difference of test electrode, at which the equilibrium of an electrode reaction is established, less that of a reference electrode. The subscript ref is omitted if the reference electrode is mentioned in the text (V)
U_o	U_{rev} under standard conditions (with respect to the mentioned reference electrode) (V)
V	volume (cm^3)
V_{ap}	apparent volume for porous structure (cm^3)
V_ν	specific volume concerning pores of type ν (cm^3/g)
$\mathbf{v}$	velocity vector (cm/sec)
w	work (Joule/Mol) or (erg/Mol)
Z	impedance per unit pore length (Ω cm)
z	number of faradays carried if reaction proceeds so that the number of moles of each product formed is equal to the stoichiometric coefficient in the reaction (dimensionless)
Γ_j	surface excess of species j (Mol/cm^2)
γ	surface tension (erg/cm)
δ_j	thickness of diffusion boundary layer of species j (cm)
η	overvoltage of an electrode reaction (V)
θ_j	electrode coverage with species j (dimensionless)
$1/\kappa$	penetration depth (cm)
$1/\kappa_p$	effective penetration depth in the continuum model of a porous structure (cm)
$\mu_j(\alpha)$	chemical potential of species j in the phase α (kcal/Mol)
$\bar{\mu}_j(\alpha)$	electrochemical potential of species j in the phase α (kcal/Mol)
ρ	density (g/cm^3)
σ_j	specific conductivity of species j ($\Omega^{-1}\,cm^{-1}$)
σ_P	effective conductivity in a porous structure ($\Omega^{-1}\,cm^{-1}$)
τ	transition time (sec)
Φ_α	inner electric potential of phase α (V)
φ	average potential in the continuum model of a porous structure (V)
ω	angular velocity (sec^{-1})

I. Introduction

1. Definition and Description of a Fuel Cell

The conversion of chemical energy of fuels into electric energy requires three steps in the process that is utilized in many power plants today:

A) Production of heat by burning the fuel.

B) Conversion of heat into mechanical energy.

C) Conversion of mechanical energy into electric energy.

For combustion engines, steps A and B are combined in the well known way. The efficiency of step B is limited since the efficiency of a closed-cycle heat engine cannot surpass a certain value at given temperatures for the input and output of heat as derived by CARNOT on thermodynamic grounds. Total efficiencies of up to 41% have been achieved for the conversion of chemical energy into electric energy in modern units.

A different approach was taken in the attempt to convert chemical energy into electric energy by electrochemical reactions with continuous supply of the reactants in a galvanic cell. In contrast to batteries, an electrical recharging of the cell after a certain period of use is not required. A device for this type of energy conversion is called "Fuel Cell". Some of the features that make fuel cells attractive are cited below:

a) In comparison to heat engines (processes A, B, and C), moving parts are only present in pumps or cooling fans of fuel cell batteries consisting of several cells.

b) Since conversion of heat into mechanical energy is not involved, the theoretically achievable efficiency is larger than the combined efficiency of processes A, B, and C.

c) Negligible air pollution.

d) Weight advantage over motor-generator sets and for longer periods of use over conventional batteries.

Schematically, a fuel cell may be represented as a system of two electrodes separated by electrolyte (see Fig. 1). Liquid or solid electrolytes are used in different types of cells. Fuel is supplied to the anode and oxygen or air to the cathode. The two electrodes are connected by a resistive load. The electrochemical oxidation of the fuel at the anode produces electrons. The electrons flow through the external circuit to the cathode on which oxygen is reduced. The ionic and neutral species that participate in the electrochemical reactions are different for the various systems. As an example, the net reactions of the hydrogen-oxygen cell in acid solutions are given below:

(1) $$\tfrac{1}{2}H_2 = H^+ + e^-,$$

(2) $$\tfrac{1}{4}O_2 + H^+ + e^- = \tfrac{1}{2}H_2O.$$

The simultaneous occurrence of the electrochemical reactions 1 and 2 is equivalent to the chemical oxidation of molecular hydrogen by oxygen:

(3) $$H_2 + \tfrac{1}{2}O_2 = H_2O.$$

Reaction 3 is called [1] a cell reaction.

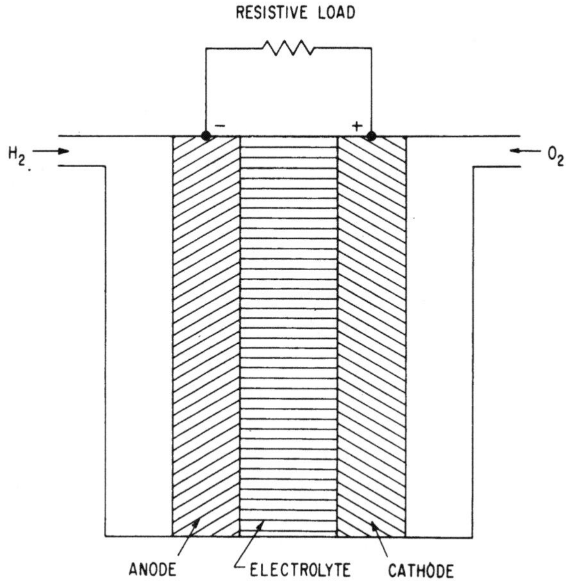

Fig. 1. Schematic diagram of a hydrogen-oxygen fuel cell with porous electrodes

2. Classification of Fuel Cells

Fuel cells that are developed at present may be classified in different ways. A classification according to the temperature of operation is used in this book. Since fuel cells with aqueous electrolytes have been operated [2] up to 300 °C, the following division of the temperature range is made:

a) Low-temperature cells operating at temperatures $\vartheta < 300$ °C.

b) High-temperature cells working at $\vartheta > 300$ °C.

Two different technological approaches are reflected by this classification. The main problem of low-temperature cells consists in finding efficient electrocatalysts so that the rates of the electrochemical reactions are still satisfactory at low polarization. Since the reaction rates increase with temperature, the role of the electrocatalyst is not so critical in the high-temperature cells. Other problems like corrosion and conductivity of the electrolyte become pertinent for the construction of a practical unit.

Another classification may be based on the type of electrolyte used in the cells. The following types are under evaluation in different laboratories:

a) Aqueous solutions [3] for temperatures below 300 °C.
b) Organic electrolytes [4].
c) Ion exchange membranes for low-temperature cells [5].
d) Fused salts at high temperatures [6].
e) Solid electrolytes for high-temperature cells [7].

The largest research effort has been put into cells belonging to categories a, c, d, and e of the second classification. Fundamental problems of these cells are discussed in this book.

3. Historical Development

The concept of the fuel cell dates [8] back to the beginning of the nineteenth century. The feasibility of a fuel cell was demonstrated by Grove [9] who employed oxygen and hydrogen as fuels. MOND and LANGER [10] reported the decline of the reactivity of platinum black in electrolytes with time and extended the life of the cell by storing the electrolyte in a porous nonconducting material in their corroboration of Grove's work. SIEGL [11] reduced the cost of the cell by supporting platinum on carbon particles. Fused electrolytes were used by BECQUEREL [12] and JABLOCKOFF [13]. OSTWALD [14] suggested replacing heat engines by electrochemical cells in which carbon is oxidized to CO_2 and oxygen is reduced. The recognition of point b of the preceding section 1 spurred the experimental efforts of HABER [15], NERNST [16], and BAUER [17] to construct a practical cell with solid electrolytes. The anodic oxidation of liquid organic fuels was studied extensively by MÜLLER and coworkers [18, 19] on different metals in aqueous electrolytes. Since it was concluded in this period that the problem is not amenable to a practical solution, the interest in electrochemical fuel cells remained small.

The situation changed after the second world war with the publication of the results of DAVTYAN [20] on solid electrolytes for high-temperature fuel cells with carbonaceous fuels and on the hydrogen-oxygen cell in the high-temperature and low-temperature range with alkaline electrolyte. The search for efficient power supplies of light weight for application in many fields, especially in the satellite technology, gave a new stimulus to the development and research on fuel cells.

BACON [21] and his coworkers had directed their efforts since 1932 to the development of the hydrogen-oxygen fuel cell with alkaline electrolytes at elevated temperatures and pressures. The use of gas diffusion electrodes [22, 23], in which the fuel gas is fed to one side while the other side is in contact with the aqueous electrolyte, played an important role in the progress reported with the (so-called) DSK (double skeleton catalyst) electrode [24] of RANEY nickel in a sintered nickel skeleton for the hydrogen electrode and of RANEY silver for the oxygen electrode in low-temperature cells with alkaline electrolytes. KETELAAR and BROERS [25] achieved practical levels of current in high-temperature cells with molten carbonates in which the electrodes are pressed against a porous magnesium oxide matrix containing the electrolyte. Carbon diffusion electrodes with catalysts added for better performance were brought into operation by KORDESCH [26]. Ion exchange membranes were used successfully in the work of GRUBB and NIEDRACH [27] on the hydrogen-oxygen fuel cell at low temperatures. The design of the Teflon-bonded platinum black electrode by NIEDRACH and ALFORD [28] represents an important step in the development of electrodes. Another attempt is the production of electrode structures consisting of porous plastic on which thin metal films are deposited as current collector and electrocatalyst [29].

The technological aspects of the new approaches towards a practical fuel cell are adequately covered in recent books [3,24,29–34]. In contrast, the emphasis of this book is on the fundamental electrochemical problems of fuel cells.

References

1. Recommendations of the Commission on Electrochemical Nomenclature and Definitions of C. I. T. C. E., R. DEFAY, N. IBL, E. LEVART, G. MILAZZO, G. VALENSI, and P. VAN RYSSELBERGHE: J. Electroanalyt. Chem. **7,** 417 (1964).
2. EISENBERG, M.: Proc. 18th Annual Power Sources Conf., Atlantic City, N. J., 1964, p. 20.
3. ADAMS, A. M., F. T. BACON, and R. G. H. WATSON: Fuel Cells, Chap. 4. New York: W. Mitchell, Jr., ed., Academic Press 1963.
4. See MAY, P. S.: Proc. 18th Annual Power Sources Conf., Atlantic City, N. J., 1964, p. 1.
5. NIEDRACH, L. W., and W. T. GRUBB: Fuel Cells, Chap. 6. N. Y.: W. Mitchell Jr., ed., Academic Press 1963.
6. PEATTICE, C. G.: Proc. IEEE **51,** 801 (1963).
7. BAUR, E., and H. PREIS: Z. Elektrochem. **43,** 727 (1937).

8. Davy, H.: Nicholson's J. nat. Phil. **144** (1802).
9. Grove, W. R.: Phil. Mag. **14,** 127 (1839).
10. Mond, L., and C. Langer: Proc. Roy. Soc. **46,** 296 (1889).
11. Siegl, K.: Elektrotech. Z. **34,** 1317 (1913).
12. Becquerel, A. C.: Traite d'éléctricité, Paris, **I,** 1833 (1855).
13. Jablockoff, P.: Compt. Rend. **85,** 1052 (1877).
14. Ostwald, W.: Z. Elektrochem. **1**, 122 (1894).
15. Haber, F., and A. Moser: Z. Elektrochem. **11**, 593 (1905).
16. Nernst, W.: German Patents **259,** 241 (1912).
17. Baur, E., and H. Ehrenberg: Z. Elektrochem. **18,** 1002 (1912).
18. Müller, E.: Z. Elektrochem. **29,** 264 (1923).
19. —, and S. Takegami: Z. Elektrochem. **34,** 704 (1928).
20. Davtyan, O. K.: Izvest. Akad. Nauk S.S.S.R., Otdel. Tekh. Nauk, 1946, 107, 215.
21. Bacon, F. T.: Beama J. **61**, 6 (1954).
22. Schmid, A.: Die Diffusionselektrode. Stuttgart: Enke 1923.
23. — Helv. Chim. Acta **7,** 370 (1924).
24. Justi, E., M. Pilkuhn, W. Scheibe u. A. Winsel: Hochbelastbare Wasserstoffdiffusionselektroden für Betrieb bei Umgebungstemperatur und Niederdruck, Abh. Mainzer Akad. Nr. **8.** Wiesbaden: Steiner 1959.
25. Broers, G. H. J., and J. A. A. Ketelaar: Ind. Eng. Chem. **52**, 303 (1960).
26. Kordesch, K.: Fuel Cells. Chap. 8. New York: W. Mitchell, Jr., ed., Academic Press 1963.
27. Grubb, W. T., and L. W. Niedrach: J. Electrochem. Soc. **107,** 131 (1960).
28. Niedrach, L. W., and H. R. Alford: J. Electrochem. Soc. **112,** 117 (1965).
29. Williams, K. R.: An Introduction to Fuel Cells. New York: American Elsevier 1966.
30. Eisenberg, M.: Advances in Electrochemistry and Electrochemical Engineering, Vol. **2.** New York: Ch. W. Tobias, ed., Interscience Publishers 1962.
31. Vielstich, W.: Brennstoffelemente. Weinheim: Verlag Chemie, 1965.
32. Baker, B. S., ed.: Hydrocarbon Fuel Cell Technology. New York: Academic Press 1965.
33. Grubb, W. T., and L. W. Niedrach: Direct Energy Conversion, Chap. 2. New York: G. W. Sutton, ed., McGraw-Hill Book Company 1966.
34. Liebhafsky, H. A., and E. J. Cairns: Fuel Cells and Fuel Batteries. New York: John Wiley and Sons, 1968.

II. General Aspects

1. Thermodynamic Considerations and Definitions

Let us consider the cell in Fig. 2 that represents a closed thermodynamic system $\gamma\alpha\beta\beta'\alpha'\delta$. Electrochemical reactions occur in the interfacial regions α/β and α'/β' between metallic phases and electrolyte. These regions have chemical, electric, and hydrodynamic properties different from those existing in the depths of the phases. Electrons are transferred across the interfaces α/γ and δ/α' between metallic phases. The electrode α is chosen as anode, electrode α' as cathode. The external circuit connecting the terminals 1 and 2 is shown as a wire for simplicity. The following reactions take place at the interfaces in the case of the hydrogen-oxygen cell:

(1) $$\tfrac{1}{2}H_2(\beta) = H^+(\beta) + e^-(\alpha),$$

(2) $$e^-(\alpha) = e^-(\gamma),$$

(3) $$e^-(\delta) = e^-(\alpha'),$$

(4) $$\tfrac{1}{4}O_2(\beta') + H^+(\beta') + e^-(\alpha') = \tfrac{1}{2}H_2O(\beta').$$

The phases in which reactants and products are located are given in parentheses. It is assumed in the thermodynamic considerations that the

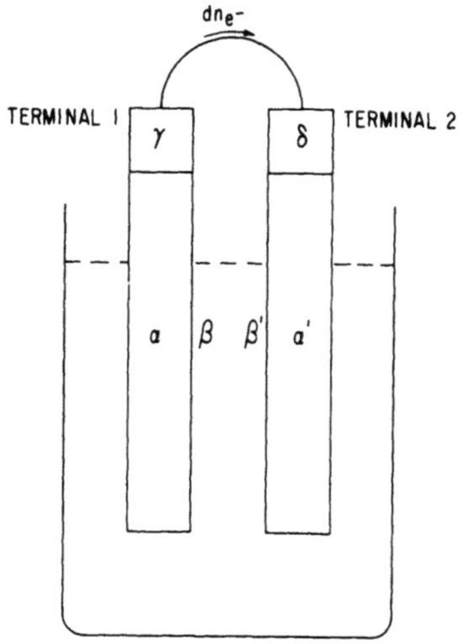

Fig. 2. Schematic diagram of a galvanic cell

concentration of reactants and products is constant between the planes β and β′ that are the boundaries of the interfacial regions towards the solution. When going from metallic phase to electrolyte, the interfacial region consists of the HELMHOLTZ double layer (transfer layer [1]), the diffuse double layer (diffuse layer [1]) and the diffusion layer. Electrons take part in the electrode reaction in the HELMHOLTZ layer. The charge density is different from zero in the diffuse double layer whether or not a current is flowing. The diffusion layer is the part of the electrolyte in which, when a current flows, the composition is different from that of the bulk.

Addition of reactions 1 to 4 yields the cell reaction:

$$\tfrac{1}{2}H_2(\beta)+\tfrac{1}{4}O_2(\beta)+e^-(\delta)=\tfrac{1}{2}H_2O(\beta)+e^-(\gamma). \tag{5}$$

The contribution of the electron transfer between metallic phases disappears from reaction 5 if the metals γ and δ are identical. The arrangement that δ is identical with γ and α′ represents a special case of this identity. If an infinitesimal number dn_{e^-} of moles of electrons is transferred from phase γ to phase δ in the time interval dt by the external circuit the current I is:

$$I=F\frac{dn_{e^-}}{dt} \tag{6}$$

Here F denotes the faraday. If only one electrochemical reaction is considered, the current is counted positive for an anodic reaction and negative for a cathodic one. The current density i is the ratio between I and apparent surface area A.

The work done by the surroundings on the system is for the infinitesimal process:

$$dw=(\bar{\mu}_{e^-}(\delta)-\bar{\mu}_{e^-}(\gamma))dn_{e^-}=(\mu_{e^-}(\delta)-\mu_{e^-}(\gamma)-(\Phi_\delta-\Phi_\gamma)F)dn_{e^-}. \tag{7}$$

Here $\bar{\mu}_{e^-}(\gamma)$ designates the electrochemical potential of the electron in the phase γ, and $\mu_{e^-}(\gamma)$ is its chemical potential. The Φ 's are the inner electric potentials [1,2] of phases γ and δ. Let us assume that γ and δ are identical metals. Then the expression for the work reduces to the form for electric work only:

$$dw=-F(\Phi_\delta-\Phi_\gamma)dn_{e^-}. \tag{8}$$

Under reversible conditions the work is equal to the change ΔG of the chemical free energy of the cell reaction at constant temperature T and constant pressure p:

$$w_{rev}=\Delta G=-zF(\Phi_\delta-\Phi_\gamma)_{rev}. \tag{9}$$

Here z is the number of faradays carried from phase γ to phase δ when the reaction proceeds so that the number of moles of each product formed is equal to the stoichiometric coefficient in the reaction written. It is said [1] that the reaction has proceeded by a unit of advancement. The difference $(\Phi_\delta - \Phi_\gamma)_{rev}$ is usually called electromotive force E of the cell. It has been suggested [1] "chemical tension of the cell" should be used instead. E represents the maximum value that $(\Phi_\delta - \Phi_\gamma)$ may achieve. The electromotive force of the hydrogen-oxygen cell is according to Eq. 9:

$$E = -\frac{\Delta G}{2F} = -\frac{\Delta G_0}{2F} - \frac{RT}{2F} \ln \frac{a_{H2O}}{a_{H_2} \cdot a_{O_2}^{1/2}}. \tag{10}$$

Here, ΔG_0 denotes the standard free energy change of the cell reaction, a_j is the activity of the species j. The value of $-\frac{\Delta G_0}{2F}$ is equal to 1.23 V at 25 °C in Eq. 10.

The potential difference $(\Phi_\gamma - \Phi_\delta)$ between terminals 1 and 2 of the cell is usually called cell voltage or cell potential. Replacement by „electric tension of the cell" has been recommended [1]. It may be written as a sum of terms:

$$\Phi_\gamma - \Phi_\delta = (\Phi_\gamma - \Phi_\alpha) + (\Phi_\alpha - \Phi_\beta) + (\Phi_\beta - \Phi_{\beta'}) + (\Phi_{\beta'} - \Phi_{\alpha'}) + (\Phi_{\alpha'} - \Phi_\delta). \tag{11}$$

The term $(\Phi_\beta - \Phi_{\beta'})$ is the ohmic potential difference ($I R_{el}$ drop) which is completely located in the electrolytic phase. It is the only measurable potential difference on the right side of Eq. 11. The other terms represent GALVANI potential differences (GALVANI tensions [1]). The metal-solution potential difference $(\Phi_\alpha - \Phi_\beta)$ is usually designated as electrode potential. Replacement by "electric tension of the electrode" has been suggested [1].

Under equilibrium conditions each of the four unmeasurable potential differences on the right side of Eq. 11 may be expressed in terms of the chemical potentials of the reactants and products involved in the electrochemical reactions at the corresponding interface. The expressions for the hydrogen-oxygen cell are:

$$\Phi_\beta - \Phi_\alpha = \frac{1}{F}\left(\frac{1}{2}\mu_{0,H_2} - \mu_{0,H^+} - \mu_{e^-}(\alpha)\right) + \frac{RT}{F} \cdot \ln \frac{p_{H_2}^{1/2}}{a_{H^+}}, \tag{12}$$

$$\Phi_\alpha - \Phi_\gamma = \frac{1}{F}(\mu_{e^-}(\alpha) - \mu_{e^-}(\gamma)), \tag{13}$$

$$\Phi_\delta - \Phi_{\alpha'} = \frac{1}{F}(\mu_{e^-}(\delta) - \mu_{e^-}(\alpha')), \tag{14}$$

$$\Phi_{\alpha'} - \Phi_{\beta'} = \frac{1}{F}\left(\tfrac{1}{4}\mu_{0,O_2} + \mu_{0,H^+} + \mu_{e^-}(\alpha') - \tfrac{1}{2}\mu_{0,H_2O}\right) + \frac{RT}{F} \cdot \ln \frac{a_{H^+} p_{O_2}^{1/4}}{\left(\frac{p_S}{p_{0,S}}\right)^{1/2}}. \tag{15}$$

The activities of the gases O_2 and H_2 have been replaced by their pressures and the activity of water by the ratio between the vapor pressure of the solvent in the solution (p_S) and that of the pure solvent ($p_{0,S}$). The subscript 0 denotes the standard chemical potentials (value of chemical potential under standard conditions).

The symbol U_{ref} is used to designate the electrode-solution potential difference of the working electrode less that of a reference electrode:

$$U_{ref} = (\Phi_{test\,el} - \Phi_S) - (\Phi_{ref\,el} - \Phi_S) = (\Phi_{test\,el} - \Phi_{ref\,el}). \tag{16}$$

It is assumed that the contribution of potential differences due to liquid junctions and the contribution of the $I R_{el}$ drop are properly eliminated in the experimental determination of the relative electrode potential U_{ref}. Frequently the word "relative" is omitted. The designation "relative electric tension" has been suggested [1] for U_{ref}. The polarization of an electrode is equal to the value of U_{ref} with current flow less the value of U_{ref} in the absence of a current. When, at a given current density, a metallic electrode is the site of a definite and unique reaction, its overvoltage η (overtension [2]) at a given instant is:

$$\eta = (\Phi_M - \Phi_S) - (\Phi_M - \Phi_S)_{rev} = U_{ref} - U_{rev,\,ref}. \tag{17}$$

$U_{rev,ref}$ is the value of U_{ref} at electrochemical equilibrium of this electrode reaction. The electrode possesses it at equilibrium and for the same chemical state of the phases as that corresponding to the instant at which the measurement is made. The subscript ref is not added for simplicity in the subsequent chapters if a hydrogen electrode in the same electrolyte as the test electrode serves as the reference electrode.

2. Efficiency of Galvanic Cells

Under reversible conditions, the external electric work w_{rev} performed by the cell is at constant T and constant p:

$$w_{rev} = -\Delta G. \tag{18}$$

The changes in free energy G, enthalpy H, and entropy S are correlated:

$$\Delta G = \Delta H - T \cdot \Delta S. \tag{19}$$

It follows that the maximum efficiency of a galvanic cell is $w_{rev}/(-\Delta G) = 1$ in terms of ΔG and equal to

$$\frac{w_{rev}}{\Delta H} = 1 - T \cdot \frac{\Delta S}{\Delta H}. \tag{20}$$

in terms of ΔH.

Under load (irreversible conditions) the corresponding efficiencies are smaller because of polarization effects on the electrodes and ohmic drops. A quantitative derivation was given for an infinitesimal process on the basis of irreversible thermodynamics by van RYSSELBERGHE [3,4].

(21) $$\frac{dw}{H' \cdot d\xi} = \frac{G'}{H'} - \frac{zF}{H'}(\eta_a + |\eta_c| + R_{el}\, i)\,.$$

Here ξ is the degree of advancement of the cell reaction; $H' = -\frac{\partial H}{\partial \xi}$ and $G' = -\frac{\partial G}{\partial \xi}$ are instantaneous values of the heat and of the chemical affinity of the cell reaction; η_a is the positive overvoltage of the anode and η_c is the negative overvoltage of the cathode; R_{el} designates the internal resistance which may include ohmic resistances in the electrodes besides the electrolytic resistance between the electrodes. The quantities H' and G' are obtainable from enthalpy and free energy data while η_a, η_c, and R_{el} have to be determined from non-equilibrium measurements.

If the resistive load (see Fig. 1) is decreased, larger currents are drawn from the fuel cell. This leads to an increase of the terms in brackets on the right side of Eq. 21. The instantaneous efficiency of fuel cells becomes smaller with decreasing load. Losses due to polarization are least for the hydrogen-oxygen cell since reaction 1 occurs with small overvoltage on platinum-metal electrocatalysts. As an example, Eq. 21 is applied here to estimate efficiencies on the basis of the performance curve for the hydrogen-oxygen fuel cell in Fig. 5 of reference 5. The first term G'/H', which represents the maximum achievable efficiency, is equal to 0.83 at 25 °C. The efficiency decreases to about 0.50 at 100 mA/cm^2 and to about 0.41 at 200 mA/cm^2.

3. Basic Requirements for a Practical Fuel Cell

The construction of fuel cells that are competitive power sources on the market depends upon a satisfactory compromise between five basic requirements:

a) Losses due to polarization and internal resistance of the cell (see Eq. 21) should be small.

b) Electrodes should be resistant to corrosion.

c) The electrolyte should be invariant.

d) The costs for electrocatalysts used in the construction of fuel cells should be low.

e) Price of fuels should be reasonable.

Let us consider the hydrogen-oxygen cell with an aqueous electrolyte as an example. Platinum metals (Pt, Ir, Rh, Pd, Ru), binary alloys of platinum metals and a restricted number of binary alloys of a platinum metal with another metal fulfill conditions a and b simultaneously in acid electrolytes. However, prices for cells with these electrocatalysts are high. Application is restricted to special projects for which the high price of fuel cell batteries with good reliability is acceptable. In alkaline electrolytes, nickel can be used as catalyst in the hydrogen electrode and lithiated nickel oxide or silver for the oxygen electrode. Efficiencies comparable to those in acid cells with platinum metal catalysts have been achieved. Thus, the alkaline low-temperature cell will become an attractive power source for general purposes as soon as requirement e has been solved for hydrogen as fuel.

The situation is more difficult when the anodic oxidation of organic fuels is desired. Oxygen containing fuels (for instance CO, CH_3OH) can be oxidized on metals of group Ib of the periodic table such as gold and silver, mixed oxides, and spinels. However, the efficiency is the best for platinum and platinum alloys. For the direct oxidation of hydrocarbons, only platinum has proved useful [6] in low-temperature cells so far.

In general, losses due to polarization are less in high-temperature fuel cells than in low-temperature cells. Satisfactory electrocatalysis is a minor problem. Corrosion in the liquid phase has to be overcome in fuel cells with aqueous electrolytes and carbonate electrolytes while corrosion of the electrocatalyst in the gaseous phase of the fuels is one of the major problems of the fuel cells with solid electrolyte.

4. Electrolytes

The specific conductivity of electrolytes in fuel cells should be large so that ohmic drops located in the solution between anode and cathode are small. It is desirable that most of the ionic current in the cell results from the transport of the ion which participates in the electrochemical reaction. Its transference number should be close to unity so that the polarization due to concentration changes of this ion is as small as possible.

Electrolytes in low-temperature fuel cells are mainly strong acids such as sulfuric acid or phosphoric acid solutions, or strong alkali such as sodium or potassium hydroxide solutions. The migrating species are H^+ or OH^-. Concentrated solutions allow operating temperatures above the boiling point of water. Phosphoric acid solutions have been used up to 250 °C, potassium hydroxide solutions up to 300 °C. Special precautions have to be taken in the cell design for the removal of reaction products if these products change the composition of the electrolyte. For example, the formation of water in the hydrogen-oxygen cell would tend to dilute the electrolyte.

Fuel cells that have ion-exchange membranes as electrolytes represent another group of low-temperature cells. Membranes containing sulfonic acid groups linked directly to a polymer network were used [7] sucessfully. The groups are not leached from the system by the production of water which is rejected because of the limited capacity of the membranes for water. An aqueous electrolyte is contained between two membranes in other types [8] of ion-exchange membrane fuel cells. Each of the membranes supports an electrode on one side.

Fused alkali carbonates are mainly used in high-temperature cells as liquid electrolytes. Operating temperatures are between 350 °C to 700 °C. The ionic current is due to the transport of the carbonate ion CO_3^{--}. The oxygen required for the anodic oxidation of fuels is supplied by this ion, for example:

$$\tfrac{1}{2}H_2+\tfrac{1}{2}CO_3^{--}=\tfrac{1}{2}H_2O+\tfrac{1}{2}CO_2+e^- . \tag{22}$$

Depletion of carbonate and possible solidification of the electrolyte is avoided by producing CO_3^{--} at the cathode, for example:

$$\tfrac{1}{4}O_2+\tfrac{1}{2}CO_2+e^-=\tfrac{1}{2}CO_3^{--} . \tag{23}$$

A portion of the anode-off gas can be recycled to the cathode for this purpose too.

Ceramic electrolytes in which the ionic current results from the transport of the oxide ion O^{--} are employed in high-temperature cells with solid electrolyte. Operating temperatures are about 1000 °C and higher in cells with good performance. Details of the conduction mechanism will be discussed later.

References

1. Recommendations of the Commission on Electrochemical Nomenclature and Definitions of C.I.T.C.E., R. Defay, N. Ibl., E. Levart, G. Milazzo, G. Valensi, and P. van Rysselberghe: J. Electroanalyt. Chem. **7,** 417 (1964).
2. Lange, E., and K. P. Miscenko: Z. Physik. Chem. **149,** 1 (1930).
3. van Rysselberghe, P.: Electrochemical Affinity. Paris: Hermann 1955.
4. — Electrochim. Acta **11,** 125 (1966).
5. Grubb, W. T., and L. W. Niedrach: Advanced Energy Conversion Systems, Chap. 2. G. Sutton, ed., McGraw-Hill 1966.
6. — Proc. 16th Annual Power Sources Conf., Atlantic City, N. J., 1962, p. 31.
7. Schanz, J. L., and E. K. Bullock: ARS Space Power Systems Conf., Santa Monica, Calif., ARS paper No. **2561—62** (1962).
8. Lurie, R. M., C. Berger, and R. J. Shuman: Fuel Cells, Vol. **2**, Chap. 11. G. J. Young, ed., New York: Reinhold 1963.

III. Mass Transport Processes

Transport processes are involved when a current is passed through a galvanic cell. Ions and neutral species that participate in the electrochemical reactions at the anode or cathode have to be transported to the respective electrode surfaces. The basic concepts of transport processes are briefly outlined in this chapter. The reader is referred for a comprehensive treatment to the book on physicochemical hydrodynamics by LEVICH [1]. Phenomena resulting from transport processes in porous electrodes of fuel cells are covered in chapter XVI.

1. Concept of the Nernst Diffusion Layer

It was shown by NERNST [1] and coworkers [2] that the rate of electrochemical reactions is given by Eq. 1 if transport processes are rate-determining at steady state.

$$I = \frac{zFDA}{\delta}\,({}_{s}c - {}_{b}c)\,. \tag{1}$$

Here ${}_{b}c$ and ${}_{s}c$ denote the concentration of the reacting species in the bulk of the electrolyte and adjacent to the electrode surface respectively. NERNST [1] assumed that the species diffuse through a region of constant width δ and that the concentration profile can be determined by linear interpolation in this thin layer of liquid. The quantity δ which is called the thickness of the NERNST diffusion layer has values between 10^{-2} and 10^{-4} cm under regular conditions of stirring.

NERNST's theory of the diffusion layer has played an important role in the development of the kinetics of electrochemical processes. However, Eq. 1 has to be considered an empirical one when applied to an electrolytic solution in motion. Experimental studies [3] demonstrated that liquid motion can be observed at distances of about 10^{-5} cm from solid surfaces. NERNST'S assumption that the liquid is static within a layer 10 to 1000 times thicker is not in agreement with the empirical evidence. In addition, neither the thickness of the diffusion layer nor its dependence upon the stirring rate is calculable from NERNST's theory.

2. Convective Diffusion

The transport of a solute in a moving liquid or gas is due to molecular diffusion as a result of concentration changes and to the entrainment and motion of solute particles by the moving phase. The combination of these two processes is called convective diffusion. Let us assume that the liquid is incompressible and that the motion is steady and laminar. The diffusional flux $\mathbf{j}_d$ written as a vector is proportional to the gradient of the chemical potential.

$$\mathbf{j}_d = -\beta \operatorname{grad} \mu = -\beta \left(\frac{\partial \mu}{\partial c}\right)_{T,p} \operatorname{grad} c \,. \tag{2}$$

It is supposed that a temperature gradient does not exist and that the mass flow resulting from the pressure gradient is negligible in comparison with the flow due to the concentration gradient. If the diffusion coefficient D is defined by Eq. 3

$$D = \beta \left(\frac{\partial \mu}{\partial c}\right)_{T,p} \tag{3}$$

the diffusional flux can be expressed:

$$\mathbf{j}_d = -D \operatorname{grad} c \,. \tag{4}$$

The diffusion coefficient is constant and independent of the composition of the liquid at small concentrations.

A flow of matter

$$\mathbf{j}_c = c\mathbf{v} \tag{5}$$

travels with the liquid and passes in one second across $1\,\mathrm{cm}^2$ of an imaginary plane constructed in the $x-y$, $x-z$, or $y-z$ plane in addition to the diffusional flux. Here $\mathbf{v}$ is the vector representing the velocity of the liquid or gas. The total flux is:

$$\mathbf{j} = c\mathbf{v} - D \operatorname{grad} c \,. \tag{6}$$

Eq. 6 is strictly valid for neutral species. If the transport of ions is involved, the flux due to a potential gradient has to be negligible in comparison to the diffusional and convective flux. The fundamental Eq. 7 that follows from the law of conservation of matter

$$\frac{\partial c}{\partial t} = -\operatorname{div} \mathbf{j} \tag{7}$$

can be transformed into Eq. 8 if the concentration dependence of D is neglected:

$$(8) \qquad \frac{\partial c}{\partial t} + v_x \frac{\partial c}{\partial x} + v_y \frac{\partial c}{\partial y} + v_z \frac{\partial c}{\partial z} = D\left(\frac{\partial^2 c}{\partial x^2} + \frac{\partial^2 c}{\partial y^2} + \frac{\partial^2 c}{\partial z^2}\right).$$

The term $\partial c/\partial t$ in Eq. 8 is zero for steady-state processes, which are of primary interest in this book.

The convective diffusion equation can be transformed to a dimensionless form by introducing a characteristic length L along which the major change of concentration occurs, and a characteristic flow velocity V_0:

$$(9) \qquad \mathrm{V_x} \frac{\partial \mathrm{C}}{\partial \mathrm{X}} + \mathrm{V_y} \frac{\partial \mathrm{C}}{\partial \mathrm{Y}} + \mathrm{V_z} \frac{\partial \mathrm{C}}{\partial \mathrm{Z}} = \frac{D}{V_0 L}\left(\frac{\partial^2 \mathrm{C}}{\partial \mathrm{X}^2} + \frac{\partial^2 \mathrm{C}}{\partial \mathrm{Y}^2} + \frac{\partial^2 \mathrm{C}}{\partial \mathrm{Z}^2}\right).$$

Here $\mathrm{V_x} = \frac{v_x}{V_0}$, $\mathrm{V_y} = \frac{v_y}{V_0}$, $\mathrm{V_z} = \frac{v_z}{V_0}$, $\mathrm{C} = \frac{c}{{}_bc}$, $\mathrm{X} = \frac{x}{L}$, $\mathrm{Y} = \frac{y}{L}$, $\mathrm{Z} = \frac{z}{L}$.

The dimensionless quantity

$$\mathrm{Pe} = \frac{V_0 L}{D}$$

is the Peclet number.

The left side of Eq. 9 represents the convective transport of matter. The right side describes molecular diffusion. When the Peclet number is small the terms on the right side of Eq. 9 can be set approximately equal to zero. The concentration gradient is largely determined by molecular diffusion. This situation exists at sufficiently small velocities and in regions of small dimensions. When the Peclet number is large, the concentration distribution results mainly from convective transport. The preceding reasoning assumes that all the dimensionless terms in Eq. 9 besides Pe are approximately unity. Since the diffusion coefficients of molecules and ions in aqueous solutions are of the order of magnitude of $D \approx 10^{-5}\,\mathrm{cm^2/sec}$, the transport of matter by convection predominates over molecular diffusion in liquids even at low velocities. Transport of solute in liquids differs strongly in this respect from transport in gases where the diffusion coefficients of molecules and ions are much larger than $10^{-5}\,\mathrm{cm^2/sec}$.

The Eq. 10

$$(10) \qquad \mathrm{V_x} \frac{\partial \mathrm{C}}{\partial \mathrm{X}} + \mathrm{V_y} \frac{\partial \mathrm{C}}{\partial \mathrm{Y}} + \mathrm{V_z} \frac{\partial \mathrm{C}}{\partial \mathrm{Z}} = 0$$

has the solution $\mathrm{C} = 1$ in the bulk of the liquid. However, this solution is not valid in the vicinity of the surface on which a heterogeneous reaction takes place. This becomes clearly apparent when ${}_{s}c=0$ is required by a large rate of the reaction. Therefore a thin layer of liquid in which the concentration terms due to diffusion in Eq. 9 cannot be neglected in comparison to the convective terms must exist near the surface. The order of magnitude of the thickness δ of the diffusion boundary layer was evaluated [1]:

$$\delta = \frac{\delta_0}{\Pr^{1/3}}. \tag{11}$$

Here δ_0 is the distance from the surface of a plate where v_x attains a value equal to 90% of the velocity V_0, the x direction being taken as normal to the surface.

$$\delta_0 = 5.2\left(\frac{\nu x}{V_0}\right)^{1/2}. \tag{12}$$

Pr designates the Prandtl number and ν the kinematic viscosity.

$$\Pr = \frac{\nu}{D}. \tag{13}$$

The concentration gradient in the diffusion boundary layer may be approximated by a linear profile. Thus the expression for the diffusional flux takes the same form as NERNST's theory (see Eq. 1). However, the dependence of δ upon D, ν, and V_0 is known.

The boundary condition at a plane surface may be written in a dimensionless form:

$$\frac{zDFA}{Lk\,{}_{b}c^{m-1}}\left(\frac{\partial C}{\partial X}\right)_{X=0} - {}_{s}C^{m} = 0. \tag{14}$$

Here $k\,{}_{b}c^{m}$ is the rate of the heterogeneous reaction in terms of current. The rate constant k is defined in a similar way as in sections 1 and 2 of chapter IV. The reaction is of the m th order with respect to c. The value of the dimensionless constant K as defined by Eq. 15

$$K = \frac{zFDA}{Lk\,{}_{b}c^{m-1}} \tag{15}$$

determines the relative significance of the rate of transport and the rate of the heterogeneous reaction. If $K \ll 1$, the boundary condition 14 becomes:

$${}_{s}C^{m} \approx 0. \tag{16}$$

The net rate of the electrochemical process is given by the maximum rate of mass transport. The transport processes are rate-determining. Conversely, $K \gg 1$ implies

$$\left(\frac{\partial C}{\partial X}\right)_{X=0} \approx 0 . \tag{17}$$

The concentration is constant in the entire electrolytic solution. The net rate is governed by the rate of the heterogeneous reaction and given by $k_b c^m$.

3. Migration and Convective Diffusion

The total flux of ions of the species n in a stirred electrolyte is:

$$\mathbf{j}_n = c_n \mathbf{v} - D_n \operatorname{grad} c_n + \frac{D_n z_n c_n F}{RT} \mathbf{E} . \tag{18}$$

The last term on the right side represents the flux due to migration of ions in the electric field which is characterized by the vector **E**. It follows from Eq. 7 at steady state:

$$\begin{aligned} v_x \frac{\partial c_n}{\partial x} + v_y \frac{\partial c_n}{\partial y} + v_z \frac{\partial c_n}{\partial z} &= D_n \left(\frac{\partial^2 c_n}{\partial x^2} + \frac{\partial^2 c_n}{\partial y^2} + \frac{\partial^2 c_n}{\partial z^2} \right) \\ &\quad - \frac{D_n z_n F}{RT} \left(\frac{\partial c_n E_x}{\partial x} + \frac{\partial c_n E_y}{\partial y} + \frac{\partial c_n E_z}{\partial z} \right) . \end{aligned} \tag{19}$$

The system of equations 19 has to be solved for the given electrolytic solution under the condition of approximate neutrality in the bulk.

$$\sum_n z_n c_n \approx 0 . \tag{20}$$

Two special cases were discussed in detail by LEVICH [1] since the solution in a general form is not feasible.

a) Binary electrolyte ($z_1 c_1 - z_2 c_2 = 0$).

b) An excess of foreign electrolyte the ions of which do not participate in the electrode reaction is present.

The set of equations 19 can be combined into one equation in case a:

$$v_x \frac{\partial c}{\partial x} + v_y \frac{\partial c}{\partial y} + v_z \frac{\partial c}{\partial z} = D_{\text{eff}} \left(\frac{\partial^2 c}{\partial x^2} + \frac{\partial^2 c}{\partial y^2} + \frac{\partial^2 c}{\partial z^2} \right) \tag{21}$$

with

(22) $$c = \frac{c_1}{z_2} = \frac{c_2}{z_1}$$

and

(23) $$D_{\text{eff}} = \frac{D_1 D_2 (z_1 + z_2)}{z_1 D_1 + z_2 D_2}.$$

The concentration c is not an explicit function of the electric field. However, Eq. 21, which has the form of the equation for convective diffusion, contains an effective diffusion coefficient D_{eff}. The problem can be approached in two steps. First the concentration distribution is derived for the given boundary conditions. The subsequent integration of Eq. 24

(24) $$\frac{\mathbf{i}}{F z_1 z_2} = -(D_1 - D_2)\operatorname{grad} c + \frac{F(D_1 z_1 + D_2 z_2)}{RT} c\mathbf{E}$$

yields the potential distribution inside the electrolyte.

In case b it can be shown [1] that the convective diffusion of ions occurs in the same way as that of neutral molecules if effects of the order c_1/c_3 are neglected. Here c_1 designates the concentration of the reacting ions, c_3 is the concentration of the foreign electrolyte. The presence of foreign electrolyte reduces the role of migration.

References

1. Levich, V. G.: Physicochemical Hydrodynamics, Prentice Hall, Inc., Englewood Cliffs, N. J., 1962.
2. Nernst, W.: Z. physik. Chem. **47,** 52 (1904).
3. Brunner, E.: Z. physik. Chem. **47,** 57 (1904).
4. Fage, A., and H. C. H. Townend: Proc. Roy. Soc. London **A 135,** 656 (1932).

IV. Kinetics of Electrode Reactions on Homogeneous Surfaces and Influence of Electrode Material

A review of the effect of the nature of the electrode on the kinetics of electrode reactions was given by PARSONS [1] recently. The different types of electrode reactions may be grouped between two limiting cases:

a) Weak interaction between reactants and electrode.

b) Strong interaction.

Simple redox systems belong to limiting case a, electrocatalytic reactions to case b. PARSONS pointed out that a distinction has to be made between primary and secondary effects. The influence of the electrode material on the free energy of formation of intermediates, including transition states, reflects primary effects. The modification of the structure of the double layer at the interface with the electrode material may in turn affect the electrode reaction (secondary effects). Primary effects are mainly discussed in this chapter.

1. Single Step Reaction

Let us consider the reaction

$$B^{m} + e^{-} = B^{m-1} \tag{1}$$

which involves the passing of a transition state B^{+} that interacts with the electrode surface, for example, through transiently adsorbed species. This is illustrated by the conventional plot of potential energy versus nuclear figuration in Fig. 3a. The minimum of the parabolic curve on the right side corresponds to the stable configuration of the species B^{m}, the minimum of the left curve to that of B^{m-1}. The intersection of the two curves represents the transition state B^{+}. If k_1 and k_2 represent the rate constants in terms of current, the rate of reaction 1 may be written:

$$I = k_1 a_{B^{m-1}} - k_2 a_{B^{m}} . \tag{2}$$

The rate constants depend upon $(\Phi_M - \Phi_S)$. Odd numbers in the subscripts of the rate constants are used in this book to denote an anodic process while even numbers designate a cathodic process.

Each of the two rate constants is related to the standard free energy of formation $\Delta G_1^{\ddagger}$ or $\Delta G_2^{\ddagger}$ of the transition state:

$$k_j = \frac{ekT}{h} \exp\left(-\frac{\Delta G_j^{\ddagger}}{RT}\right). \tag{3}$$

$\Delta G_1^{\ddagger}$ and $\Delta G_2^{\ddagger}$ are called energies of activation. They cannot be determined experimentally. The electron charge is designated by e, k is Boltzmann's constant, and h Planck's constant. It is assumed that the transition state has a charge $(m-q)$ with $0<q<1$ and that it is located

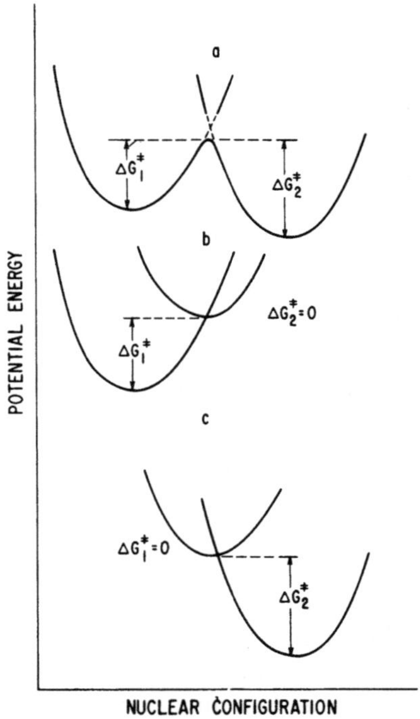

Fig. 3. Schematic diagrams of potential energy versus nuclear configuration for a redox process.
a) Regular situation, b) Cathodic barrierless process, c) Anodic barrierless process

within the inner region of the double layer where the potential is equal to $\Phi_S+\zeta(\Phi_M-\Phi_S)$. The potential of the outer Helmholtz plane [2] of the double layer is put equal to Φ_S, disregarding a potential drop in the diffuse double layer for simplicity. The polarization of the transition state is neglected in a first approximation. The standard free energies may be expressed in terms of standard electrochemical potentials, for example:

(4) $$\Delta G_2^{\ddagger} = \bar{\mu}_{0,\ddagger} - \bar{\mu}_{0,B^m} - q\bar{\mu}_{0,e^-}$$
or

(5) $$\Delta G_2^{\ddagger} = \mu_{0,\ddagger} - \mu_{0,B^m} - q\mu_{0,e^-} + F(q+(m-q)\zeta)(\Phi_M - \Phi_S).$$

A similar expression applies if the reaction occurs in the opposite direction:

(6) $$\Delta G_1^{\ddagger} = \mu_{0,\ddagger} + (1-q)\mu_{0,e^-} - \mu_{0,B^{m-1}} - F(1-q-(m-q)\zeta)(\Phi_M - \Phi_S).$$

Introducing U according to Eq. II, 16, Eq. 5 and 6 may be rewritten:

(7) $$\Delta G_2^{\ddagger} = \Delta G_{2,\text{ref}}^{\ddagger} + F(q+(m-q)\zeta)U,$$

(8) $$\Delta G_1^{\ddagger} = \Delta G_{1,\text{ref}}^{\ddagger} - F(1-q-(m-q)\zeta)U$$
with

(9) $$\Delta G_{2,\text{ref}}^{\ddagger} = \mu_{0,\ddagger} - \mu_{0,B^m} - q\mu_{0,e^-} + F(q+(m-q)\zeta)(\Phi_M - \Phi_S)_{\text{ref}},$$

$$\Delta G_{1,\text{ref}}^{\ddagger} = \mu_{0,\ddagger} + (1-q)\mu_{0,e^-} - \mu_{0,B^{m-1}} - F(1-q-(m-q)\zeta)(\Phi_M - \Phi_S)_{\text{ref}}.$$
(10)

The comparison of the rate of a given electrode reaction on different electrodes requires the choice of a suitable standard state for the comparison. Two standards are in use at present:

c) $(\Phi_M - \Phi_S)_0$ corresponding to the equilibrium of the electrode reaction under standard conditions of unit activity of reactants

(11) $$(\Phi_M - \Phi_S)_0 = \frac{1}{F}(\mu_{0,B^m} + \mu_{0,e^-} - \mu_{0,B^{m-1}}).$$

d) $(\Phi_M - \Phi_S)_z$ corresponding to the point of zero charge [2] at which the equal and opposite charges on the solid and solution are both zero.

Eq. 5 may be rewritten for the two states of comparison:

$$\Delta G_{2,0}^{\ddagger} = \mu_{0,\ddagger} - (1-q)\mu_{0,B^m} - q\mu_{0,B^{m-1}} + (m-q)\zeta(\mu_{0,B^m} + \mu_{0,e^-} - \mu_{0,B^{m-1}}).$$

(13) $$\Delta G_{2,z}^{\ddagger} = \mu_{0,\ddagger} - \mu_{0,B^m} - q\mu_{0,e^-} + F(q+(m-q)\zeta)(\Phi_M - \Phi_S)_z.$$

The parameters $\mu_{0,\ddagger}$, μ_{0,e^-}, $(m-q)\zeta$ in Eq. 12 and Eq. 13 and $(\Phi_M - \Phi_S)_z$ in Eq. 13 depend upon the electrode material. The equations reduce to those corresponding to weak interaction between the transition state and the electrode surface for $\zeta \to 0$.

(14) $$\Delta G_{2,0}^{\ddagger} = \mu_{0,\ddagger} - (1-q)\mu_{0,B^m} - q\mu_{0,B^{m-1}},$$

(15) $$\Delta G_{2,z}^{\ddagger} = \mu_{0,\ddagger} - \mu_{0,B^m} - q\mu_{0,e^-} + Fq(\Phi_M - \Phi_S)_z.$$

It follows from Eq. 14 that the standard free energy of activation is independent of the nature of the electrode material in a first approximation, since $\mu_{0,\neq}$ and q are unaffected by the nature of the substance to which the transition state is weakly bound. A detailed discussion is given in references 3 to 5. Experimental evidence presented in Table I of reference 1 seems to confirm this conclusion. The exchange current density

(16)
$$i_0 = \frac{k_1 a_{\mathrm{Bm-1}}}{A} = \frac{k_2 a_{\mathrm{Bm}}}{A}$$

of each of three redox systems does not vary much on different metals at $(\Phi_M - \Phi_S)_{rev}$. Measurements of the rates at the point of zero charge cannot give this information since $(\Phi_M - \Phi_S)_z$ in Eq. 15 varies with the electrode material. The variation occurs whether the transition state is weakly or strongly bonded. Comparison of the exchange current densities at $(\Phi_M - \Phi_S)_{rev}$ after correction for possible secondary effects on the basis of models for the double layer is generally recommended. [2]

As discussed by KRISHTALIK [6], there exist two limiting cases for redox processes involving an activation barrier. For a cathodic process, the positive activation energy $\Delta G_2^{\ddagger}$ decreases continuously with cathodic polarization when η becomes more negative. This follows from Eq. 7 if $\Delta G^{\ddagger}_{2,\mathrm{ref}}$ is replaced by $\Delta G^{\ddagger}_{2,rev}$ and U by η. However, $\Delta G_2^{\ddagger}$ cannot be less than zero. If $\Delta G_2^{\ddagger} = 0$ is reached at η_c, the rate of the cathodic process does not increase further with polarization at $\eta < \eta_c$. Eq. 8 may also be rewritten in terms of $\Delta G^{\ddagger}_{1,rev}$ and η. Let us assume that $\Delta G_1^{\ddagger}$ becomes zero at η_a. Then the rate of the anodic process does not increase with η at $\eta > \eta_a$. A redox process is called "barrierless" if it occurs at either $\Delta G_1^{\ddagger} = 0$ or $\Delta G_2^{\ddagger} = 0$. Both cases are illustrated in Fig. 3b and 3c. With increasing cathodic polarization the curve on the right side is shifted upwards until the intersection coincides with its minimum (Fig. 3b). With increasing anodic polarization, the right parabolic curve will eventually intersect the left curve at its minimum (Fig. 3c).

2. Two Step Reaction with Adsorbed Intermediate

Let us consider two examples which are of importance for the discussion of the mechanism of the hydrogen electrode in acid electrolytes:

(17)
$$H^+ + e^- = H_{ad},$$

(18)
$$H_{ad} + H_{ad} = H_2 .$$

Adsorbed hydrogen atoms formed by the Volmer reaction [7] combine to H_2 molecules on the surface in the Tafel [8] reaction. In the second mechanism, the Volmer reaction occurs simultaneously with the Heyrovsky [9] reaction:

(19) $$H^+ + H_{ad} + e^- = H_2 .$$

Hydrogen molecules are formed by discharge of H^+ ions on H atoms.

Under equilibrium conditions, $(\Phi_M - \Phi_S)_{rev}$ is determined by Eq. II, 12 for both mechanisms. It will vary with the electrode material because of $\mu_{0,e^-}(\alpha)$. The chemical potential of H atoms is the same for different electrode materials:

(20) $$\mu_{0,H} + RT \ln a_H = \tfrac{1}{2}\mu_{0,H_2} + \frac{RT}{2} \ln a_{H_2} .$$

The activity a_H is a function of the coverage θ of the electrode surface with H atoms. The coverage will depend upon the nature of the electrode material because of $\mu_{0,H}$. The standard free energy $\Delta G_{0,H}$ of hydrogen adsorption is related to $\mu_{0,H}$:

(21) $$\Delta G_{0,H} = 2\mu_{0,H} - \mu_{0,H_2} .$$

The standard states are $a_H = 1$ and $a_{H_2} = 1$. The reader is referred to the discussion of hydrogen adsorption on platinum metals in chapter VI as example for the influence of the electrode material upon $\Delta G_{0,H}$.

The standard free energies of activation of the cathodic partial currents of reactions 17 and 19 may be written in the form of Eq. 5. For simplicity the factor $(q + (1-q)\zeta)$ is put equal to $(1-\alpha)$ for reaction 17 and equal to $(1-\beta)$ for reaction 19. Similar expressions that contain α and β as factors hold for the standard free energies of activation for the anodic direction. The parameters α and β are called transfer coefficients. It is assumed that the adsorption of H atoms follows a Langmuir isotherm while adsorption of H_2 molecules is negligible. The effect of the strength of adsorption is reflected entirely by the equilibrium coverage θ_0 at $p_{H_2} = 1$ atm. Introducing η, the rates of reactions 17 and 19 are:

(22) $$I_V = k_1 \theta \exp \frac{\alpha \eta F}{RT} - k_2 (1-\theta) a_H + \exp \frac{-(1-\alpha)\eta F}{RT} ,$$

(23) $$I_H = k_3 (1-\theta) a_{H_2} \exp \frac{\beta \eta F}{RT} - k_4 \theta a_H + \exp \frac{-(1-\beta)\eta F}{RT} .$$

For simplicity of writing, the rate constants in Eq. 22 and 23 are introduced here in a slightly different way than in the preceding section.

They satisfy Eq. 3 if $\Delta G_j^{\ddagger}$ is replaced by $\Delta G_{j,rev}^{\ddagger}$. The rate of the Tafel reaction may be expressed:

(24) $$I_T = k_5(1-\theta)^2 a_{H_2} - k_6\theta^2 .$$

The rate constants are given by Eq. 3 with the pre-exponential factor $\frac{2ekT}{h}$ instead of $\frac{ekT}{h}$. The energies of activation may be expressed [10]:

(25) $$\Delta G_5^{\ddagger} = \Delta G_0 - \gamma \Delta G_{0,H} ,$$

(26) $$\Delta G_6^{\ddagger} = \Delta G_0 + (1-\gamma)\Delta G_{0,H} .$$

Here γ is a constant and ΔG_0 is the value of $\Delta G_5^{\ddagger}$ and $\Delta G_6^{\ddagger}$ at $\Delta G_{0,H}=0$. Since reaction 18 is not an electrochemical reaction, the free energies of activation $\Delta G_5^{\ddagger}$ and $\Delta G_6^{\ddagger}$ are accessible to an experimental determination. Both energies depend linearly upon the standard free energy of hydrogen adsorption.

In equilibrium, the partial anodic current is equal to the partial cathodic current of the same reaction:

(27) $$I_{0,V}(\theta) = k_1\theta \exp\frac{\alpha\eta F}{RT} = k_2(1-\theta)\, a_{H^+} \exp\frac{-(1-\alpha)\eta F}{RT} = k_1^{1-\alpha} k_2^{\alpha} \theta^{1-\alpha}(1-\theta)^{\alpha} a_{H^+}^{\alpha} ,$$

(28) $$I_{0,H}(\theta, a_{H_2}) = k_3(1-\theta) a_{H_2} \exp\frac{\beta\eta F}{RT} = k_4\theta\, a_{H^+} \exp\frac{-(1-\beta)\eta F}{RT} = k_3^{1-\beta} k_4^{\beta}\, \theta^{\beta}(1-\theta)^{1-\beta} a_{H_2}^{1-\beta} a_{H^+}^{\beta} ,$$

(29) $$I_{0,T}(a_{H_2}) = k_5(1-\theta)^2 a_{H_2} = k_6\theta^2 = \frac{2ekT}{h}\exp\frac{-\Delta G_0}{RT} \frac{a_{H_2}\exp\frac{-(1-\gamma)\Delta G_{0,H}}{RT}}{\left(1 + a_{H_2}\exp\frac{-\Delta G_{0,H}}{RT}\right)^2}$$

$I_{0,V}(\theta)$, $I_{0,H}(\theta, a_{H_2})$, and $I_{0,T}(a_{H_2})$ are the exchange currents of reactions 17 to 19.

As discussed [11, 12] the exchange current $I_{0,V}(\theta)$ passes through a maximum with increasing coverage. The maximum appears at $\theta = 1-\alpha$ ($\theta = \frac{1}{2}$ for $\alpha = \frac{1}{2}$). The dependence of $I_{0,H}(\theta, a_{H_2})$ upon θ is similar. However, the influence of a_{H_2} has to be considered [13] in the discussion of the coverage dependence on platinum metals.

The exchange currents corresponding to $a_{H_2}=1$ ($p_{H_2} \approx 1$ atm) and equilibrium coverage θ_0 will be designated by $I_{0,V}$, $I_{0,H}$, and $I_{0,T}$. They should be determined in the conventional way [14] from the slope of steady-state current-potential curves through the origin. This is feasible if one of the three processes is rate-determining and if $\eta=0$ is established by the electrode at $a_{H_2}=1$. A correction [14] for the influence of mass transport processes may be necessary in the computation of the exchange current from the slope of the $I-U$ curve through $\eta=0$. However, there are conditions under which the exchange currents can be measured on electrodes that do not function as a hydrogen electrode. These conditions will be discussed subsequently.

Introducing $I_{0,V}$, $I_{0,H}$, and $I_{0,T}$ the equations 22 to 24 may be rewritten:

$$(30)\qquad I_V = I_{0,V}\left(\frac{\theta}{\theta_0}\exp\frac{\alpha\eta F}{RT} - \frac{1-\theta}{1-\theta_0}\,\frac{a_{H^+}}{{}_b a_{H^+}}\exp\frac{-(1-\alpha)\eta F}{RT}\right),$$

$$(31)\qquad I_H = I_{0,H}\left(\frac{1-\theta}{1-\theta_0}\,\frac{p_{H_2}}{{}_b p_{H_2}}\exp\frac{\beta\eta F}{RT} - \frac{\theta}{\theta_0}\,\frac{a_{H^+}}{{}_b a_{H^+}}\exp\frac{-(1-\beta)\eta F}{RT}\right),$$

$$(32)\qquad I_T = I_{0,T}\left(\left(\frac{1-\theta}{1-\theta_0}\right)^2\frac{p_{H_2}}{{}_b p_{H_2}} - \left(\frac{\theta}{\theta_0}\right)^2\right).$$

The ratio $a_{H_2}/{}_b a_{H_2}$ of the activities for H_2 has been replaced by the ratio of the respective pressures.

As long as the condition $|I_j/I_{0,j}| \ll 1$ is fulfilled, the electrode reaction j may be treated as if it were in equilibrium. The exchange currents $I_{0,V}$, $I_{0,H}$, and $I_{0,T}$ are a measure of the reactivity of the electrode material with respect to the VOLMER, HEYROVSKY or TAFEL reaction. The electrode functions as hydrogen electrode at $p_{H_2}=1$ atm if either $I_{0,V}$, $I_{0,T} \gg |I_R|$ or $I_{0,V}, I_{0,H} \gg |I_R|$. Here I_R designates the total residual current resulting from oxidation and reduction of impurities, anodic metal dissolution, and cathodic reduction of molecular oxygen.

The kinetic expression for the Tafel reaction leads to an anodic limiting current

$$(33)\qquad I_{l,T} = I_{0,T}\,\frac{p_{H_2}}{{}_b p_{H_2}}$$

if $\theta, \theta_0 \ll 1$ for $I/I_{0,T} \gg 1$. A cathodic limiting current

$$(34)\qquad I_{l,T} = -\frac{I_{0,T}}{\theta_0^2}$$

results if θ tends towards 1 for $-\dfrac{I_T}{I_{0,T}} \gg 1$.

Hydrogen evolution occurs by the Volmer-Tafel mechanism if $I_{0,\mathrm{H}} \ll I_{0,\mathrm{T}}$. Steady-state conditions are:

$$I = I_{0,\mathrm{V}}\left(\frac{\theta}{\theta_0}\exp\frac{\alpha\eta F}{RT} - \frac{1-\theta}{1-\theta_0}\,\frac{a_{\mathrm{H}^+}}{{}_{\mathrm{b}}a_{\mathrm{H}^+}}\exp\frac{-(1-\alpha)\eta F}{RT}\right)$$

(35)

$$= I_{0,\mathrm{T}}\left(\left(\frac{1-\theta}{1-\theta_0}\right)^2 \frac{p_{\mathrm{H}_2}}{{}_{\mathrm{b}}p_{\mathrm{H}_2}} - \left(\frac{\theta}{\theta_0}\right)^2\right).$$

Four limiting cases may be distinguished for $|I_{\mathrm{R}}| < 0.1 \cdot |I|$:

a) $$0.1 \cdot |I| < I_{0,\mathrm{T}},\ I_{0,\mathrm{V}}.$$

The overvoltage is due to concentration changes of H_2 and H^+ in the diffusion layer. The limiting diffusion current densities are:

(36) $$i_{\mathrm{d,H}^+} = -\frac{D_{\mathrm{H}^+} F}{\delta_{\mathrm{H}^+}}\,{}_{\mathrm{b}}c_{\mathrm{H}^+},$$

(37) $$i_{\mathrm{d,H}_2} = \frac{2D_{\mathrm{H}_2} F}{\delta_{\mathrm{H}_2}}\,{}_{\mathrm{b}}c_{\mathrm{H}_2}.$$

The D's designate the diffusion coefficients, and δ is the thickness of the respective diffusion layer. The correlation between overvoltage and current density is:

(38) $$\eta = \frac{-RT}{2F}\ln\left(1 - \frac{i}{i_{\mathrm{d,H}_2}}\right) + \frac{RT}{F}\ln\left(1 - \frac{i}{i_{\mathrm{d,H}^+}}\right).$$

Hydrogen overvoltage is independent of electrode material.

b) $$I_{0,\mathrm{V}} \ll I_{0,\mathrm{T}}$$
$$I_{0,\mathrm{V}} < 0.1 \cdot |I| < I_{0,\mathrm{T}}.$$

The Volmer reaction is rate-determining. Equilibrium between H atoms and H_2 remains established. It follows from Eq. 30:

(39) $$I = -I_{0,\mathrm{V}}\frac{1-\theta}{1-\theta_0}\,\frac{a_{\mathrm{H}^+}}{{}_{\mathrm{b}}a_{\mathrm{H}^+}}\exp\frac{-(1-\alpha)\eta F}{RT}.$$

The relationship between η and I may be written in the form of Tafel's law

(40) $$-\eta = \mathrm{a} + \mathrm{b}\ln(-I)$$

for a cathodic process if $\frac{1-\theta}{1-\theta_0} \approx 1$ and $\frac{a_{\mathrm{H}^+}}{{}_{\mathrm{b}}a_{\mathrm{H}^+}} \approx 1$ in the studied range. This holds for either weak interaction ($\theta_0 \ll 1, \Delta G_{0,\mathrm{H}} > 0$) between H atoms and electrode or strong interaction ($\theta_0 > 0.9, \Delta G_{0,\mathrm{H}} < -10$ kcal/M).

Extrapolation of the Tafel line with the slope $b = \frac{RT}{(1-\alpha)F}$ to $\eta = 0$ allows the determination of $I_{0,V}$ from $a = \frac{RT}{(1-\alpha)F} \ln I_{0,V}$. The electrode does not have to function as hydrogen electrode.

c)
$$I_{0,T} \ll I_{0,V}$$
$$I_{0,T} < 0.1 \cdot |I| < I_{0,V}.$$

The Tafel reaction represents the rate-determining step. Combination of Eq. 32, with the equation resulting from the equilibrium of the Volmer reaction

(41)
$$\eta = \frac{RT}{F} \ln \left(\frac{1-\theta}{\theta} \cdot \frac{\theta_0}{1-\theta_0} \frac{a_{H^+}}{{}_b a_{H^+}} \right)$$

gives for $a_{H^+}/{}_b a_{H^+} = 1$:

(42)
$$\eta = \frac{RT}{2F} \ln \left(1 - \frac{I}{I_{0,T} \left(\frac{1-\theta}{1-\theta_0} \right)^2} \right).$$

A Tafel line with the slope $b = RT/2F$ is only to be expected [12] for weak interaction between H atoms and electrode. Extrapolation of the Tafel line to $\eta = 0$ allows $I_{0,T}$ to be determined whether the electrode establishes $\eta = 0$ at $p_{H_2} = 1$ atm or not. Condition c is not compatible with strong interaction because $-I$ cannot become larger than $-\frac{I_{0,T}}{\theta_0^2}$.

d)
$$I_{0,T}, I_{0,V} < |I_R|.$$

The Potential difference $(\phi_M - \phi_S)$ is determined [15] at $p_{H_2} = 1$ atm by the simultaneous occurrence of two different electrode reactions. The anodic current of the one reaction is equal to the absolute value of the cathodic current of the second reaction. An electrode that is the site of several electrochemical reactions with distinct values of U_{rev} is called a poly-electrode. The steady-state condition 35 represents the relationship between θ and I. The anodic partial currents are negligible. The coverage increases with η. Only weak interaction is compatible with condition d. A Tafel line with slope $b = \frac{RT}{(1-\alpha)F}$ is found for $\frac{a_{H^+}}{{}_b a_{H^+}} = 1$. Extrapolation of the Tafel line to $\eta = 0$ allows to **determine** $I_{0,V}$.

A similar distinction may be made at $|I_R| < 0.1 \cdot |I|$ if hydrogen evolution occurs by the Heyrovsky mechanism ($I_{0,T} \ll I_{0,H}$). The steady-state condition is:

$$I = 2I_{0,V}\left(\frac{\theta}{\theta_0}\exp\frac{\alpha\eta F}{RT} - \frac{1-\theta}{1-\theta_0}\frac{a_{H^+}}{{}_b a_{H^+}}\exp\frac{-(1-\alpha)\eta F}{RT}\right)$$

(43)

$$= 2I_{0,H}\left(\frac{1-\theta}{1-\theta_0}\frac{p_{H_2}}{{}_b p_{H_2}}\exp\frac{\beta\eta F}{RT} - \frac{\theta}{\theta_0}\frac{a_{H^+}}{{}_b a_{H^+}}\exp\frac{-(1-\beta)\eta F}{RT}\right).$$

e) $$0.1 \cdot |I| < I_{0,V}, I_{0,H}.$$

This case is identical with case a. As demonstrated [16], one of the six rate constants in Eq. 22 to 24 may be expressed as a function of the other five.

f) $$I_{0,V} \ll I_{0,H}$$

$$I_{0,V} < 0.1 \cdot |I| < I_{0,H}.$$

If $I_{0,V}$ is replaced by $2I_{0,V}$, Eq. 39 holds. The variables θ, η, p_{H_2} and a_{H^+} are correlated:

(44) $$\eta = \frac{RT}{F}\ln\left(\frac{1-\theta}{\theta}\cdot\frac{\theta}{1-\theta_0}\cdot\frac{{}_b a_{H^+}}{a_{H^+}}\frac{p_{H_2}}{{}_b p_{H_2}}\right).$$

A Tafel line with slope $b = \frac{RT}{(1-\alpha)F}$ will be found at $\frac{a_{H^+}}{{}_b a_{H^+}} = 1$ for weak interaction. Extrapolation yields $a = \frac{RT}{(1-\alpha)F}\ln 2I_{0,V}$. The discussion of the case of strong interaction is more difficult because of the influence of p_{H_2}.

g) $$I_{0,H} \ll I_{0,V}$$

$$I_{0,H} < 0.1 \cdot |I| < I_{0,V}.$$

The Heyrovsky reaction is rate-determining. The coverage follows from Eq. 43. It is:

(45) $$I = -2I_{0,H}\frac{\theta}{\theta_0}\frac{a_{H^+}}{{}_b a_{H^+}}\exp\frac{-(1-\beta)\eta F}{RT}.$$

Tafel's law is only fulfilled for strong interaction. Extrapolation of the Tafel line with the slope $b = \frac{RT}{(1-\beta)F}$ yields $a = \frac{RT}{(1-\beta)F}\ln 2I_{0,H}$ at $\eta = 0$.

h) $$I_{0,T}, I_{0,H} < |I_R|.$$

This case is similar to case d. The coverage is a function of η:

(46) $$\theta = \frac{I_{0,V}}{I_{0,H}} \frac{1}{\frac{1-\theta_0}{\theta_0}\exp\frac{\eta F}{RT} + \exp\frac{(\alpha-1-\beta)\eta F}{RT}}.$$

The coverage will increase with η if $\frac{1-\theta_0}{\theta_0} \gg \exp\frac{(\alpha-2-\beta)\eta F}{RT}$ in the range studied. It will decrease if $\frac{1-\theta_0}{\theta_0} \ll \exp\frac{(\alpha-2-\beta)\eta F}{RT}$. Only the case of weak interaction leads to a Tafel line with slope $b = \frac{RT}{(1-\alpha)F}$. The exchange current $I_{0,V}$ can be obtained by extrapolation.

Summarizing the preceding considerations it may be said that $I_{0,V}$ is accessible to an experimental determination in cases b, d, f, h and $I_{0,T}$ in case c for weak interaction between H atoms and electrode. For strong interaction, $I_{0,V}$ can be determined in case b and $I_{0,H}$ in case g. It is assumed that η does not approach values for which one of the two discharge reactions becomes barrierless.

3. Successive Electrode Reactions with One Rate-Determining Step

An electrochemical reaction may be a sequence of several steps involving one or more intermediates. If a particular step is rate-determining, the standard free energies $\Delta G_1^{\ddagger}$ and $\Delta G_2^{\ddagger}$ can be computed. The method that is commonly employed is to assume quasi-equilibrium for all the preceding steps and thus to obtain the concentration of the intermediates as a function of potential (compare cases c and f in section 2 of this chapter). In an alternate derivation PARSONS [17] treated the passage of the representative point of a system over a series of activation barriers. The state P immediately before the highest barrier and the state Q immediately after it have to be considered besides the initial and final states. This treatment [17–19] which is equivalent [18] to the quasi-equilibrium analysis is discussed subsequently.

The completion of the net reaction

(47) $$xX + yY^{m} + ne^{-} = wW + zZ^{(ym-n)/z}$$

requires the formation and decomposition of ν identical activated complexes. The stoichiometric number ν represents the number of times the rate-determining step takes place when the net reaction occurs once. If the activated complex is formed from X and Y^m the rate-determining step may be written:

$$\frac{xX}{v}+\frac{yY^m}{v}+\frac{ne^-}{v}=\frac{wW}{v}+\frac{zZ^{(ym-n)/z}}{v}. \tag{48}$$

Let us assume that the transition state has a charge $(ym-nq)/v$ with $0<q<1$ and that it is located within the inner region of the double layer where the potential is equal to $\phi_S+\zeta_1(\phi_M-\phi_S)$. The potential of the outer Helmholtz plane is made equal to ϕ_S for simplicity. The location of the state P or Q is characterized by $\phi_S+\zeta_2(\phi_M-\phi_S)$ or $\phi_S+\zeta_3(\phi_M-\phi_S)$ respectively. The generalization of Eq. 5 and Eq. 6 yields:

$$\Delta G_2^{\ddagger}=\mu_{0,\ddagger}-\frac{x}{v}\mu_{0,X}-\frac{y}{v}\mu_{0,Y^m}-\frac{nq}{v}\mu_{0,e^-}+\frac{nF}{v}\left(q+\zeta_1\left(y\cdot\frac{m}{n}-q\right)-\zeta_2 y\frac{m}{n}\right)(\phi_M-\phi_S), \tag{49}$$

$$\Delta G_1^{\ddagger}=\mu_{0,\ddagger}+\frac{n}{v}(1-q)\mu_{0,e^-}-\frac{w}{v}\mu_{0,W}-\frac{z}{v}\mu_{0,Z^{(ym-n)/z}}-\frac{nF}{v}\left(1-q-\zeta_1\left(y\frac{m}{n}-q\right)+\zeta_3\left(y\frac{m}{n}-1\right)\right)(\phi_M-\phi_S). \tag{50}$$

The magnitude of ζ_j $(j=1,2,3)$ depends upon the interaction of the electrode with the intermediate state j. In a first approximation, the electrode material does not influence the standard free energies of activation in the case of weak interaction $(\zeta_1=\zeta_2=\zeta_3=0)$. If ζ_2 or ζ_3 are very small, the interaction of state P or Q can be disregarded. If strong interaction exists for all three states, the assumption $\zeta_2=\zeta_2=\zeta_3$ is reasonable. Then Eq. 49 and Eq. 50 simplify to:

$$\Delta G_2^{\ddagger}=\Delta G_{2,\text{chem}}^{\ddagger}+\frac{nFq}{v}(1-\zeta_1)(\phi_M-\phi_S), \tag{51}$$

$$\Delta G_1^{\ddagger}=\Delta G_{1,\text{chem}}^{\ddagger}-\frac{nF}{v}(1-q)(1-\zeta_1)(\phi_M-\phi_S). \tag{52}$$

Here $\Delta G_{2,\text{chem}}^{\ddagger}$ and $\Delta G_{1,\text{chem}}^{\ddagger}$ contain the terms which are independent of $(\phi_M-\phi_S)$.

Tafel relations between U and I result if one of the two partial currents in the generalized form of Eq. 2

$$I=k_1 a_W^{w/v} a_{Z^{(ym-n)/z}}^{z/v}-k_2 a_X^{x/v} a_{Y^m}^{y/v} \tag{53}$$

is much smaller than the other and if the activities in the remaining term are independent of I. The cathodic (b_c) and the anodic (b_a) slopes of the Tafel line are:

$$(54)\qquad b_c = \frac{\nu R T}{\mathrm{n} F} \frac{2.303}{\mathrm{q}+\zeta_1\left(\mathrm{y}\frac{\mathrm{m}}{\mathrm{n}} - \mathrm{q}\right) - \zeta_2 \mathrm{y}\frac{\mathrm{m}}{\mathrm{n}}},$$

$$(55)\qquad b_a = \frac{\nu R T}{\mathrm{n} F} \frac{2.303}{(1-\mathrm{q})-\zeta_1\left(\mathrm{y}\cdot\frac{\mathrm{m}}{\mathrm{n}} - \mathrm{q}\right) + \zeta_3\left(\mathrm{y}\cdot\frac{\mathrm{m}}{\mathrm{n}} - 1\right)}.$$

The Tafel slopes depend strongly upon the assumptions of the interaction between the electrode and the intermediate states. Simple expressions result if the interaction is either weak or strong for all three states. The sum of the cathodic and anodic transfer coefficients α_c and α_a

$$(56)\qquad \alpha_c + \alpha_a = \frac{2.303\,\nu R T}{\mathrm{n} F}\left(\frac{1}{b_c} + \frac{1}{b_a}\right) = 1 - \zeta_2 \mathrm{y}\frac{\mathrm{m}}{\mathrm{n}} + \zeta_3\left(\mathrm{y}\frac{\mathrm{m}}{\mathrm{n}} - 1\right)$$

becomes equal to 1 only for weak interaction. The properties of the rate-determining step in successive reactions differ from those of a single step reaction.

It was pointed out by GNANAMUTHU and PETROCELLI [19] in their treatment of different mechanisms for the reduction of molecular oxygen in acid electrolytes

$$(57)\qquad O_2 + 4H^+ + 4e^- = H_2O$$

and in alkaline solutions

$$(58)\qquad O_2 + 2H_2O + 4e^- = 4OH^-$$

that the intermediates of the rate-determining step have to be considered adsorbed ($\zeta_1 = \zeta_2 = \zeta_3$) in many cases. In addition the rate-determining step has to be of the form in Eq. 48. The respective activities have to be independent of I in the Tafel region and the coverage with any type of oxygen layer that does not participate in reaction 57 or 58 has to be constant. The usefulness of b_c and b_a as a diagnostic criterion in the distinction of different mechanisms of the oxygen electrode is severely limited by these conditions.

4. Some Features of Mechanisms Involving the Simultaneous Formation of Different Products

Electrochemical reactions in which several products are formed simultaneously differ in some essential aspects from the mechanisms discussed in sections 1 to 3 of this chapter. These features are discussed subsequently. The anodic oxidation of methanol in acid solutions is treated as an example for the simultaneous occurrence of several anodic processes under a) to f). The oxidation [20] of formic acid by molecular oxygen in aqueous electrolytes serves as an example for the overlapping of anodic and cathodic processes under g) and h).

a) Number of Electrons Involved in the Net Reaction

Under steady-state conditions, the number of electrons that are involved in the net reaction 1 and in the sequences of reactions 17 and 18 or 17 and 19 is a whole number. If n_{CH_3OH} moles of CH_3OH are completely oxidized to CO_2 at constant current on platinum, the required number of coulombs is equal [21] to 6 $n_{CH_3OH}F$. The net reaction for the complete oxidation of a certain amount of methanol is a four-electron process.

The situation is different when the instantaneous efficiency of CO_2 production is considered. As it is well known, formaldehyde and formic acid are intermediate products in the anodic oxidation of methanol. Since CH_2O and HCOOH are found in the bulk of the solution, the instantaneous oxidation may be described by the scheme [22]:

(59) 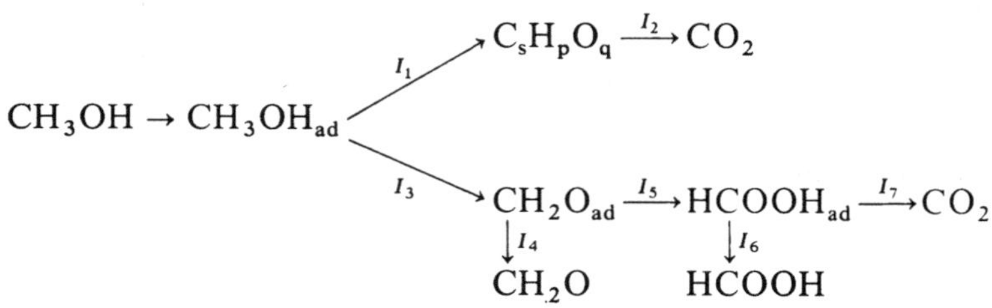

Either chemisorbed carbonaceous species of the net composition $C_sH_pO_q$ or CH_2O_{ad} are formed anodically from adsorbed methanol molecules. The diffusion of CH_2O and HCOOH into the bulk of the solution is counted in electrical units. A distinction between two different paths [22] for CO_2 production is difficult at small coverages with $C_sH_pO_q$. The net current is expressed for steady-state conditions by Eq. 60:

(60) $$I = I_1 + I_2 + I_3 + I_5 + I_7 .$$

The current of CO_2 production is:

(61) $$I_{CO_2} = I - (I_4 + I_6) = I_2 + I_2 + I_5 + 2I_7 .$$

The production of one CO_2 molecule requires six electrons in the limiting case that $I_4 \ll I_5$ and $I_6 \ll I_7$. If I_4 and I_6 are not negligible, the instantaneous number of electrons of the net reaction will be smaller than 6 and need not be an integer.

The experimental results of the author shown in Fig. 4 illustrate this situation. Methanol was oxidized in 0.5 M H_2SO_4 at constant current on platinized platinum (50 cm^2 apparent surface area) at room temperature. The initial bulk concentrations of methanol are given in the caption to Fig. 4. The initial bulk concentrations of formaldehyde and formic

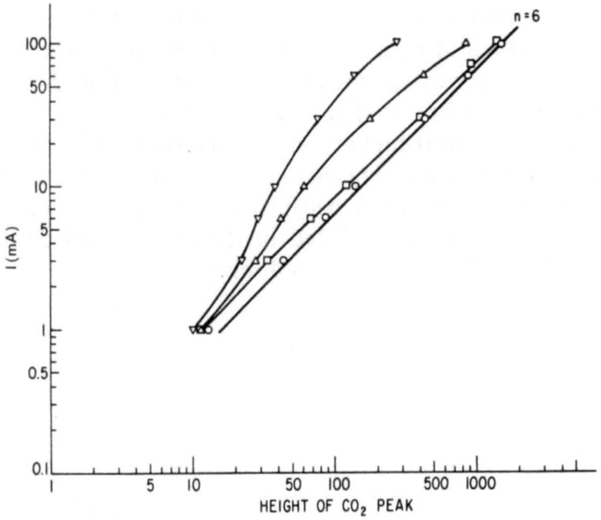

Fig. 4. CO_2 production during methanol oxidation on platinized platinum in 0.5 M H_2SO_4 as a function of current at room temperature. Straight line represents the CO_2 production for a six-electron process.
○ 0.01 M CH_3OH, □ 0.1 M CH_3OH, △ 1.0 M CH_3OH, ▽ 5.0 M CH_3OH

acid were zero. The measurements were started at 10 mA. The production of carbon dioxide was determined by gas chromatography from the CO_2 content of helium by which the solution was stirred at a constant flow rate of 0.5 cm^3/sec. The reading was taken when the increase in the height of the CO_2 peak with time became very small, the time being counted from the switching-on of a given current. Calibration [24] of the gas chromatograph yielded the solid line corresponding to a six-electron process. The plots of current versus peak height in Fig. 4 demonstrate that methanol oxidation is nearly a six-electron process at ${}_{b}c_{CH_3OH}=0.01$ M under the given conditions. The influence of I_4 and I_5 increases considerably with ${}_{b}c_{CH_3OH}$ at constant current. It depends upon U.

The values of U measured with respect to a hydrogen electrode in the same solution were determined as a function of I for the points in Fig. 4. The results are plotted in Fig. 5. There is little difference in the values of U at a given I although the efficiency for CO_2 production varies considerably.

b) Meaning of Steady-State Conditions

Steady-state conditions are said to be achieved for simple redox processes if $dU/dt \approx 0$ at constant I or $dI/dt \approx 0$ at constant U. It was observed during the measurements presented in Fig. 5 that dU/dt became negligible after about 15 minutes when the current was changed from one value to the next higher one. The increase of the height of the CO_2 peak with time became very small after about 60 minutes. The latter time is longer than 15 minutes because a stationary concentration of dissolved CO_2 has to be established in the 200 cm^3 of electrolyte at each current. This is due to the high solubility of CO_2.

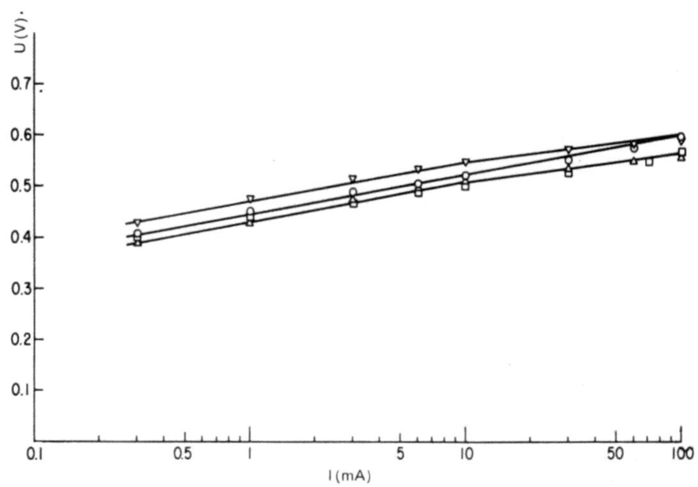

Fig. 5. Dependence of potential upon current during the steady-state oxidation of methanol in 0.5 M H_2SO_4.
○ 0.01 M CH_3OH, □ 0.1 M CH_3OH, △ 1.0 M CH_3OH, ▽ 5.0 M CH_3OH

The bulk concentrations ${}_{b}c_{CH_2O}$ and ${}_{b}c_{HCOOH}$ continue to increase after 60 minutes at ${}_{b}c_{CH_3OH} > 0.01$ M since $I_4 + I_5 > 0$. Thus a true steady state with respect to the bulk concentrations has not been reached in 60 minutes. However, the result that $I_4 + I_6$ is practically constant after 60 minutes implies that the concentrations c_{CH_2O} and c_{HCOOH} at the electrode surface tend towards certain stationary values and that $c_{CH_2O} \gg {}_{b}c_{CH_2O}$ and $c_{HCOOH} \gg {}_{b}c_{HCOOH}$ for each current under the prevailing conditions. The concentrations c_{CH_2O} and c_{HCOOH} depend upon the rate of stirring, since the diffusion layer thickness is a function of the stirring rate.

c) Rate-Determining Step

The five electrochemical reactions in the oxidation scheme 59 are net reactions. Each of the net reactions consists of different steps of which one or, possibly, a combination of two are rate-determining. In order to find the rate-determining steps, it is necessary to determine the partial currents I_j separately as function of c_{CH_3OH}, c_{CH_2O}, c_{HCOOH} and of the coverage with chemisorbed carbonaceous species at different potentials. This has not yet been done for reaction scheme 59. It is only meaningful to speak about one rate-determining step of the net process in the limiting case where $I_4 \ll I_5$ and $I_6 \ll I_7$.

d) Overvoltage

Since several electrode reactions with distinct values of U_{rev} occur simultaneously, the measured value of U is a mixed one (mixed electric tension [25]). The values of U_{rev} may be computed from thermodynamic data for the net reactions.

(62) $$CH_3OH = CH_2O + H_2, \qquad U_0 = 0.14\,V,$$

(63) $$CH_2O + H_2O = HCOOH + H_2, \qquad U_0 = 0.13\,V,$$

(64) $$HCOOH = CO_2 + H_2, \qquad U_0 = 0.25\,V,$$

(65) $$CH_3OH + H_2O = CO_2 + 3H_2, \qquad U_0 = 0.02\,V.$$

Here U_0 designates the value of U_{rev} under standard conditions at 25 °C. Overvoltages can be evaluated as difference $U - U_{rev}$ for the anodic reactions corresponding to reactions 62 to 64. If $I_4 \ll I_5$ and $I_6 \ll I_7$ an overvoltage for the net process 65 may be defined.

e) Meaning of the Tafel Slope

The plot of U versus log I in Fig. 5 demonstrates that the data can be represented by Tafel lines between 0.03 mA and 10 mA. The slope of the Tafel lines is nearly the same (about 75 mV) for the four bulk concentrations. This result might be regarded as evidence that the same step is rate-determining. However, such an interpretation is in contradiction to the data in Fig. 4. The drastic change in the efficiency of CO_2 production with the methanol bulk concentration at constant potential is not reflected by the Tafel slope. The results in Fig. 4 and Fig. 5 illustrate that the origin of an experimental Tafel relationship is not easily interpretable in the case of complex reactions.

f) Dependence of the Net Rate Upon pH

The slope of the Tafel line of methanol oxidation on smooth platinum was found [26, 27] not to change with pH in a first approximation. The $I-U$ curve is shifted just as the equilibrium potential of the hydrogen electrode. Measurements of methanol oxidation made by the author on platinized platinum in 0.05 M H_2SO_4 at 25 °C revealed similar efficiencies for CO_2 production as the measurements in 0.5 M H_2SO_4. These results suggest strongly that the pH dependence cannot be used as an unambiguous criterion for a rate-determining step in complex reactions.

g) Origin of the Open-Circuit Potential for Overlapping Anodic and Cathodic Processes

Formic acid may be oxidized by molecular oxygen in acid electrolytes

(66) $$HCOOH+\tfrac{1}{2}O_2=CO_2+H_2O$$

on platinum as a result of the simultaneous occurrence of an anodic reaction

(67) $$HCOOH=CO_2+2H^++2e^-$$

and a cathodic reaction

(68) $$O_2+4H^++4e^-=2H_2O.$$

When the rate I_a of reaction 67 becomes equal to the rate I_c of reaction 68 on the same electrode at open circuit, steady-state conditions have been achieved:

(69) $$I_a(U_{\mathrm{Op.}}, {}_bc_{\mathrm{HCOOH}}, \theta_{C_sH_pO_q})=I_c(U_{\mathrm{Op.}}, {}_bp_{O_2}, \theta_{C_sH_pO_q}).$$

Assuming mass transport processes as negligible for simplicity, the rate I_a controlled by the rate-determining step of net reaction 67 will depend upon the relative electrode potential, the bulk concentration of formic acid, and the coverage with chemisorbed carbonaceous species. A similar statement holds for I_c. The value of the open-circuit potential U_{0p} is determined by Eq. 69. On the potential scale it is located [15] between the U_0 value of reaction 67 and 68 under standard conditions.

h) Polarization

If the electrode potential is different from U_{0p} when an anodic and a cathodic reaction overlap, a net current I flows:

$$I = I_a(U, {}_b c_{HCOOH}, \theta_{C_sH_pO_q}) - I_c(U, {}_b p_{O_2}, \theta_{C_sH_pO_q}). \tag{70}$$

The current is anodic for $U > U_{0p}$ and cathodic for $U < U_{0p}$. The relative electrode potential U established under current is a mixed potential as defined by WAGNER and TRAUD [15].

i) Effect of Electrode Material on the Products of a Branched Reaction

An attempt was made by PARSONS [28] recently to describe the effect of electrode material on a simple branched electrochemical reaction and at the same time to allow for the participation of adsorbed intermediates and inhibitors. An idealized system was assumed. A reactant produces an adsorbed species in a one-electron oxidation reaction. The adsorption of the intermediate is Langmuirian. The intermediate is involved in two parallel reactions leading to two different products. The exchange currents of the three partial reactions were expressed as functions of the reactants and the coverage in a similar way as for the hydrogen evolution reactions in section 2 of this chapter. The exchange currents of the partial reactions were sketched as a function of the standard free energy of adsorption ΔG_0 of the intermediate at equilibrium. The information on the behavior at finite overvoltages of the one-electron reaction can be represented in a three-dimensional diagram with the coordinates of log I, η, and ΔG_0 on the basis of the kinetic equations. The effect of the electrode material under working conditions is found by taking a section at constant overvoltage. As pointed out by PARSONS [28], the predictions depend strongly on the assumptions about the detailed reaction mechanism and the relative rates of the partial reactions.

5. Correlation between Hydrogen Overvoltage and Free Energy of Hydrogen Adsorption

The attempts to correlate hydrogen overvoltage with other physical or chemical properties of metal electrodes are a classical example of the endeavor to understand the effect of the nature of the electrode material on an electrochemical reaction. The free energy of hydrogen adsorption appears to be the one property of metals for which a direct correlation with hydrogen overvoltage might be expected. It is a property of the surface in contrast to bulk properties like d-bond character, lattice constant, melting point, etc. The dependence of hydrogen overvoltage upon $\Delta G_{0,H}$ has been discussed [6, 10, 29—46] extensively.

In general, the discussions assume a homogeneous surface. However, the qualitative predictions [6, 10, 34] are also valid for heterogeneous surfaces since isotherms for heterogeneous surfaces differ [10] only over a range of medium coverage from isotherms for homogeneous surfaces. It follows from Eq. 27 to 29 that the three exchange currents will be small both for weak interaction between hydrogen atoms and metal ($\theta_0 \ll 1, \Delta G_{0,H} > 0$) and for strong interaction ($\theta_0 \approx 1, \Delta G_{0,H} < -10$ kcal/M). Going from weak to strong interaction, each of the exchange currents should pass through a maximum at intermediate values of $\Delta G_{0,H}$.

A large amount of experimental data concerning the correlation between hydrogen overvoltage and other properties of metal electrodes has been complied [36] recently. The exchange current density i_0 of the rate-determining step of hydrogen evolution was computed from the Tafel constants a and b for different metals from data in the literature. If the $I-U$ curve consisted of two Tafel regions, a and b were determined from the part at less cathodic overvoltage. The logarithm of i_0 is plotted versus the atomic number of the material in Fig. 6. Although the magnitude of i_0 may fluctuate considerably for the same metal, "volcano" type curves are obtained for each of the long periods. The results in Fig. 6 can be correlated to the free energy of hydrogen adsorption in the gas phase. Metals that adsorb hydrogen strongly are located on the left side of each of the periods. Metals with weak hydrogen adsorption are on the right side. The so-called d-metals, which exhibit a large reactivity for catalytic reactions in general, are positioned in the middle. The experimental results in Fig. 6 seem to confirm the theoretical prediction [6, 10, 34] of a maximum for the exchange currents at intermediate values of $\Delta G_{0,H}$. However, such a conclusion is based on the assumption that $\Delta G_{0,H}$ has nearly the same value for a given metal in the gas phase and in the electrolytic solution. The validity and limitations of this assumption are discussed subsequently.

As described in chapter VI, approximate values of $\Delta G_{0,H}$ may be determined experimentally for Rh, Ru, and Os. For Pt, Pd, and Ir this may be done for those H atoms that participate in the reactions at $\eta = 0$. In sulfuric acid solutions the $\Delta G_{0,H}$ values fall in the range between 0 and -10 kcal/M designating intermediate interaction between H atoms and metal. Hydrogen adsorption is small on Au [37, 38] and Ag [39] and negligible on Hg [40], Tl [41], Pb [41], and Cd [41] in electrolytic solutions. Thus the increase of i_0 from weak to intermediate interaction is in good agreement with the theoretical predictions for the middle curve and for the right curve in Fig. 6. However, experimental evidence is lacking that the decrease of i_0 from Pt, Ir, Os to Re, W and from Pd,

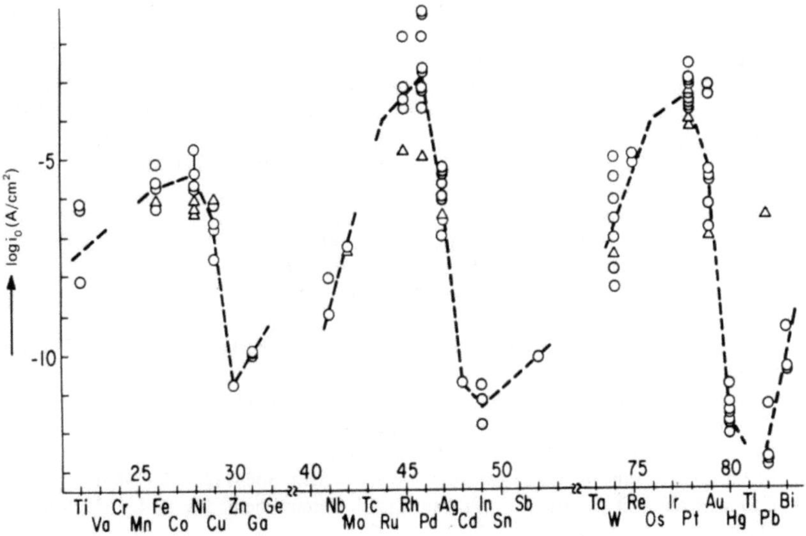

Fig. 6. Plot of the logarithm of the exchange current density of hydrogen evolution versus the atomic number of the metals

Rh, Ru to Mo, Nb is due to the transition from intermediate to strong interaction. The metals that adsorb hydrogen strongly in the gas phase exhibit a large affinity for the formation of oxide layers too. The presence of aqueous electrolyte and the tendency to oxide layer formation is likely to make the affinity for hydrogen adsorption different in the gas phase and in the electrolyte. An example for which experimental evidence exists is nickel, which functions as hydrogen electrode in alkaline solutions. The equilibrium coverage θ_0 at $\eta = 0$ is small [42, 43] while hydrogen adsorption is detectable [44, 45] at $p_{H_2} = 10^{-10}$ atm in the gas phase. Adsorption of OH and, possibly, formation of thin films of $Ni(OH)_2$ interferes [43] with hydrogen adsorption. It is considered likely that iron electrodes behave in a similar way to nickel electrodes. The decrease in the initial capacitance determined [42] in M KOH from the decay of U after interruption of the cathodic polarization from about $300\,\mu F/cm^2$ at $\eta = -0.4\,V$ to about $150\,\mu F/cm^2$ at $-0.55\,V$ may be due to a decrease in the adsorption [46] of OH radicals rather than to a decrease of electrochemically active H atoms.

References

1. Parsons, R.: Surface Sci. **2,** 418 (1964).
2. — Advances in Electrochemistry, Vol. **1**, Chap. 1. P. Delahay, ed., New York: Interscience 1961.
3. Marcus, R. A.: Can. J. Chem. **37,** 155 (1959).
4. Hush, N. S.: J. Chem. Phys. **28,** 962 (1958), Trans. Faraday Soc. **57,** 557 (1961).
5. Levich, V. G., and R. R. Dogonadze: Coll. Czech. Chem. Commun. **26,** 193 (1961).
6. Krishtalik, L. I.: Zhur. Fiz. Khim. **33,** 1715 (1959).
7. Erdey-Gruz, T., and M. Volmer: Z. physik. Chem. **A 150,** 203 (1930).
8. Tafel, J.: Z. physik. Chem. **50,** 641 (1905).
9. Heyrovsky, J.: Rec. Trav. Chim. Pays-Bas **46**, 582 (1927).
10. Parsons, R.: Trans. Faraday Soc. **54,** 1053 (1958).
11. Eucken, A., and B. Weblus: Z. Elektrochem. **55,** 114 (1951).
12. Breiter, M., and R. Clamroth: ibid. **58,** 493 (1954).
13. — Electrochim. Acta **7,** 25 (1962).
14. Vetter, K. J.: Elektrochemische Kinetik, Chap. 2E. Berlin-Göttingen-Heidelberg: Springer 1961.
15. Wagner, C., and W. Traud: Z. Elektrochem. **44,** 391 (1938).
16. Frumkin, A. N.: Zhur. Fiz. Khim. **31,** 1875 (1957).
17. Parsons, R.: Trans. Faraday Soc. **47,** 1332 (1951).
18. Fraser, G. H., and R. G. Barradas: J. Electrochem. Soc. **112,** 462 (1965).
19. Gnanamuthu, D. S., and J. V. Petrocelli: J. Electrochem. Soc. **114,** 1036 (1967).
20. Schwabe, K.: Z. Elektrochem. **61,** 744 (1957).
21. Vielstich, W.: Chem. Ing. Techn. **35,** 362 (1963).
22. Breiter, M. W.: Discuss. Faraday Soc., **45**, 79 (1968).
23. Andrew, M. R.: Electrochim. Acta **11,** 1425 (1966).
24. Breiter, M. W.: J. Electroanalyt. Chem. **14,** 407 (1967).
25. Defay, R., N. Ibl, E. Levart, G. Milazzo, G. Valensi, and P. van Rysselberghe: J. Electroanalyt. Chem. **7,** 417 (1964).
26. Bagotzky, V. S., and Yu. B. Vasiliev: Electrochim. Acta **9,** 869 (1964).
27. Khasova, O. A., Yu. B. Vasiliev, and V. S. Bagotzky: Izv. Akad. Nauk. S.S.S.R. Otd. Khim. 1539, 1778 (1965).
28. Parsons, R.: Discussions Faraday Soc. **45**, 40 (1968).
29. Bonhoeffer, K. F.: Z. physik. Chem. **A 113,** 199 (1924).
30. Horiuti, J., and M. Polanyi: Acta Physicochim. USSR **2,** 505 (1935).
31. Rüetschi, P., and P. Delahay: J. Chem. Phys. **23,** 195 (1955).
32. Temkin, M., and A. Frumkin: Zhur. Fiz. Khim. **29**, 1513 (1955).
33. Conway, B. E., and J. O'M. Bockris: J. Chem. Phys. **26,** 532 (1957).
34. Gerischer, H.: Bull. Soc. Chim. Belg. **67,** 506 (1958).

35. Breiter, M.: Mitteilungsblatt der Chemischen Gesellschaft in der DDR, Sonderheft **1959,** p. 9.
36. Kita, H.: J. Electrochem. Soc. **113,** 1095 (1966) and subsequent discussions.
37. Breiter, M., C. A. Knorr, and W. Völkl: Z. Elektrochem. **59,** 681 (1955).
38. Will, F. G., and C. A. Knorr: Z. Elektrochem. **64,** 270 (1960).
39. Gerischer, H., and W. Mehl: Z. Elektrochem. **59,** 1049 (1955).
40. Frumkin, A. N.: Advances in Electrochemistry, Vol. **1**, Chap. 2. P. Delahay, ed., New York: Interscience 1961.
41. — Advances in Electrochemistry, Vol. **3**, Chap. 5. P. Delahay, ed., New York: Interscience 1963.
42. Past, W. E., and S. A. Iofa: Zhur. Fiz. Khim. **33**, 1230 (1959).
43. Weininger, J. L., and M. W. Breiter: J. Electrochem. Soc. **111,** 707 (1964).
44. Suhrmann, R.: Z. Elektrochem. **60,** 804 (1956).
45. Broeder, J. J., L. L. van Reijen, W. M. H. Sachtler, and G. C. A. Schuit: Z. Elektrochem. **60,** 838 (1956).
46. Franke, K., C. A. Knorr, and M. Breiter: Z. Elektrochem. **63,** 226 (1959).

V. Electrode Reactions on Heterogeneous Surfaces

The two basic premises of the discussion of electrode kinetics in the preceding chapter are:

a) Each of the rate constants has the same value on any site of the surface.

b) The current distribution is uniform across the surface.

Condition a is fulfilled in a strict sense for liquid metal electrodes like mercury, gallium, and amalgams only. A plane surface cut from a single crystal in the direction of one of the nominal faces approximates condition a. However, the exposure of other faces cannot be eliminated completely because of the limited precision of the cutting which is usually not better than $\pm 1°$ with respect to the theoretical orientation.

Condition b can be met for homogeneous surfaces by the design of the cell, for instance by planar arrangement of test and counter electrode. Polarization [1—4] tends to produce a secondary current distribution that is less non-uniform than the primary distribution in the absence of polarization. If certain parameters that are characteristic [1—4] for the arrangement of the electrodes, the conductivity of the electrolyte, and the polarization are chosen properly, the current distribution can be considered uniform on a homogeneous surface.

Condition a is not fulfilled for real surfaces of polycrystalline material used commonly in the study of electrocatalytic reactions. Each of the rate constants may cover a large spectrum of values when passing across the surface. If the spectrum is large, the current distribution need no longer be uniform even when the conditions of uniformity for a homogeneous surface are met.

1. Structure and Composition of Surfaces of Solids

If a plane surface is prepared by cutting from a piece of solid metal, the surface consists of a large number of small grains. The grains are separated by grain boundaries. Both the grains and the grain boundaries represent cross sections through the crystallites and the adjacent grain boundaries of the non-uniform metal in the direction of the cut. Similarly, numerous grains are exposed on cold-rolled sheets and drawn wires. On a finer scale there are steps on each of the grains. Irregularities in the steps are called kinks.

Chemical and structural inhomogeneities that exist inside the solid metal are also present on the surface. Structural defects include lattice vacancies, interstitial atoms, and dislocations besides the above-mentioned grain boundaries. The free energy of formation of lattice vacancies is sufficiently small [5] for a small concentration to be stable at ambient temperatures. In contrast, the free energies of formation of the other defects are large and positive. The crystal would not contain them if it had been formed under equilibrium conditions. The chemical potential of the atoms at the latter defects is larger than that of the atoms in the bulk. It is generally assumed that these atoms are more reactive than the atoms in the bulk. The great reactivity of rough metal deposits and of compounds like Raney nickel may result from a large number of defects.

Chemical inhomogeneities are due to the segregation of particles of different composition at dislocations and grain boundaries during the production of the material or during heat treatment. Although chemical segregation cannot take place in a perfectly pure metal, there can be a non-uniform distribution of vacancies, as demonstrated [6] by soft regions around grain boundaries on the surface. The components of an alloy may interact to a different extent with the various defects. Associations are likely to occur in the vicinity of defects. Diffusion of substitional constituents can be expected over distances of 10 to 100 Å in the vicinity of defects at room temperature even for alloys with high melting points. Chemical inhomogeneities on the surface represent centers of different reactivity.

2. Current Distribution on Heterogeneous Surfaces

Let us assume that a single-step reaction, for instance, reaction IV, 1 occurs on a heterogeneous surface. The proportionality that is found for many electrocatalytic reactions between I and A for heterogeneous surfaces of solid metals suggests that the sites on which each of the rate constants k_1 and k_2 has the same value are uniformly distributed. The smallest apparent area A for which this statement still holds depends upon the structure and composition of the surface in question. In a crude approximation this area may be equal to the average grain size. If there is a finite number of types of sites with rate constants $k_{1,v}$ and $k_{2,v}$, the rate of the reaction in electric units is:

(1) $$I = \sum_{v}(k_{1,v} a_{B^{m-1}} - k_{2,v} a_{B^m}) .$$

The distribution of the k's can be expressed as a periodic function of appropriate coordinates, for instance x and y for a plane surface.

The rate constants satisfy Eq. IV, 3 if $\Delta G_j^{\ddagger}$ is replaced by $\Delta G_{j,v}^{\ddagger}$. Different values of the rate constant result from a variation of $\mu_{0,\ddagger}$, ζ, and $(\phi_M - \phi_S)$ according to Eq. IV, 5 and IV, 6. It follows from the discussion in section 1 of chapter IV that only reactions with strong interaction between the transition state and the surface have to be considered. If $(\phi_M - \phi_S)$ is not constant in the case of weak interaction it can only be due to a non-uniform current distribution.

In order to obtain a solution of the problem, the Poisson equation has to be solved with Eq. 1 as the boundary condition at the test electrode and a second, possibly simpler, boundary condition at the counter electrode. If concentration changes are involved, the equations for mass transport have also to be taken into account [7]. In this case the activities in Eq. 1 are replaced according to Eq. 2.

$$a_j = f_j c_j . \tag{2}$$

Here f_j designates the individual activity coefficient of the species j. Assuming that f_j does not depend upon c_j because of a suitable composition of the electrolyte, the product of activity coefficient and rate constant is frequently combined to a new rate constant with the unit A/M. The computation outlined depends strongly upon the assumption of the distribution function for the rate constants across the surface.

The distribution of I and U was computed [8, 9] for simple periodic variations of one kinetic parameter in the x direction of a plane electrode of infinite size. A dependence of the kinetic parameter on the y direction was not considered. The $I-U$ relation at the test electrode was approximated by a linear one. If the deviation of the kinetic parameter from an average value is assumed [9] to be small, the magnitude of the other characteristic parameters can be derived for which either I is a periodic function of x at constant U, or U is a periodic function of x at constant I. The predictions made on the basis of the simple models [8, 9] for heterogeneous surfaces are in agreement with general observations on metal deposition and on the electrochemical etching of metals during anodic dissolution. The reader is referred to the original papers for the detailed analysis.

3. Approximate Kinetic Expressions for Electrocatalytic Reactions on Heterogeneous Surfaces

The approach differs from the more rigorous approach discussed in section 2 of this chapter. Models for heterogeneous surfaces are introduced for which the effect of the heterogeneity is expressed solely in the activity terms that depend upon the coverage with the adsorbed intermediate. No influence of the heterogeneity upon I and U is assumed. The expression [11] "uniformly heterogeneous" describes the situation adequately. Historically the kinetic equations were derived in connection with the Volmer reaction. The subsequent discussion follows the same pattern since the Volmer reaction on platinum is the example studied thoroughly. A slight overlapping between the content of this section and that of chapter VI cannot be avoided.

Kinetic equations were derived by TEMKIN [12] for electrocatalytical reactions occurring on a surface for which the heat of adsorption W of the intermediate varies in a linear fashion between W_1 and W_2:

$$W = W_1 - (W_1 - W_2)\mathrm{s}. \tag{3}$$

The parameter s has values between 0 and 1 across the surface. The part of the surface for which the heat of adsorption is greater than or equal to W is designated by s. The Volmer reaction (compare Eq. IV, 17 and IV, 22) can be written [12] in a range of medium coverage θ with hydrogen atoms:

$$\begin{aligned} I_V = {} & k_1 \exp\left(\mathrm{f}\theta + \frac{\beta(\phi_M - \phi_S)F}{RT}\right) \\ & - k_2 a_{H^+} \exp\left(-\mathrm{f}\theta - \frac{(1-\beta)(\phi_M - \phi_S)F}{RT}\right). \end{aligned} \tag{4}$$

Here f denotes a constant. At equilibrium Eq. 4 leads to a linear dependence of θ upon $(\phi_M - \phi_S)$:

$$\theta = \frac{\mathrm{f}}{2} \ln \frac{k_2 a_{H^+}}{k_1} + \frac{F(\phi_M - \phi_S)}{2RT\mathrm{f}}. \tag{5}$$

If $(\phi_M - \phi_S)$ is replaced by η by transforming Eq. 5 in the usual way, the coverage can be expressed as a function of the hydrogen pressure p_{H_2}. The so-called Temkin isotherm results.

Eq. 5 was used [13] to interpret the nearly linear relation between θ and η observed on platinized platinum in hydrochloric acid and hydrobromic acid solutions when the adsorbed hydrogen is removed at constant current (anodic charging curve).

A generalization of Eq. 4 is given by Eq. 6:

$$\begin{aligned} I_V = {} & k_1 \theta \exp\left(\mathrm{f}\theta + \frac{\beta(\phi_M - \phi_S)F}{RT}\right) \\ & - k_2(1-\theta) a_{H^+} \exp\left(-\mathrm{f}\theta - \frac{(1-\beta)(\phi_M - \phi_S)F}{RT}\right). \end{aligned} \tag{6}$$

At equilibrium, expression 6 leads to a relation equivalent to the FRUMKIN [14] isotherm:

$$\frac{\theta}{1-\theta}\exp(f\theta)=\frac{k_2 a_{H^+}}{k_1}\exp\left(\frac{(\phi_M-\phi_S)F}{RT}\right). \tag{7}$$

The Langmuir isotherm results for $f=0$.

A different type of uniform heterogeneity was suggested by EUCKEN and WEBLUS [15]. It was assumed that two types of sites exist for hydrogen adsorption on platinum and that Eq. 6 holds for each type with different values of k_1, k_2, and f. Since f was found to be small [15], the suggestion is practically equivalent to the Langmuirian type of adsorption. An application is demonstrated by the results in Fig. 7.

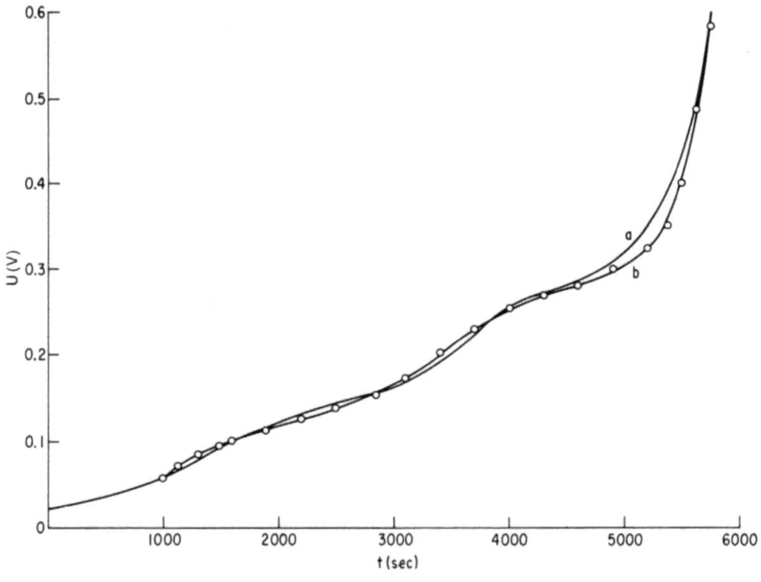

Fig. 7. Comparison between experimental (a) and computed (b) charging curves on platinized platinum in 0.5 M H_2SO_4

The classical $U-t$ curve [13] for the equilibrium charging of a platinized platinum electrode at constant current in 0.5 M H_2SO_4 at room temperature is represented by curve a. The experimental curve was approximated by curve b computed [16] with

$$F\Gamma_1=0.27\,\text{coul/cm}^2 \quad \text{and} \quad \frac{k_1}{k_2 a_{H^+}}=1.16\cdot 10^2$$

for one type of site, and

$$F\Gamma_2 = 0.22\,\text{coul/cm}^2 \quad \text{and} \quad \frac{k_1}{k_2 a_{H^+}} = 4.4 \cdot 10^4$$

for the other type. Here $F\Gamma_1$ is the charge required for the formation of a monolayer of H atoms on the sites of type 1, and $F\Gamma_2$ is that for type-2 sites. The computation was started at 0.06 V to avoid the influence of H_2 accumulated in the diffusion layer and of H atoms dissolved in the platinum. Since the charging curve was started at about 0.03 V, these two contributions were not eliminated experimentally. It is suggested that the two effects lead to the appearance of the initial portion of the charging curve with a curvature different [16] from that in the subsequent parts of the $U-t$ curve in the hydrogen region. The agreement between curves a and b is good above 0.06 V if the initial portion is disregarded for the reason mentioned.

Detailed kinetic treatments of similar problems were recently published by SRINIVASAN and GILEADI [17] and by HALE and GREEF [18].

References

1. KASPER, C.: Trans. Amer. Electrochem. Soc. **78,** 131 (1940).
2. WAGNER, C.: J. Electrochem. Soc. **98,** 116 (1951).
3. DROSSBACH, P.: Z. Elektrochem. **56,** 23 und 599 (1952).
4. BREITER, M., and Th. GUGGENBERGER: Z. Elektrochem. **62,** 859 (1958).
5. LOMER, W. M.: Progr. Metal Phys. **8,** 255 (1959).
6. AUST, K. T., A. J. PEAT, and J. H. WESTBROOK: Acta Met. **14,** 1469 (1966).
7. ASADA, K., F. HINE, S. YOSHIZAWA, and S. OKADA: J. Electrochem. Soc. **107,** 242 (1960).
8. DROSSBACH, P., and R. MEGGLE: Z. Elektrochem. **61,** 415 (1957).
9. WAGNER, C.: Electrochem. Acta **12,** 131 (1967).
10. LEVICH, V. G.: Physicochemical Hydrodynamics, Chap. VI. Englewood Cliffs, N. J.: Prentice Hall Inc. 1962.
11. FRUMKIN, A. N.: Advances in Electrochemistry, Vol. **3**, Chap. 5. P. Delahay, ed., New York: Interscience 1963.
12. TEMKIN, M.: Zhur. Fiz. Khim. **15,** 296 (1941).
13. SLYGIN, A., and A. FRUMKIN: Acta Physicochim. U.R.S.S. **3,** 791 (1935).
14. FRUMKIN, A. N.: Z. Physik **35,** 792 (1926).
15. EUCKEN, A., and B. WEBLUS: Z. Elektrochem. **55,** 114 (1951).
16. BREITER, W. M.: Trans. Faraday Soc. **62,** 2887 (1966).
17. SRINIVASAN, S., and E. GILEADI: Electrochim. Acta **11,** 321 (1966).
18. HALE, J. M., and R. GREEF: Electrochim. Acta **12,** 1409 (1967).

VI. Characterization of the Surface of Platinum Metals and Platinum Metal Alloys by Hydrogen Adsorption and Comparison of the Results with other Techniques

The properties of the surfaces of platinum metals can be studied by the cathodic deposition or the anodic removal of the adsorbed layer of H atoms in a simple way. Since the platinum metals and their alloys play an important role in electrocatalysis, the essential aspects of hydrogen adsorption under different experimental conditions are discussed subsequently.

1. Electrochemical Determination of Isotherms of Hydrogen Adsorption

A certain amount of hydrogen atoms, measured as Q_H in coul/cm^2 of apparent surface area, is adsorbed at stationary values of the hydrogen overvoltage η on the electrode. Charging curves [1—19], impedance measurements [20—27], and the potentiostatic technique [28—34] of applying a periodic voltage of triangular shape or a sawtooth voltage have been used for the study of hydrogen adsorption on platinum [1—6, 8—10, 12, 16—18, 20—23, 25—28, 30—34], iridium [7, 9, 19, 24, 29, 32], rhodium [9, 11, 19, 24, 29], palladium [13], ruthenium [14, 19], and osmium [15, 19].

The principle of the potentiostatic technique is described in connection with Fig. 8. A periodic voltage of triangular shape was applied to the test electrode. The amplitude of the sweep was chosen such that U changed between 0.05 V and 1.5 V. The electrolyte was saturated with an inert gas. The adsorbed hydrogen is removed during the anodic sweep in the so-called hydrogen region [3] according to the Volmer reaction. An oxygen layer is formed [28, 29] in the oxygen region:

$$H_2O = O_{ad} + 2H^+ + 2e^- . \quad (1)$$

Eq. 1 holds for acid electrolytes. A similar reaction can be formulated for alkaline solutions. The oxygen layer, which is considered here for simplicity to consist of O_{ad} only, is reduced first during the cathodic sweep. Then the hydrogen layer forms. The Volmer reaction and the corresponding reaction in the oxygen region occur in readily distinguishable potential ranges at sweep rates between 0.01 V/sec and 10 V/sec in acid and alkaline electrolytes during the anodic sweep,

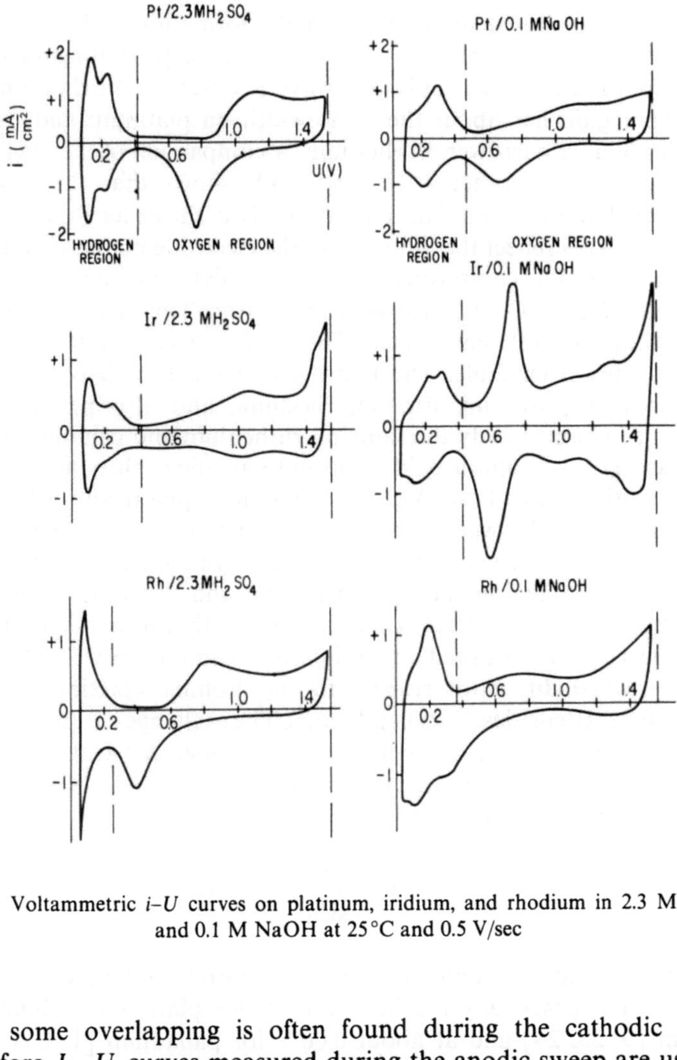

Fig. 8. Voltammetric *i–U* curves on platinum, iridium, and rhodium in 2.3 M H_2SO_4 and 0.1 M NaOH at 25 °C and 0.5 V/sec

while some overlapping is often found during the cathodic sweep. Therefore $I-U$ curves measured during the anodic sweep are used for a further evaluation only. Curves obtained at 0.5 V/sec on smooth wires of platinum, iridium and rhodium in 2.3 M H_2SO_4 and in 0.1 M NaOH at 25 °C are plotted in Fig. 8. The $i-U$ curves are readily reproducible and do not change their shape with time after a few initial cycles. The initial cycles serve [35] to remove impurities from the surface. Impurities do not diffuse back to the electrode surface to any appreciable extent in the time of half a cycle because their volume concentration is small in pure electrolytes. Hydrogen adsorption is investigated on a clean surface.

The $i-U$ curves demonstrate that the formation and the removal of the hydrogen layer occur in two distinct waves on platinum and iridium in acid solutions, whereas only one wave is observed on rhodium. The hydrogen region has about the same width on platinum and iridium. It is considerably narrower on rhodium. A comparison of the area under the $i-U$ curves between 0.2 V and 0.4 V shows that the adsorbed amount of H atoms at 0.2 V decreases rapidly in the order Pt, Ir, and Rh. The $i-U$ curves reflect the influence of the electrode material on hydrogen adsorption. The composition of the electrolyte influences [3, 4, 7, 11, 13—15, 20, 22, 25, 33, 36] hydrogen adsorption. This is a primary effect since $\Delta G_{0,H}$ has different values. The $i-U$ curves for 0.1 M NaOH in Fig. 8 are an example. The hydrogen region broadens in alkaline electrolytes on platinum, iridium, rhodium, and also palladium. In contrast it remains nearly the same on ruthenium and osmium [14, 15] when passing from sulfuric acid solution to alkaline solutions.

The equilibrium of the Volmer reaction is practically established at sufficiently small sweep rates on platinum, iridium, and rhodium. Each of the $i-U$ curves in Fig. 8 is nearly symmetrical to the U axis in the hydrogen region during the anodic and cathodic sweep. The condition $|i| \ll i_{0,V}$ is fulfilled. Charging curves that were taken [14, 15] on electrolytically deposited ruthenium and osmium in 0.05 M H_2SO_4 indicate reversibility with respect to the Volmer reaction at much smaller current densities ($|i| < 0.1\,\mu A/cm^2$). The exchange current densities of the Volmer reaction are considerably smaller for ruthenium and osmium than for the other four platinum metals.

Let us define the coverage θ:

$$\theta = \frac{Q_H}{{}_sQ_H}. \tag{2}$$

Here ${}_sQ_H$ designates the amount of H atoms corresponding to saturation coverage. It is practically reached at $\eta = 0$ for platinum, iridium, and rhodium [9, 23, 24] and at about 0.06 V for palladium [13]. A small error is introduced by reversing the sweep at 0.05 V. However, the reversal at 0.05 V has the advantage of making [37] the contribution from H atoms dissolved in platinum (and presumably in iridium and rhodium) and from H_2 molecules in the diffusion layer negligible at sweep rates $v > 0.1$ V/sec. Corrections, the magnitude of which depends upon the sweep rate, have to be made for these effects if the potential is reversed at $U < 0.05$ V. If $I \ll I_{0,V}$ the coverage can be obtained by stepwise integration of the anodic parts of the $I-U$ curves, or directly from anodic charging curves as a characteristic function of the electrode potential on the platinum metals.

If the condition of reversibility of the Volmer reaction is not completely fulfilled, as for the measurements on Ru [14] and Os [15], an approximate $U-t$ curve may be constructed in the middle between the anodic and the subsequent cathodic charging curve. The cathodic curve is measured by reversing the current in the double layer region and maintaining its amplitude. A differentiation between adsorbed hydrogen and hydrogen dissolved in palladium is possible [13, 38] on the basis of the experimental observation that certain portions of the charging curve do not depend upon the composition of the electrolyte. These portions are therefore attributed to bulk phenomena like dissolution of hydrogen and formation of the β phase. An appropriate correction for the simultaneous charging of the double layer has to be applied [3, 28, 29] in all cases.

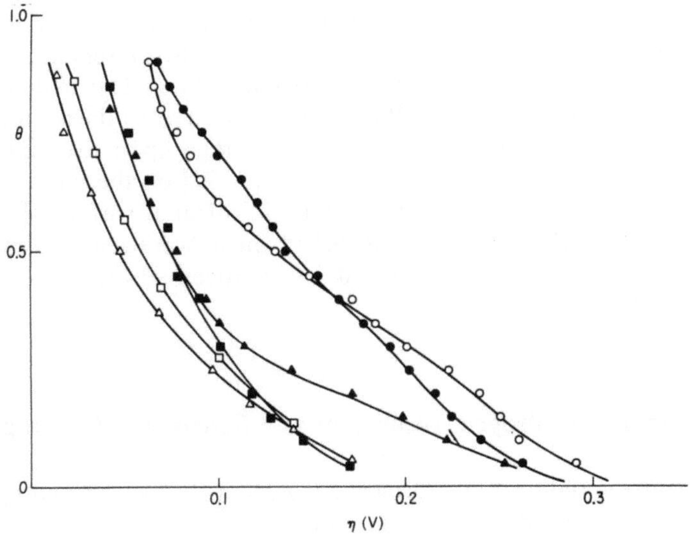

Fig. 9. Hydrogen coverage θ as a function of η on platinum (●), iridium (▲), rhodium (■) in 2.3 M H_2SO_4 at 30 °C, on palladium (○) in 0.5 M H_2SO_4 at 23 °C, and on ruthenium (□) and osmium (△) in 0.05 M H_2SO_4 at room temperature

The $\theta-\eta$ curves for Pt, Ir, Rh in 2.3 M H_2SO_4 at 30 °C and for palladium in 0.05 M H_2SO_4 at 23 °C in Fig. 9 were taken from reference 32 and reference 13 respectively. The curves for Ru and Os in 0.05 M H_2SO_4 at room temperature were obtained in the way outlined above from the data in references 14 and 15 under the assumption of 1.7 coul and 1.4 coul respectively as the charge for the monolayer formation. The coverage as a function of overvoltage represents an adsorption isotherm in a semilogarithmic plot. The pressure p_{H_2} which produces the same coverage at equilibrium can be computed [3] with the aid of Nernst's equation, taking into consideration the influence of the water vapor pressure at higher temperatures. Thus isotherms for hydrogen adsorption are obtainable from electrochemical measurements in a simple way. The temperature range is limited by the thermal stability of the electrolyte.

The effect of the electrode material on hydrogen adsorption is clearly reflected by the $\theta-\eta$ curves in Fig. 9. The coverage is very small at 0.2 V on Rh, Ru, and Os. It increases from 0.14 for Ir to 0.25 for platinum and 0.30 for Pd. At 0.1 V the coverage becomes larger in the order Os, Ru, Rh, Ir, Pd, and Pt. Platinum and palladium display very similar affinity for hydrogen adsorption. The waves of the $i-U$ curves on Pt and Ir in Fig. 8 are barely recognizable as humps in the $\theta-\eta$ curves in Fig. 9 because of the smoothing effect of the integration. As demonstrated by curve a in Fig. 7, these humps are also present in charging curves [3] on platinized platinum in 0.5 M H_2SO_4. The middle part of the $\theta-\eta$ curves on Pt and Pd can be considered as linear in a first approximation. A linear dependence of coverage upon the logarithm of p_{H_2} follows from Temkin's assumptions [3, 39] on the uniform heterogeneity of the surface or from the assumption of repulsive forces [39, 40] between the species which are adsorbed on a homogeneous surface. The latter assumption is not favored for hydrogen adsorption on platinum metals.

2. Heat of Hydrogen Adsorption as a Function of Coverage

The differential heat of hydrogen adsorption W_d has been computed from $\theta-\eta$ curves determined at different temperatures on Pt [16, 32—34], Ir [19, 32], Rh [19, 32], and Pd [13] according to Eq. 3:

$$W_d = R\,T^2\left(\frac{\partial \ln p_{H_2}}{\partial T}\right)_\theta \qquad (3)$$

The measurements with the sweep technique yielded W_d values that were practically independent of temperature for Pt and Ir between $-10°$ and 80 °C, and for Rh between 30° and 70 °C in 2.3 M H_2SO_4. Similarily W_d was found [13] to be constant at a given θ between 0° and 43 °C on Pd in 0.5 M H_2SO_4.

Fig. 10 represents W_d as a function of θ on palladium, platinum, iridium, and rhodium. In general the heat of adsorption decreases with increasing coverage, as was found in the gas phase for different metals by calorimetric determination [41] of W_d. The W_d values on Pt, Ir, and Rh are more than 10 kcal/M lower than that of palladium at a given coverage. A difference of this magnitude between Pt and Pd appears somewhat large, considering the similarity of the isotherms on

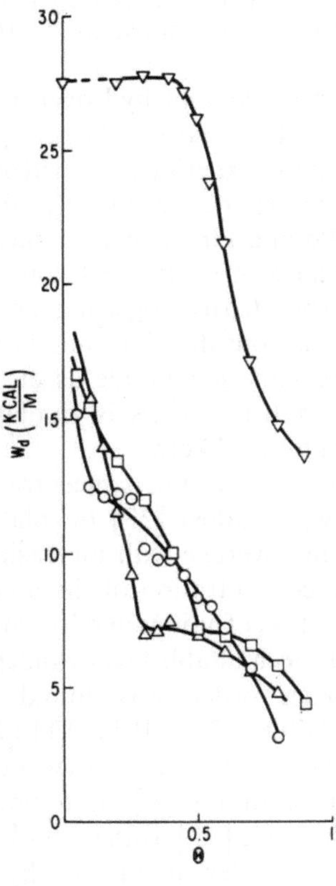

Fig. 10. Differential heat of hydrogen adsorption as a function of hydrogen coverage on palladium (▽), platinum (□), iridium (△), and rhodium (○)

the two metals. More experimental work on palladium is required to resolve this discrepancy.

The $W_d - \theta$ curves on Pd, Pt, and Ir consist of two main parts. On palladium W_d is about 27.5 kcal/M between $\theta = 0$ and $\theta = 0.4$ and decreases rapidly with θ for $\theta > 0.5$. In contrast W_d shows a rapid decrease with θ between 0 and 0.48 on platinum and between 0 and 0.28 on iridium. A more gradual decrease follows on both metals.

The first part is so narrow on rhodium that no distinction will be made. Adsorbed H atoms were called [42] "firmly bonded" in the first part and "weakly bonded" in the second by analogy with results in the gas phase [43—45]. The appearance of two major regions is directly related to the occurrence of current peaks in $I-U$ curves and humps in charging curves. It was suggested [42] that the strongly bonded hydrogen corresponds to the irreversibly adsorbed hydrogen in the gas phase while the weakly bonded hydrogen is equivalent to the reversibly adsorbed hydrogen.

It has been assumed so far that hydrogen is adsorbed in the form of atoms in both regions. The relatively large W_d values of the strongly bonded hydrogen represent experimental evidence for the correctness of this assumption in the first part of the $W_d-\theta$ curves on Pd, Pt, and Ir. The W_d values on Pd and kinetic considerations [42] based on the interpretation of impedance measurements on platinum favor atomic adsorption of the weakly bonded hydrogen in the second region. Hydrogen adsorption on rhodium is considered atomic because the $W_d-\theta$ curve covers W_d values of the same magnitude as on platinum and iridium. Hydrogen adsorption on Ru and Os is similar to that of rhodium. Therefore atomic adsorption is likely.

The influence of the nature and concentration of the electrolyte on the $W_d-\theta$ curves was studied [33] on platinum. The differential heat decreases at a given coverage with increasing adsorbability of the anion at constant bulk concentration of the anions. This effect is due to the competitive adsorption of anions and hydrogen atoms. The specific adsorption of anions has been established by independent measurements [46—49]. The following order was found for W_d: 0.5 M NaOH > 1 M $HClO_4$ > 2.3 M H_2SO_4 > 3 M HCl > 0.3 M HBr. A similar effect of the anions on the shape of charging curves was discussed for platinized platinum [3], smooth platinum [4] and electrolytically deposited layers of iridium [7], rhodium [11], ruthenium [14], and osmium [15]. An interpretation of the greater affinity for hydrogen adsorption on platinized platinum in alkaline electrolytes was already advanced by Slygin and Frumkin [3]. The metal is negatively charged [46] with respect to the electrolyte. The presence of cations in the immediate vicinity of the surface strengthens the bond of hydrogen atoms which are polarized so that the negative end of the dipole points into the solution. A similar interpretation is suggested for palladium, iridium, and rhodium.

A comparison between $W_d-\theta$ curves obtained on Pt by calorimetric measurements [50] at 7 °C in the gas phase and by electrochemical measurements [32, 51] in 2.3 M H_2SO_4, 0.5 M H_2SO_4 or 0.5 M D_2SO_4 is shown in Fig. 11. The $W_d-\theta$ curves determined in 0.5 M H_2SO_4 and 0.5 M D_2SO_4 coincide within the accuracy of the measurements. The different location of the $W_d-\theta$ curves in 0.5 M H_2SO_4 and 2.3 M H_2SO_4 may be caused by the specific adsorption [47] of SO_4^{--}. Specific adsorption of anions lowers [33, 42] the heat of hydrogen adsorption more for the strongly bonded than for the weakly bonded atoms. The presence of

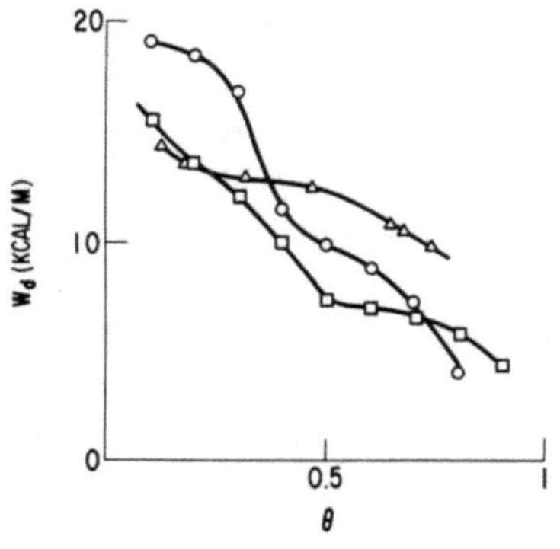

Fig. 11. Heat of hydrogen adsorption as a function of coverage. ○ : 0.5 M H_2SO_4 or 0.5 M D_2SO_4, □ : 2.3 M H_2SO_4, △ : gas phase

electrolyte makes the difference in the heats of adsorption of the strongly and weakly bonded hydrogen greater in solutions than in the gas phase. The W_d values at a given coverage are comparable for sulfuric acid solutions and the gas phase.

3. Langmuir Approximation of the Isotherms of Hydrogen Adsorption

Two overlapping Langmuir isotherms with different values of $\Delta G_{0,H}$ were used [21] to approximate the whole isotherm of hydrogen adsorption on platinum in sulfuric acid solution. The approximation looks satisfactory at room temperature on the scale of the plot given in reference 21. Recently the applicability of the Langmuir isotherm to the adsorption of the weakly and strongly bonded hydrogen was tested [52] by comparing experimental and computed $i-U$ curves on smooth platinum in 0.5 M H_2SO_4 at temperatures between 0° and 70 °C. The values of the parameters needed for the computation were taken from the experimental data. The Langmuir isotherm was found to yield acceptable results for $\theta < 20$ °C and to be a rough approximation for 20 °C $\leq \theta \leq$ 70 °C. A better fit was achieved between 20° and 70 °C by the Frumkin isotherm [53] in which a change of the free energy of adsorption with coverage is taken into account:

$$R\,T \ln p_{H_2} = \Delta G_{0,F} - 4R\,T\beta\theta + 2R\,T \ln \frac{\theta}{1-\theta}. \tag{4}$$

The positive β values that gave the best fit were smaller than 1 and varied with temperature.

The fact that a Langmuir approximation works at all appears contradictory to the continuous decrease of W_d with θ, demonstrating the existence of sites with a wide spectrum of W_d values. However, the formal application of the Langmuir isotherm

$$R\,T \ln p_{H_2} = \Delta G_{0,L} + 2R\,T \ln \frac{\theta}{1-\theta} \tag{5}$$

requires only that $\Delta G_{0,L}$ is independent of θ. If $\Delta G_{0,L}$ changes slightly with θ, as found, Eq. 5 holds approximately. The values of $\Delta G_{0,L}$ were determined from the adsorption isotherms in Fig. 9 at $\theta = 0.5$. Here θ designates the total coverage for Rh, Ru, and Os and the coverage of the weakly bonded hydrogen on Pd, Pt, and Ir. The $\Delta G_{0,L}$ values are put together in Table 1.

Table 1. *Free energy* $\Delta G_{0,L}$ *of hydrogen adsorption on* Rh, Ru, *and* Os *and of the weakly bonded* H *atoms on* Pd, Pt, *and* Ir *according to the Langmuir approximation*

Metal	Pt	Pd	Ir	Rh	Ru	Os
Total coverage	0.72	0.70	0.64	0.5	0.5	0.5
$\Delta G_{0,L}$ (kcal/M)	−4.5	−3.8	−2.9	−3.4	−2.8	−2.2

As mentioned in section 5 of chapter IV, the $\Delta G_{0,L}$ values fall within the range that is characteristic for intermediate interaction between metal and hydrogen atoms.

It was suggested [42, 54] that $\Delta G_{0,L}$ is approximately independent of θ because of a compensation effect. The heat W_d decreases continuously with θ. If the entropy of adsorption ΔS becomes less negative with coverage because the bonding of the H atoms is less strongly localized, $\Delta G_{0,L}$ need not change much with coverage. This interpretation is supported by a nearly linear dependence between $W_d(\theta \to 0)$ and $-\Delta S(\theta \to 0)$. Sets of these two quantities were determined at small coverage from isotherms in different electrolytes. The reader is referred to references 42 and 54 for a detailed discussion.

4. Influence of Surface Structure on Hydrogen Adsorption at Platinum

Hydrogen adsorption was studied by WILL [34] on the three main faces (100), (111), and (110) of platinum single crystal electrodes in 4 M H_2SO_4 at different temperatures with the sweep technique. The $i-U$ curves in Fig. 12 were obtained at 25 °C and 0.1 V/sec during the 260th periodic sweep. The potential sweep covered the range between 0.05 V and 1.55 V. Each of the three curves exhibits two large waves during the anodic sweep, as on polycrystalline electrodes. While the ratio of the left current maximum i_1 (weakly bonded hydrogen atoms)

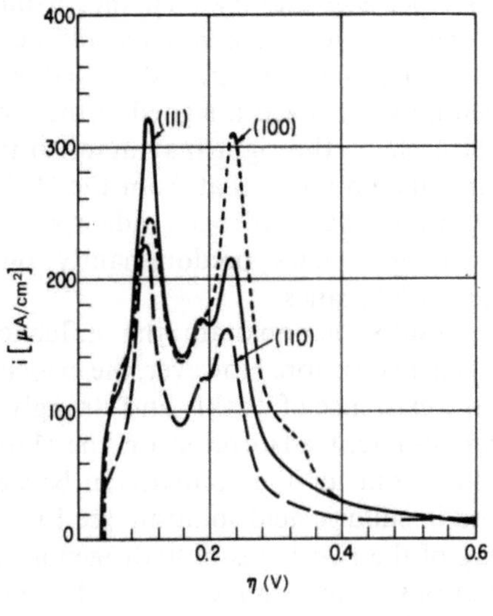

Fig. 12. Voltammetric i–U curves on the three main faces of platinum single crystals in 4 M H_2SO_4 at 25°C and 0.1 V/sec in the hydrogen region

and the right maximum i_2 (strongly bonded hydrogen) is about 1.2 on polycrystalline electrodes, it has the value of 0.8 for the (100) face, 1.5 for the (111) face, and 1.5 for the (110) face after many cycles. In addition a small third current maximum is present at about 0.19 V during the initial sweeps (see also section 7 of this chapter). After ten sweeps the ratio i_1/i_2 was 0.79 for the (100) face, 2.1 for the (111) face, and 2.9 for the (110) face.

The heat of adsorption was determined [34] for the three faces after 260 cycles. It was independent of temperature between 0° and 80 °C for the (100) and (111) faces and between 0° and 40 °C for the (110) face.

The general shape of the $W_d - \theta$ curves is similar to that of the curves on polycrystalline platinum in Fig. 11. Large differences exist in the W_d values at a given coverage. The heat W_d decreases in the order (100)>(111)>(110) for $\theta \leq 0.8$.

The number of platinum atoms in the surface planes is $1.5 \cdot 10^{15}$ per cm^2 for the (111) face, $1.3 \cdot 10^{15}$ for the (100) face, and $0.92 \cdot 10^{15}$ for the (110) face. This leads to theoretical values for ${}_{S}Q_H$ of 0.24 mcoul/cm^2, 0.21 mcoul/cm^2, and 0.15 mcoul/cm^2. The experimental ${}_{S}Q_H$ values are much larger on the three faces, especially after continued cycling. WILL [34] estimated that the roughness factor increased from 1.87 after ten sweeps to 2.45 after 260 sweeps on the (100) face. The initial roughness of the virgin surfaces becomes greater with the number of sweeps. Each of the nominal faces exposes several crystal planes in different proportions.

The fact that the current maxima have different heights for the faces but occur at nearly the same potentials led WILL to a tentative assignment: The left maximum was assigned to the removal of hydrogen atoms from the (110) plane, the right maximum to that from the (100) plane, and the third maximum to that from the (111) plane. It follows from this interpretation that hydrogen adsorption on cold-worked polycrystalline platinum occurs predominantly on crystallites that expose the (110) and (100) planes.

The preceding results demonstrate the influence of the surface structure on hydrogen adsorption. However, the evidence is not sufficient to decide whether the existence of weakly and strongly bonded hydrogen is really due to predominant adsorption on the (110) and (100) faces. It was already pointed out for the comparison between $W_d - \theta$ curves in the gas phase and in sulfuric acid solution (see Fig. 11) that the difference in the bonding of the two types of hydrogen atoms is accentuated in electrolytic solutions. The complexity of the interaction between hydrogen atoms and polycrystalline platinum in electrolytes is illustrated by the voltammetric $i - U$ curves in Fig. 13. These curves were measured [36] at 30 °C in 1 M $HClO_4$ with different additions of HCl. The sweep rate was 30 mV/sec. Small additions ($<10^{-5}$ M HCl) cause slight changes of the $i - U$ curves in the oxygen region. There are two waves in the hydrogen region as in sulfuric acid solutions. The location of the two current maxima is not much different from that of the corresponding maxima in Fig. 12. There are three current maxima in the hydrogen region at ${}_{b}c_{HCl} > 10^{-5}$ M. These peaks appear at less anodic U values with increasing ${}_{b}c_{HCl}$ when the potential region of hydrogen adsorption becomes narrower [3, 4]. The peaks marked I and II in Fig. 13 coincide [33] in 1 M HCl and 3 M HCl, so that only one big maximum at about 0.12 V and a small one at about 0.22 V remain. These results indicate that adsorption sites which are not located on the same face, but which

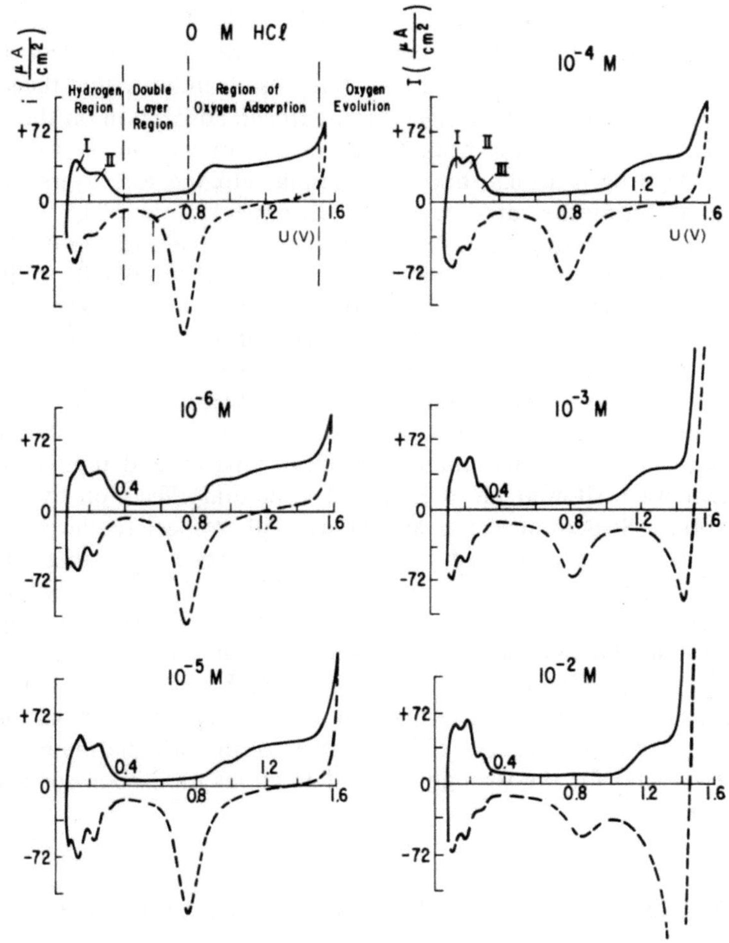

Fig. 13. Voltammetric *i–U* curves on smooth platinum in 1 M $HClO_4$ with different additions of HCl at 30 °C and 30 mV/sec

have nearly the same value of $\Delta G_{0,L}$, are grouped together as a consequence of the competitive adsorption of Cl^- ions. It is suggested that the existence of such groups is correlated to the compensation effect described in section 3 of this chapter. The strong influence of the composition of the electrolyte on hydrogen adsorption is superimposed on the structural effects and makes separation difficult.

The author's studies on the origin of the small third peak in Fig. 12 revealed several details. The small peak is not detectable at 30 mV/sec and 300 mV/sec during the cathodic sweep in 0.5 M H_2SO_4 at 30 °C if the same resolution is used during the anodic and cathodic sweep.

It appears during the anodic sweep only if the sweep is reversed at $U < 0.13$ V. The small peak becomes more pronounced with decreasing potential of reversal for $U < 0.13$ V. These results indicate that the small third peak should not be assigned to hydrogen adsorption on the (111) face. It is likely that the peak results as a combination of two small effects. Hydrogen atoms that are diffusing into the top layers of the platinum during the cathodic sweep between 0.13 V and 0.05 V return to the surface during the subsequent anodic sweep and are oxidized. Similarly hydrogen molecules formed between 0.13 V and 0.05 V diffuse back to the electrode surface during the following anodic sweep.

It was already noted by Ershler [4] that anodic charging curves have a similar shape in a given solution on smooth and platinized platinum. Smooth and platinized platinum seemed to differ mainly in the effective surface area. A more detailed study [18] of platinized platinum and Raney platinum electrodes prepared and pretreated in different ways showed the presence of structural effects on hydrogen adsorption. Structural effects are largest with respect to the sites on which the strongly bonded hydrogen is adsorbed. Charging curves taken [55] on platinum black powder in 0.5 M H_2SO_4 display the humps that are characteristic for the two types of hydrogen adsorption (compare Fig. 7). A similar statement applies to charging curves on Teflon-bonded platinum black [56] in 2.5 M H_2SO_4 and 6 M KOH at 25 °C. The fine structure of the hydrogen arrest is weak in charging curves on platinized porous carbon electrodes [57]. Relative to smooth platinum, there are fewer sites for the strongly bonded H atoms on carbon-supported platinum.

5. Determination of the Electrochemically Active Surface

It has become a standard procedure in the last ten years to take the charge ${}_SQ_H$ from charging curves recorded in the absence of fuels as a measure of the electrochemically active surface of fuel cell electrodes containing a platinum metal catalyst. Two assumptions are involved:

a) The ratio of H atoms to Pt atoms in contact with the electrolyte and not blocked by impurities is one in the absence of fuel.

b) The electrochemical reaction in question occurs on all these Pt atoms and not only on certain centers.

Assumption a appears justified in view of the experimental evidence for smooth platinum [9, 23, 28, 42] and the similarity of the shape of charging curves on smooth platinum, platinized platinum, Teflon-bonded platinum black, and platinized porous carbon. The validity of assumption b has to be checked for each of the electrochemical reactions. The subsequent discussions of reactions in aqueous low-temperature fuel cells will show that the validity of assumption b is the exception rather than the rule. Thus ${}_{S}Q_{H}$ represents a measure of the maximum surface area available to the net reactions. In spite of this restriction, a comparison of ${}_{S}Q_{H}$ values indicates the efficient method of preparing porous electrodes with a maximum ${}_{S}Q_{H}$ for a given loading with catalyst. As discussed in section 6 of this chapter, the measurement of ${}_{S}Q_{H}$ in the absence of fuel and ${}_{S}Q'_{H}$ in the presence of carbonaceous fuel is an important method for the determination of the surface coverage with carbonaceous species.

If the material which supports the catalyst is electrochemically active, for instance, carbon, a correction for the charge consumed on the support has to be applied. Since the determination of the charge consumption on carbon is not accurate [57], a method that uses the

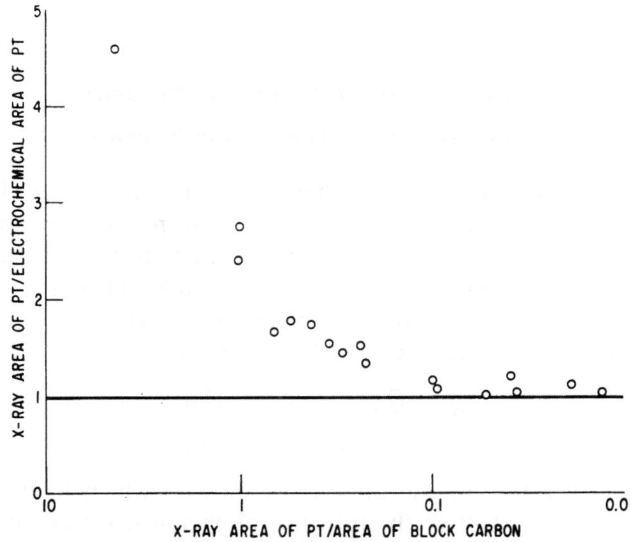

Fig. 14. Ratio of surface area from X ray diffraction to area from the electrochemical determination as a function of platinum loading of the carbon support

hydrogen and the oxygen arrest in combination was developed [57]. A reproducibility of 10% or better was achieved as long as the charge consumption on the support did not exceed 25% of the total charge of the arrest. The average size of platinum crystallites on carbon was determined [57] for different electrodes from X-ray diffraction data. The area of a crystallite was calculated on the assumption that the crystallites were cubes with one face covered by contact with the support.

The available platinum area was also measured with the aid of the electrochemical technique. The results are compared in Fig. 14. Agreement is poor if the ratio of platinum to carbon surface exceeds 0.1. Good agreement exists for ratios between 0.1 and 0.01. It was suggested [57] that the platinum crystallites are crowded at high platinum to carbon area ratios so that much of the platinum is not in contact with the electrolyte.

Surface areas of platinum black [56, 57] and rhodium black [55] powders were determined by krypton adsorption, using the BET method, and by the hydrogen arrest from charging curves. The agreement of the areas obtained by the two techniques for the same sample was good. Satisfactory agreement was also reported [58] for surface areas measured on platinized platinum. These results confirm the correctness of assumption a.

6. Hydrogen Adsorption in the Presence of Chemisorbed Carbonaceous Species

Let us designate by ${}_{S}Q_{H}$ the charge of a monolayer of H atoms adsorbed on a platinum metal electrode in a given electrolytic solution when carbonaceous species are absent. If carbonaceous species are chemisorbed, the number of sites for hydrogen adsorption is decreased. The charge corresponding to the complete coverage of the remaining sites by H atoms is designated ${}_{S}Q'_{H}$. The quantity

$$\theta' = \frac{{}_{S}Q_{H} - {}_{S}Q'_{H}}{{}_{S}Q_{H}} = 1 - \frac{{}_{S}Q'_{H}}{{}_{S}Q_{H}} \qquad (6)$$

is a relative measure for the number of sites on which hydrogen adsorption does not occur in the presence of chemisorbed carbonaceous species or other species. It is also a measure for chemisorbed species if the hydrogen atoms do not react with them. The adsorption of acetic acid [59] and of nitrides [60] was studied on platinum in this way. The quantity θ' was considered [59, 60] a direct measure for the coverage with adsorbed species without experimental verification of this assumption.

The essential points concerning the simultaneous adsorption of hydrogen atoms and carbonaceous species are discussed subsequently for the simple case of carbon monoxide as blocking species in acid electrolytes. Adsorption of carbon monoxide has been studied on smooth platinum by pulse techniques [63—65] and on platinized platinum by anodic charging curves [66, 67]. Hydrogen atoms do not reduce [63, 65] the absorbed carbon monoxide at temperatures $\vartheta \leq 60\,°C$ on platinum.

The charging curve a in Fig. 15 was taken [68] from 0.07 V at 50 mA on a platinized Pt electrode with $50\,cm^2$ of apparent surface area in 0.5 M H_2SO_4 at 25 °C. The electrolyte was stirred with purified helium at $0.5\,cm^3/sec$. The charging curve has the characteristic shape with two humps in the hydrogen region and displays the practically linear dependence of potential upon time in the double layer region between about 0.4 V and 0.7 V and also in the initial part of the oxygen region for $U > 0.9\,V$. The absence of arrests in the double layer region is the

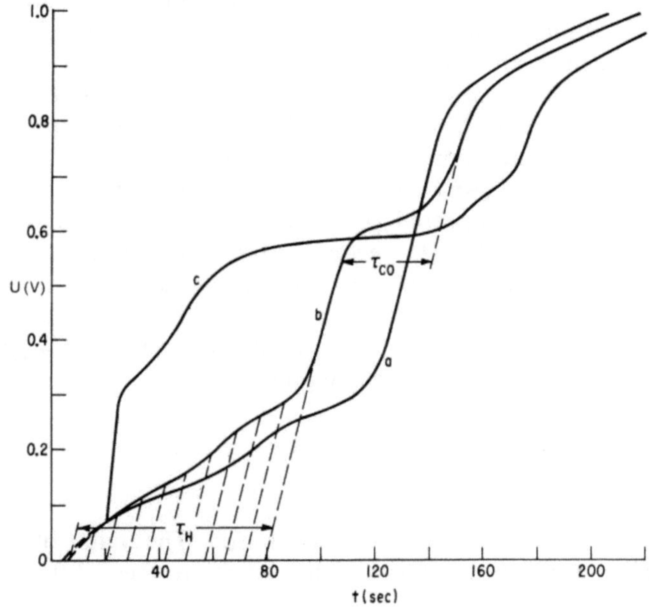

Fig. 15. Anodic charging curves on platinized platinum in 0.5 M H_2SO_4 at 25 °C and 50 mA. Curve a: no coverage of CO_{ad}, curve b: small coverage of CO_{ad} ($CO_{ad} = 0.25$), curve c: large coverage of CO_{ad} ($CO_{ad} = 1$)

criterion for the absence of chemisorbed carbonaceous species and for oxidizable species dissolved in the electrolyte. Curve a was always recorded at the beginning of every sequence of runs. The electrode was brought back to 0.1 V at the end of the measurement of curve a. Stirring with 80 % Ar + 20 % CO was carried out for a certain time. Adsorption of CO occurred with the electrode at open circuit. Subsequent stirring with pure He for 1500 sec served to remove the CO molecules from the electrolyte. Then a second charging curve was started from 0.07 V at 50 mA. The adsorbed carbon monoxide was removed by anodic oxidation to CO_2. The charging curves are not plotted above 1.0 V, although the current was maintained until 1.3 V were reached. Then the electrode stayed at open circuit for 2500 sec under He stirring until all the carbon dioxide had been removed from the electrolyte. This was verified by gas chromatography of the helium carrier gas. Another sequence could then be started.

The hydrogen branch was extrapolated to $U=0$ as indicated by dotted lines for curves a and b in Fig. 15. The graphical determination of the transition time τ_H and of the hydrogen coverage is illustrated for curve b. Curve b exhibits a short arrest in the double layer region. The arrest is due to the oxidation of CO_{ad} to CO_2. Since the charging curves a, b, and c are parallel to each other at $U>0.9$ V, it is concluded that the surface is free of CO_{ad} there. The transition time τ_{CO} for the anodic oxidation of CO_{ad} is determined as shown for curve b. An increase of $Q_{CO}=\tau_{CO}\cdot I$ is accompanied by a decrease of ${}_sQ'_H$. At longer times of stirring with 80% Ar + 20% CO, a second transition region for the oxidation of CO_{ad} appeared [67] between about 0.3 V and 0.55 V at 50 mA. Simultaneously ${}_sQ'_H$ becomes very small (see curve c). It was established [68] that the length of each of the two arrests of the oxidation of CO_{ad} did not increase with time of stirring beyond that of the corresponding arrest in curve c. Curve c is characterized by ${}_sQ'_H/{}_sQ_H=0.02$. The coverage with CO_{ad} can be computed:

$$\theta_{CO}=\frac{Q_{CO}}{{}_sQ_{CO}}. \tag{7}$$

It was ${}_sQ_{CO}=6.3$ coul during these studies. The value of ${}_sQ_{CO}$ was obtained as product of the sum of the transition times of the two CO_{ad} arrests and of the current from curve c.

The $\theta-\eta$ curves for hydrogen adsorption in the presence of CO_{ad} in Fig. 16 were constructed from the hydrogen branch of anodic charging curves. The $\theta-\eta$ curves practically coincide at $\theta \geq 0.2$ for $\theta_{CO} \leqq 0.25$. However, a decrease of the hydrogen coverage is already detectable in the region of the strongly bonded H atoms ($\theta<0.2$) at small coverage with CO_{ad}. The $\theta-\eta$ curves are shifted to the left with increasing θ_{CO} at $\theta_{CO}>0.25$. The free energy $\Delta G_{0,H}$ becomes less negative. A distinction between weakly and strongly bonded hydrogen may still be made at $\theta_{CO}=0.41$. The distinction is not feasible at $\theta_{CO} \geqq 0.52$. The experimental $\theta-\eta$ curve degenerates to a straight line (Temkin isotherm). A similar behavior was found by PODLOVCHENKO and STENIN [69] for hydrogen adsorption in the presence of chemisorbed carbonaceous species that

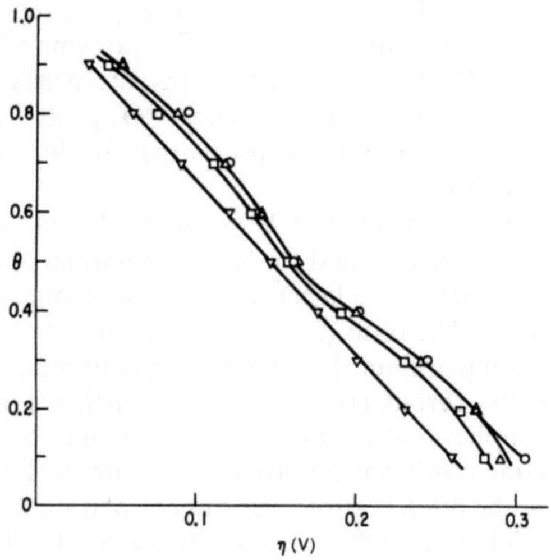

Fig. 16. Hydrogen coverage as a function of η on platinized Pt in 0.5 M H_2SO_4 at 25 °C in the presence of adsorbed carbon monoxide.
○ $Q_{CO}/_sQ_{CO}=0$, △ $Q_{CO}/_sQ_{CO}=0.25$, □ $Q_{CO}/_sQ_{CO}=0.41$, ▽ $Q_{CO}/_sQ_{CO}=0.52$

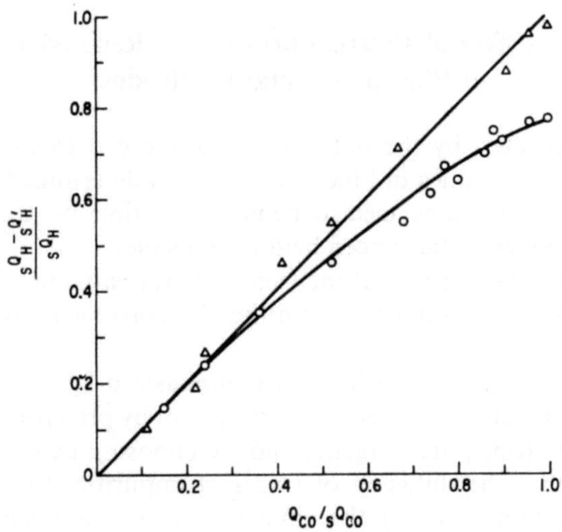

Fig. 17. Plot of θ′ as defined by Eq. VI, 6 versus coverage θ_{CO} with carbon monoxide. △ platinized platinum in 0.5 M H_2SO_4 at 25 °C, ○ smooth platinum in 1 M $HClO_4$

are produced anodically in 0.5 M H_2SO_4 at methanol bulk concentrations of about 0.5 M. While BRUMMER and CAHILL [70] reported a similar influence of CO_{ad} on hydrogen adsorption at smooth platinum in 1 M H_2SO_4 at 40 °C, they claimed the opposite behavior, namely an attractive interaction between H_{ad} and $CO_{2,r}$, at $U<0.18$ V and $\theta_{CO_{2,r}} \leqq 0.6$. Here $CO_{2,r}$ designates species that are formed [71] by the reaction of CO_2 with H_{ad}.

The quantity θ' is plotted in Fig. 17 versus θ_{CO} for platinized platinum from the data in reference 68 and for smooth platinum from the data in reference 63 (Fig. 5 there). The latter results were obtained by GILMAN in 1 M $HClO_4$ at 30 °C. The ratio ${}_sQ_{CO}/{}_sQ_H$ was 1.5 for platinized platinum and 1.12 for smooth platinum. The points for platinized platinum scatter around a straight line through the origin with an inclination of 45° against the θ axis. This implies that the quantity θ' is equal to θ_{CO} on platinized platinum. The same result was reported in reference 65 for smooth platinum. In contrast, the $\theta'-\theta_{CO}$ curves for smooth platinum in Fig. 17 deviates from the straight line for $\theta'=0.3$. It tends towards $\theta'=0.78$ for $\theta_{CO}=1$. Similar curves were found on smooth platinum in acid solutions for the dependence of θ' upon the coverage with carbonaceous species formed anodically in the presence of methanol [63] or formic acid [72]. The curvature of these curves does not disappear if θ' is not defined with respect to the total apparent surface, as here, but with respect [73, 74] to that portion which is covered with organic species at maximum adsorption.

7. Effect of Pretreatment on the Reactivity of Platinum Metal Electrodes

As demonstrated by the data in Fig. 6, there is frequently a large discrepancy in the values of kinetic parameters determined by different researchers for the same electrochemical reaction on a given metal electrode. Two factors have been held responsible:

a) Different reactivity resulting from the pretreatment.

b) Adsorption of impurities from the electrolyte leading to different results.

The influence of factor b has been emphasized again and again by BOCKRIS [75]. It can be reduced to a large extent by preparing the electrolytic solutions from pure reagents and by choosing experimental conditions such that the diffusion of traces of impurities to the electrode surface is negligible during the time of the measurement. Although factor b is important, the author thinks that most discrepancies in the values of kinetic parameters are due to factor a, especially for platinum metal electrodes. Examples of the influence of both factors are described subsequently. Although the discussion is restricted to platinum, the general conclusions apply to other platinum metals because of their similar behavior under the respective experimental conditions.

The transition of smooth platinum from a state of small reactivity to one of large reactivity was investigated [35] by the voltage sweep technique and by voltammetry with superimposed ac voltage [76]. A small ac signal is superimposed upon the sweep in the latter case and is used to determine the ohmic and capacitive component of the impedance of the interface platinum/electrolyte as a function of U at constant frequency ω. The $I-U$ curve, the R_S-U curve and the $1/\omega C_S-U$ curve of the electrode impedance in an analog series circuit can be recorded [77] simultaneously. The sweep extended from 0.2V to about 1.5V. A frequency of 10000 Hz was employed [76].

After washing with distilled water, a fresh electrode of low reactivity was used as supplied by the glass blower. The $i-U$ curve a in Fig. 18 corresponds to the 2nd cycle, curve b to the 110th cycle. The anodic sweep is represented by a solid line, the cathodic sweep by a dashed line.

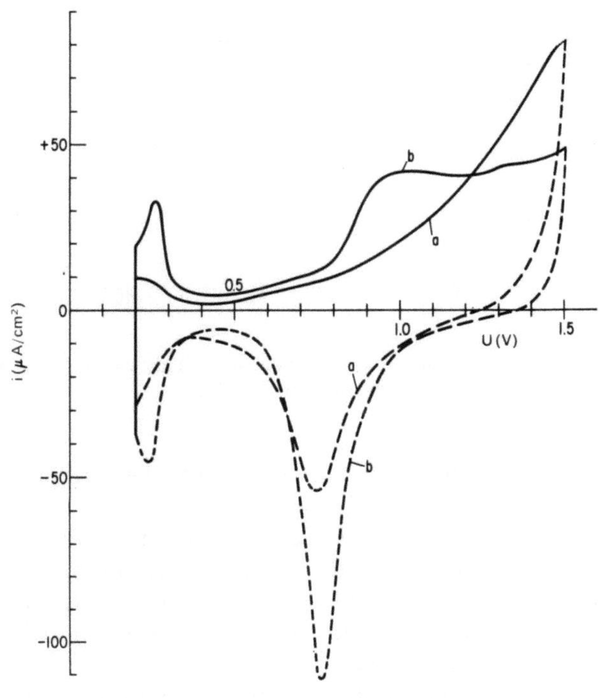

Fig. 18. Voltammetric i–U curves on smooth platinum in 0.5 M H_2SO_4 at 30 °C and 30 mV/sec during the second sweep (a) and during the 110th sweep (b)

Curve a is typical for platinum electrodes of low reactivity. Neither the hydrogen waves nor the anodic wave in the oxygen region have the shape observed on active electrodes [29] (compare curve b in Fig. 18). The current in the hydrogen region is lower at a given U value for curve a than for curve b. The formation of the oxygen layer between 0.7 V and 1.1 V appears inhibited in curve a.

The anodic charge $Q_{O,an}$ was determined by integration of the $i-U$ curves between 0.6 V and 1.5 V as a function of the number n of cycles. Similarly the charge $Q_{O,cath}$ was obtained for $U \geq 0.4$ V from the cathodic sweep. Both quantities are plotted as a function of n in Fig. 19. The transition from curve a to curve b in Fig. 18 is accompanied by a gradual decrease of $Q_{O,an}$ and an increase of $Q_{O,cath}$. The two curves in Fig. 19

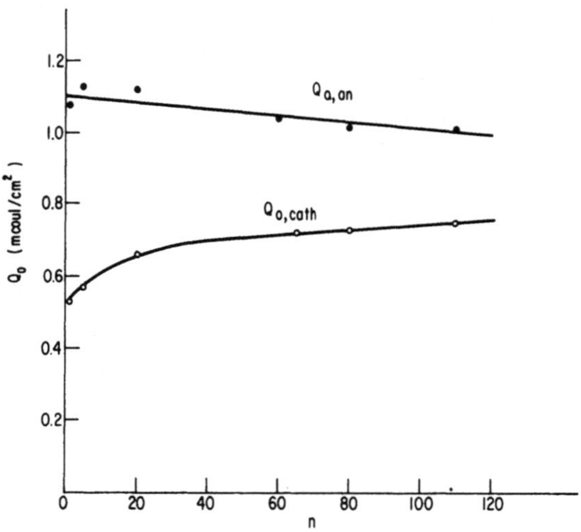

Fig. 19. Anodic charge $Q_{O,an}$ in the oxygen region and cathodic charge $Q_{O,cath}$ as a function of the number of cycles

tend to approach each other. This result may be interpreted in two different ways:

c) The charge $Q_{O,an}$ includes a contribution from the oxidation of organic impurities adhering to the surface. Since the reduction of the oxygen layer occurs mainly at $U < 0.9$ V, where the rate of oxidation of impurities is small, the quantity $Q_{O,cath}$ corresponds approximately to the charge of the oxygen layer.

d) Dissolution of oxygen atoms in the top layers of platinum leads [17] to an anodic charge which is larger than the cathodic charge. The assumption has to be postulated that the desorption of the oxygen atoms in the top layers is strongly inhibited in contrast to their rapid dissolution.

The results of the impedance measurements in Fig. 20 favor interpretation c for the first hundred cycles. The capacitance C_P in an analog parallel circuit of a resistance R_P and a capacitance C_P has low values during the initial cycles. The shape of the $C_P - U$ curve a is typical [61, 78] for the presence of adsorbed organic impurities. After the impurities have been largely oxidized, in about 100 cycles, the shape of the

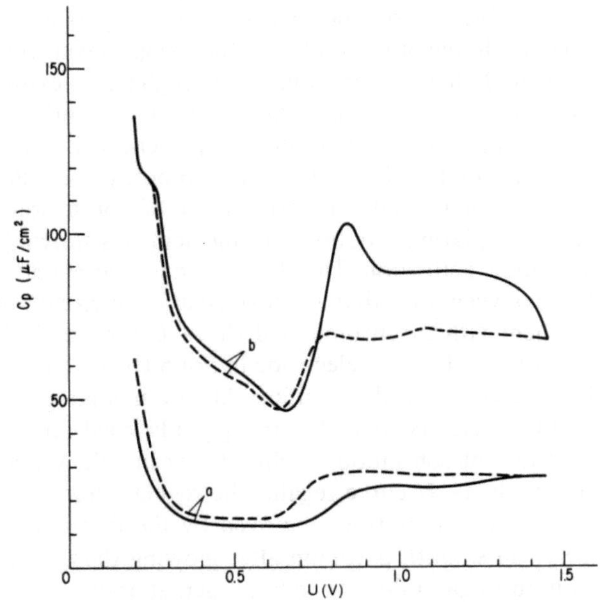

Fig. 20. Capacitance C_P at 10000 Hz as a function of U during the second cycle (a) and during the 110th cycle (b)

$C_P - U$ curve b is that of an electrode with high reactivity. The preceding results suggest that the anodic pretreatment introduced for platinum metals by KNORR and coworkers [79] removes impurities from the surface.

The result that $Q_{O,an} > Q_{O,cath}$ even after 110 cycles can no longer be explained by the oxidation of organic impurities. It is considered as evidence for a slow dissolution of oxygen atoms in the top layers of platinum or for the formation of patches of oxide during the continued cycling. Experimental evidence for the dissolution of oxygen atoms during the formation of oxygen layers on smooth platinum was already presented by OBRUCHEVA [80].

It was claimed recently by SCHULDINER and WARNER [17] that coulometric studies involving the oxidation or reduction of substances adsorbed on platinum can be seriously affected by the presence of dissolved oxygen atoms. However, the experimental evidence for this claim may be interpreted in a different way. The time for the reduction of an oxygen layer on platinum by molecular hydrogen will depend upon the distribution of the layer if H_2 removes [81] the oxygen layer on a partially covered surface, H_2 oxidation at one point being matched by the reduction of the layer at another. The reduction is considered [81] the result of the simultaneous occurrence of the H_2 oxidation on free sites and of the discharge of H^+ on top of the oxygen layer. If the surface is uniformly covered, there is a delay before enough free sites are available for the H_2 oxidation to occur at a fast rate. On the other hand, the reduction starts immediately if free sites are present. It is suggested that the oxygen layer is not as uniform after short anodic pulses (shorter than 1 msec) than after potentiostatic deposition [81, 82] of several minutes.

Pretreatment of platinum in hot chromic acid was shown [83] to be equivalent to anodic activation. The electrode has a large reactivity after reduction of the oxygen layer that was formed at open circuit. In contrast to the pretreatment with hot chromic acid, the $i-U$ curve of the first sweep after treatment of the platinum electrode in hot nitric acid does not have the shape characteristic for a clean surface. An intermediate product in the reduction of nitric acid seems to be strongly adsorbed on platinum. It takes several hours of continuous cycling between 0.05 V and 1.4 V at 30 mV/sec before the $i-U$ curve regains the regular shape.

A small amount of impurities removed by the first anodic cycle is present [83, 84] on smooth platinum after glowing the platinum in the blue flame of natural gas. Glowing in hydrogen at about 900 °C reduced [84] the value of ${}_{S}Q_{H}$ and led to the appearance of an additional arrest in the double-layer region. A spectacular poisoning of the platinum surface resulted [83] from glowing in a small yellow flame of natural gas. The poisoning did not disappear after 3 hr of cycling between 0.05 V and 1.4 V and withstood treatment in hot chromic acid. It is suggested that residues of hydrocarbons or carbon itself were burned into the surface since mechanical abrasion was required for the removal of the poisoning.

Curves of the type a in Fig. 18 were also found [78] during the initial cycles on platinum electrodes that had been pretreated anodically and left at open circuit in electrolytic solutions for days. However, it took only 10 cycles to restore the regular shape of the $i-U$ curves. Adsorption of impurities from the electrolyte is the cause of the loss of reactivity in the latter case. The comparison of C_P-U curves confirms this conclusion. The time in which the loss of reactivity occurs depends upon the amount of impurities in the electrolyte, the conditions of stirring, and the potential. It is conceivable that the ratio $Q_{O,an}/Q_{O,cath}=2$ reported in references 85 and 86 is largely due to factor c.

If a smooth platinum wire electrode was held for 1 min at 2.0 V and cycled subsequently between 0.05 V and 0.5 V at 0.45 V/sec in 2.3 M H_2SO_4, the charge ${}_{S}Q_H$ decreased [33] within 20 min to about half the value at 0.5 min at temperatures between 0° and 70° C. The $\theta-\eta$ curve determined from the $i-U$ curve after cycling for 1 min practically coincided with the curve after 20 min of cycling. The $W_d-\theta$ curves demonstrated that the value of W_d at a given θ is lower after 20 min than after 1 min. The effect is greatest for the sites of the strongly bonded hydrogen. While the decrease of ${}_{S}Q_H$ with time was mainly attributed [33] to a decrease of the surface roughness by recrystallization, it now appears to the author that partial adsorption of impurities cannot be ruled out. Probably both processes occurred simultaneously.

Area losses of carbon-supported platinum were determined [87] by electrochemical measurements and by X-ray diffraction. These losses cannot be due to adsorption of impurities since they were detected in roughly the same measure by both techniques. Recrystallization of the supported platinum was suggested [87] as the cause of the reduction of the surface area with time.

8. Hydrogen Adsorption on Binary Platinum Metal Alloys

Examples which illustrate the essential features of binary platinum metal alloys are discussed for three types of alloys in this section:

a) Alloys between two platinum metals.

b) Alloys between a platinum metal and another metal which is relatively inert in the given electrolyte.

c) Alloys between a platinum metal and another metal which has a tendency to dissolve in the electrolyte.

Ruthenium-platinum [88—91] and ruthenium-palladium [88] alloys are of particular interest among the alloys of group a since the oxidation of certain fuels occurs [88, 89] with a lower net overvoltage at the same current on alloy electrodes of a porous structure at ruthenium contents below 20% than on platinized platinum. The regions characteristic for the weakly and strongly bonded hydrogen are not recognizable in anodic charging curves [88] on composite electrodeposits with Ru:Pt=1:9 in 0.5 M H_2SO_4. The fine structure is much weaker in the charging curve on the electrodeposit than on platinum in 1 M KOH. The shape of the isotherm of hydrogen adsorption in sulfuric acid solution appears to change rather rapidly with increasing ruthenium content from that of platinum to that of ruthenium (see Fig. 9). Although the net composition of the electrodeposits was determined [92] by a radio-tracer technique, it is not certain if the nominal bulk composition of the Ru-Pt alloys was the same as that of the top layer at the surface. Different composition in the top layers may be the reason why the maximum reactivity of Ru-Pt alloys is not observed at the same bulk composition by different researchers.

Platinum-gold alloys represent an example of group b. The electrochemical properties of these alloys have been studied [93—98] extensively. They can exist [99] as homogeneous alloys or as a mixture of a platinum-rich phase α_1 and a gold-rich phase α_2. The cold-drawn wires used in these investigations [94—98] were of the latter type. It was demonstrated [96—98] that the electrocatalytic properties of the platinum-rich phase are similar to those of platinum. Hydrogen adsorption [96] and methanol oxidation [97] occur only on this phase. The

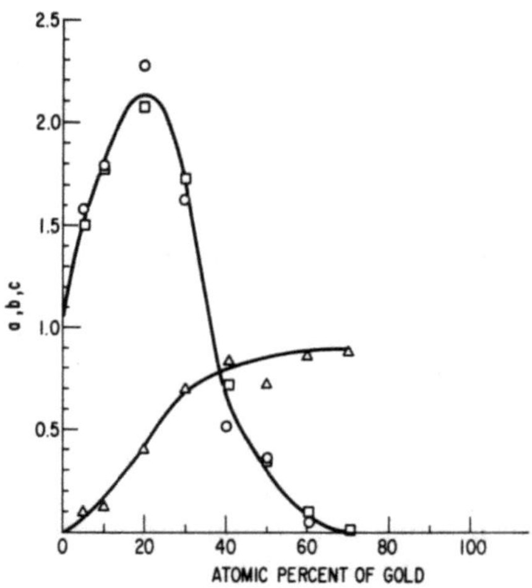

Fig. 21. Parameters a, b, c as a function of the bulk composition of Pt -Au alloys

isotherms of hydrogen adsorption coincide [96] practically for the alloys with gold contents below 30%. A distinction between weakly and strongly bonded hydrogen can be made up to this gold content.

The surface composition may be characterized by three parameters a, b, and c. Parameters a and b were determined [94] by approximating the $I-U$ curve of the alloy during the cathodic sweep by a linear combination of the corresponding curves on smooth platinum and gold between 1.5 V and 0.4 V:

(8) $$I(U) = a\,I_{Pt}(U) + b\,I_{Au}(U)\,.$$

Parameters a and b are a relative measure of the amount of each of the two phases on the surface. Parameter c is defined as:

(9) $$c = \frac{{}_{S}Q_{H}}{{}_{S}Q_{H,Pt}}\,.$$

Here ${}_{S}Q_{H}$ is the total amount of hydrogen on the alloy, ${}_{S}Q_{H,Pt}$ is that on platinum. The parameters are plotted as a function of gold content in Fig. 21. Parameter b increases with the gold content of the bulk and tends towards a limiting value of about 0.9 above 50% Au. The replacement of the phase α_1 by the phase α_2 is practically complete on the surface at a gold content of 60%. Curves a and c virtually coincide, as would be expected if hydrogen adsorption takes place on the platinum-rich phase only. Curve a displays a maximum at about 20% Au. It was suggested [94] that the decrease of the phase α_1 is overcompensated between 3 and 20% Au by an increase in roughness. Thus a parabolic shape results.

Platinum-chromium alloys belong to group c. Platinum and chromium form homogeneous alloys below 75 at.% Cr. The compound $PtCr_3$ exists at chromium contents above 75%. Three alloys (10, 30, and 52% Cr) and the compound $PtCr_3$ were studied [100] in 0.5 M H_2SO_4 at 30 °C. The samples had been annealed in argon for 1 hr. Electron microprobe scans showed the samples to be fairly homogeneous across the surface. Quantitative values of Cr and Pt were determined for the alloy with 52% Cr from the measurement of both CrK_α and PtL_α X-ray intensities and the use of theoretical correction procedures. The Cr content was found to be 42% which is less than the bulk content.

Voltammetric $i-U$ curves taken [100] at 3 mV/sec between 0 and 1.5 V are representative for the electrochemical conversion of chromium oxide layers at chromium contents above 10%. In contrast to the Pt-Au system, it is not possible to characterize the surface composition by $i-U$ curves. However, the $1/\omega C_S - U$ curves in Fig. 22 display the two characteristic hydrogen peaks [28] between 0 and 0.4 V. Here $1/\omega C_S$ is the capacitive component of the electrode impedance in an analog series circuit, measured by voltammetry with superimposed ac current at 1000 Hz. A calibration of the pseudocapacitance [20] of H atoms as a function of the number of Pt atoms on the surface is not feasible because of different roughness factors for the alloys and because of the slow dissolution of chromium oxide from the surface layers. The dissolution leads to a gradual increase of the capacitance with time. The number of Pt atoms participating in the Volmer reaction becomes larger. It is considered unlikely for this reason that the top layers consist of homogeneous crystallites as does the bulk. The linear decrease [100] of the lattice constant with chromium content between 10 and 52% represents evidence for the homogeneity of the alloys in the bulk only.

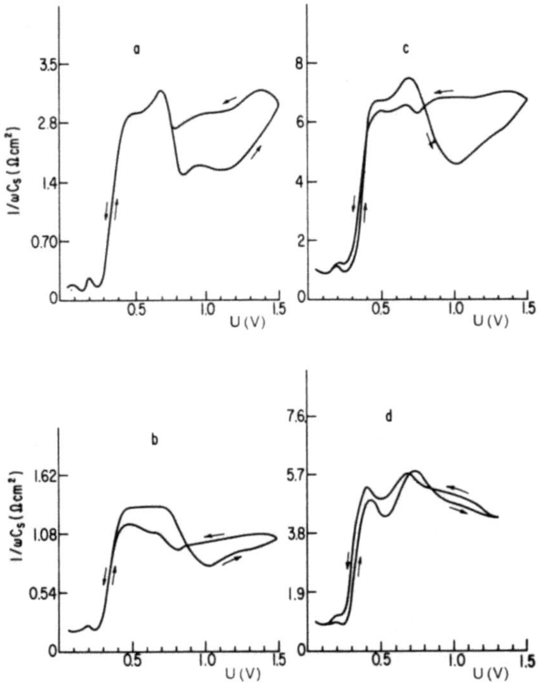

Fig. 22. Capacitive component of Pt—Cr alloys as a function of potential at 1000 Hz. a: 10 at.%, b: 30 at.%, c: 52 at.%, d: 75 at.%

The results obtained for solid alloys (Ru—Pt [101], Pt—Au [94], and Pt—Cr [100]) emphasize the necessity of studying the composition of the top layers on the surface. Different methods of production or heat treatment are likely to lead to a different surface composition even in the case where the bulk composition is the same. The effect of the pretreatment of Pt-Au and Pd-Au alloys on the rate of O_2 reduction is another example reported [102] recently. The bulk composition cannot be taken as representative for the electrocatalytic properties of alloys. Such a correlation requires a knowledge of the surface composition which should be determined simultaneously with the kinetic parameters.

Different roughness factors should also be taken into account when comparing the reactivity of alloys. A change of the roughness factor with the bulk composition is demonstrated for the Pt-Au system by curve a in Fig. 21. Similarly the surface area expressed in m^2/g was found [89] to be larger for Ru-Pt catalsyts (Ru:Pt = 5:95) than for platinum.

References

1. Bowden, F. P.: Proc. Roy. Soc. London **125A**, 446 (1929).
2. Butler, J. A. V., and G. Armstrong: Proc. Roy. Soc. London **137A**, 604 (1932).
3. Slygin, A., and A. Frumkin: Acta Physicochim. U.R.S.S. **3**, 791 (1935).
4. Ershler, B.: Acta Physicochim. U.R.S.S. **7**, 327 (1937).
5. Pearson, J. D., and J. A. V. Butler: Trans. Faraday Soc. **34**, 1163 (1938).
6. Hickling, A.: Trans. Faraday Soc. **41**, 333 (1945).
7. Obrucheva, A. D., and I. L. Burinstein: Doklady Akad. Nauk S.S.S.R. **63**, 403 (1948).
8. Salkind, Z., and B. Ershler: Zhur. Fiz. Khim. **25**, 565 (1951).
9. Breiter, M., C. A. Knorr, and W. Völkl: Z. Elektrochem. **59**, 681 (1955).
10. Giner, J.: Z. Elektrochem. **63**, 386 (1959).
11. Tyurin, Yu. M.: Doklady Akad. Nauk S.S.S.R. **126**, 827 (1959).
12. Presbrey, C. H., Jr., and S. Schuldiner: J. Electrochem. Soc. **108**, 985 (1961).
13. Vert, Zh. L., I. A. Mosevich, and I. Tverdovsky: Doklady Akad. Nauk S.S.S.R. **140**, 149 (1961).
14. Stoyanovskaya, T. N., G. P. Khomchenko, and G. D. Vovchenko: Vestnik Moscow Univ., Ser. Khim. No. **5**, 30 (1962) and No. **2**, 20 (1963).
15. Khomchenko, G. P., N. G. Ul'ko, and G. D. Vovchenko: Elektrokhim. **1**, 659 (1965).
16. Marvet, R. V., and O. A. Petry: Elektrokhim. **1**, 1225 (1965).
17. Schuldiner, S., and T. B. Warner: J. Electrochem. Soc. **112**, 212 (1965).
18. Marvet, R. V., and O. A. Petry: Elektrokhim. **4**, 591 (1967).
19. — — Elektrokhim. **3**, 1445 (1967).
20. Dolin, P., and B. Ershler: Acta Physicochim. U.R.S.S. **13**, 747 (1940).
21. Eucken, A., and B. Weblus: Z. Elektrochem. **55**, 114 (1951).
22. Wicke, A., and B. Weblus: Z. Elektrochem. **56**, 169 (1952).
23. Breiter, M., H. Kammermaier, and C. A. Knorr: Z. Elektrochem. **60**, 37 (1956).
24. — — — Z. Elektrochem. **60**, 119 (1956).
25. Birinzeva, T. P., and B. N. Kabanov: Zhur. Fiz. Khim. **33**, 844 (1959).
26. Breiter, M. W.: J. Phys. Chem. **68**, 2249 (1964).
27. — J. Phys. Chem. **68**, 2254 (1964).
28. Will, F. G., and C. A. Knorr: Z. Elektrochem. **64**, 258 (1960).
29. — — Z. Elektrochem. **64**, 270 (1960).
30. Franklin, T. C., and S. L. Cooke Jr.: J. Elektrochem. Soc. **107**, 556 (1960).
31. Kolotyrkin, Ya., and A. N. Chemodanov: Doklady Akad. Nauk S.S.S.R. **134**, 128 (1960).
32. Böld, M., and M. Breiter: Z. Elektrochem. **64**, 897 (1960).
33. Breiter, M., and B. Kennel: Z. Elektrochem. **64**, 1180 (1960).
34. Will, F. G.: J. Electrochem. Soc. **112**, 451 (1965).
35. Breiter, M. W.: Electrochim. Acta **11**, 905 (1966).
36. — Electrochim. Acta **8**, 925 (1963).

37. BEN DANIEL, D. J., and F. G. WILL: J. Electrochem. Soc. **115**, 909 (1967).
38. STETSENKO, A., and I. TVERDOVSKY: Zhur. Fiz. Khim. **26**, 647 (1952).
39. TEMKIN, M.: Zhur. Fiz. Khim. **15**, 296 (1941).
40. LANGMUIR, I.: J. Am. Chem. Soc. **54**, 2798 (1932).
41. BEEK, O.: Discussions Faraday Soc. **8**, 118 (1950).
42. BREITER, M.: Electrochim. Acta **7**, 25 (1962).
43. MIGNOLET, J. C. P.: J. Chim. Phys. **54**, 19 (1957).
44. SUHRMANN, R., G. WEDLER, and H. GENTSCH: Z. physik. Chem. N. F. **17**, 350 (1958).
45. KAVTARADSE, N.: Zhur. Fiz. Khim. **32**, 909 and 1055 (1958).
46. SLYGIN, A., A. FRUMKIN, and W. MEDWEDOWSKY: Acta Physicochim. U.R.S.S. **4**, 911 (1936).
47. BALASHOVA, N. A.: Doklady Akad. Nauk S.S.S.R. **103**, 639 (1955).
48. POPAT, P. V., and N. HACKERMAN: J. Phys. Chem. **62**, 1198 (1958).
49. OBRUCHEVA, A. D.: Zhur. Fiz. Khim. **32**, 2155 (1958).
50. ASTON, J. G.: J. Phys. Chem. **67**, 2042 (1963).
51. BREITER, M. W.: Trans. Faraday Soc. **60**, 1445 (1964).
52. — J. Electroanalyt. Chem. **8**, 449 (1964).
53. FRUMKIN, A. N.: Z. Physik **35**, 792 (1926).
54. BREITER, M. W.: Ann. N.Y. Acad. Sci. **101**, 709 (1963).
55. BOGDANOVSKI, G. A., G. P. KHOMCHENKO, and G. D. VOVCHENKO: Zhur. Fiz. Khim. **34**, 1408 (1965).
56. NIEDRACH, L. W.: J. Electrochem. Soc. **111**, 1309 (1964).
57. CONNOLLY, J. F., R. J. FLANNERY, and G. ARONOWITZ: J. Electrochem. Soc. **113**, 577 (1966).
58. JONCICH, M. J., and N. HACKERMAN: J. Electrochem. Soc. **111**, 1286 (1964).
59. OIKAWA, M., and T. MUKAIBO: J. Electrochem. Soc. Japan **20**, 568 (1952).
60. FRANKLIN, T. C., and R. SOTHERN: J. Phys. Chem. **58**, 951 (1954).
61. BREITER, M. W.: J. Electrochem. Soc. **109**, 42 (1962).
62. —, and S. GILMAN: J. Electrochem. Soc. **109**, 622 (1962).
63. GILMAN, S.: J. Phys. Chem. **67**, 78 (1963).
64. WARNER, T. B., and S. SCHULDINER: J. Electrochem. Soc. **111**, 992 (1964).
65. BRUMMER, S. B., and J. I. FORD: J. Phys. Chem. **69**, 1355 (1965).
66. PADYUKOVA, G. L., A. B. FASMAN, and D. V. SOKOLSKY: Elektrokhim. **2**, 885 (1966).
67. BREITER, M. W.: Electrochim. Acta **12**, 1213 (1967).
68. — J. Phys. Chem. **72**, 1305 (1968).
69. PODLOVCHENKO, B. I., and B. F. STENIN: Elektrokhim. **3**, 649 (1967).
70. BRUMMER, S. B., and K. CAHILL: Discussions Faraday Soc. **45**, 67 (1968).
71. GINER, J.: Electrochim. Acta **9**, 63 (1964).
72. BREITER, M. W.: Electrochim. Acta **8**, 447 (1963).
73. HASOVA, O. A., YU. B. VASILIEV, and V. S. BAGOTZKY: Elektrokhim. **1**, 84 (1965).
74. BAGOTZKY, V. S., and YU. B. VASILIEV: Elektrokhim. **11**, 1439 (1966).
75. BOCKRIS, J. O'M.: Modern Aspects of Electrochemistry, No. **1**, Chap. 4. London: Butterworths Scientific Publication 1954.
76. See DELAHAY, P.: New Instrumental Methods in Electrochemistry, Chap. 7, V. New York: Interscience Publishers 1954.
77. BREITER, M. W.: J. Electroanalyt. Chem. **7**, 38 (1964).
78. — Electrochim. Acta **7**, 533 (1962).
79. KANDLER, L., C. A. KNORR, and C. SCHWITZER: Z. physik. Chem. **A 180**, 281 (1937).
80. OBRUCHEVA, A. D.: Zhur. Fiz. Khim. **26**, 1448 (1952).
81. BREITER, M. W.: J. Electrochem. Soc. **109**, 425 (1962).
82. BECKER, M., and M. BREITER: Z. Elektrochem. **60**, 1080 (1956).
83. BREITER, M. W.: J. Electroanalyt. Chem. **8**, 230 (1964).

84. Ershler, B., G. Deborin, and A. Frumkin: Acta Physicochim. U.R.S.S. **8**, 565 (1938).
85. Vetter, K., and D. Berndt: Z. Elektrochem. **62**, 378 (1958).
86. Feldberg, S. W., C. G. Enke, and C. E. Bricker: J. Electrochem. Soc. **110**, 826 (1963).
87. Connolly, J. F., R. J. Flannery, and B. L. Meyers: J. Electrochem. Soc. **114**, 241 (1967).
88. Petry, O. A., B. I. Podlovchenko, A. N. Frumkin, and Hira Lal: J. Electroanalyt. Chem. **10**, 253 (1965).
89. Adlhart, O. J., H. Shields, and S. Pudick: Engelhard Ind. Tech. Bull. **6**, 37 (1965).
90. Brooman, E. W., and T. P. Hoar: Platinum Metals Rev. **10**, 135 (1966).
91. Schmidt, H., and W. Vielstich: Z. Analyt. Chem. **224**, 84 (1967).
92. Petry, O. A., and V. E. Kasarinov: Elektrokhim. **1**, 1389 (1965).
93. Lapteva, K. A., T. I. Borisova, and M. G. Slinko: Zhur. Fiz. Khim. **30**, 61 (1956).
94. Breiter, M. W.: J. Phys. Chem. **69**, 901 (1965).
95. — Electrochim. Acta **10**, 543 (1965).
96. — Trans. Faraday Soc. **61**, 749 (1965).
97. — J. Phys. Chem. **69**, 3377 (1965).
98. — Ber. Bunsenges. Physik. Chem. **69**, 612 (1965).
99. Darling, A. S.: Platinum Metals Rev. **6**, 60 and 106 (1962).
100. Breiter, M. W.: J. Electroanalyt. Chem. **10**, 191 (1965).
101. Entina, V. S., and O. A. Petry: Elektrokhim. **4**, 119 (1968).
102. Damjanovic, A., and V. Brusic: J. Electroanalyt. Chem. **15**, 29 (1967).

VII. Anodic Oxidation of Molecular Hydrogen at Low Temperatures

The net reactions for the H_2 oxidation in acid and alkaline electrolytes are:

(1) $$\tfrac{1}{2}H_2 = H^+ + e^- ,$$

(2) $$\tfrac{1}{2}H_2 + OH^- = H_2O + e^- .$$

Noble metals and some of their alloys are materials that offer sufficient resistance to corrosion in acid media and are good electrocatalysts at the same time. The mechanism of the anodic H_2 oxidation is discussed for noble metals and their alloys in section 1 of this chapter. The conclusions derived for acid solutions can be easily transferred to alkaline electrolytes since mass transport processes are rate-determining under regular conditions. Less expensive nickel electrodes perform satisfactorily in alkaline solutions. Section 2 deals with studies of the H_2 oxidation on different types of nickel electrodes. Results of studies of the H_2 oxidation at platinum in contact with an ion exchange membrane are described in section 3.

1. Mechanism of the H_2 Oxidation on Noble Metals and Noble Metal Alloys

It was concluded first by KNORR and coworkers [1] that diffusion of molecular hydrogen is the rate-determining step of hydrogen evolution on smooth platinum and palladium electrodes pretreated anodically. The same conclusion holds for the anodic processes given by Eq. 1 and 2. The interpretation based on steady-state measurements [1] is valid at cathodic current densities $-i \leq 1\,\text{mA/cm}^2$ in strong acid electrolytes ($i/i_{d,H^+} \ll 1$). The correctness of this interpretation has been substantiated by a number of independent techniques which allow the concentration c_{H_2} of H_2 molecules adjacent to the surface to be determined:

a) Measurements [2] with the rotating platinum disc electrode.
b) Impedance measurements [3, 4] on Pt, Ir, and Rh.
c) Voltammetry [5] with continuously changing potential on Pt.
d) Use [6] of the rotating ring-disc electrode [7].

The increase of c_{H_2} with overvoltage η was found to agree well with that computed according to Nernst's equation from the measured

η values at $-i \leq 1$ mA/cm^2. The influence of hydrogen removal from the surface by gas bubbles has to be taken into account [3, 4] at $-i > 1$ mA/cm^2. In spite of the corraborating evidence that mass transport processes are rate-determining on active platinum metal electrodes under conditions of regular stirring, a kinetic control has been postulated by SCHULDINER [8] and BOCKRIS [9] for reaction 1.

The good agreement between experimental curves and $i-U$ curves computed according to Eq. IV, 38 with the experimental values of i_{d,H_2} is demonstrated in Fig. 23. The theoretical curves are represented

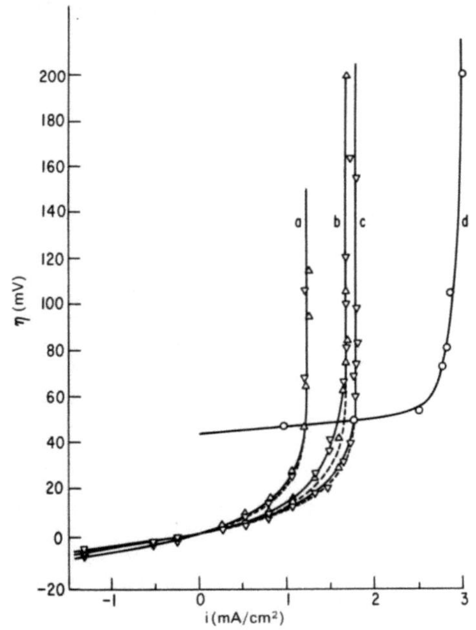

Fig. 23. Steady-state current-potential curves measured potentiostatically on smooth platinum with increasing intensity (a, b, c) of the H_2 stirring and on smooth palladium (d) in sulfuric acid solutions

by dashed lines. The $i-U$ curves on smooth platinum were taken [10] in 1 M H_2SO_4 at different stirring intensities. Triangles designate the points measured from cathodic to anodic overvoltage. Inverted triangles are used in the opposite direction. An increase in the intensity of the H_2 stirring causes a decrease of the thickness of the diffusion layer and leads to an increase of i_{d,H_2}. The slope of the $i-\eta$ curve through $\eta=0$ agreed well [10] with the theoretical value:

$$\left(\frac{\Delta\eta}{\Delta i}\right)_{i=0} = \frac{RT}{2F}\,\frac{1}{i_{d,H_2}}. \tag{3}$$

Similar results were obtained on smooth iridium and rhodium electrodes [10]. The $i-U$ curve on smooth palladium [11] in 0.5 M H_2SO_4 is characterized by circles in Fig. 23. The Pd wire was kept at 0.8 V until the current achieved the steady-state value of i_{d,H_2}. The initially larger current at 0.8 V is due to the anodic removal of hydrogen bonded in the β phase [12]. The $i-U$ curve which was measured step-wise in the cathodic direction displays [11] the limiting current density i_{d,H_2} between

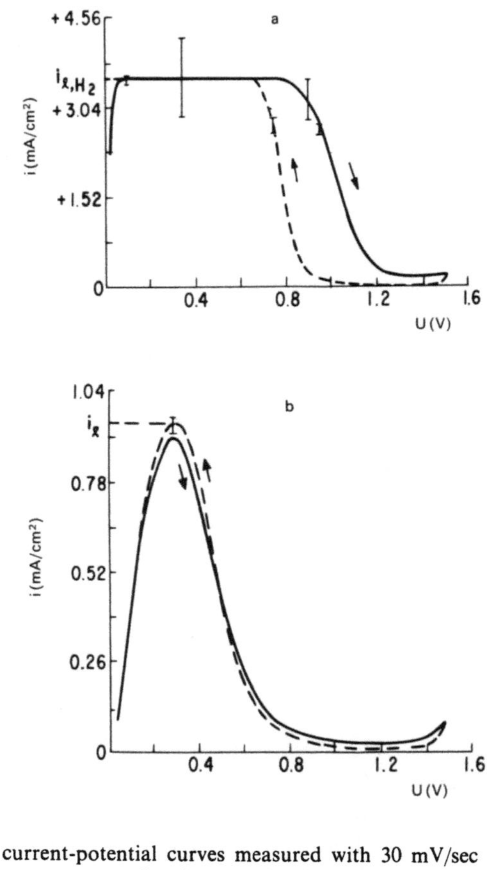

Fig. 24. Periodic current-potential curves measured with 30 mV/sec in 0.5 M H_2SO_4 under vigorous H_2 stirring.
a) smooth platinum-gold alloy (30 at. % Au), b) smooth gold

0.8 V and 0.2 V. In contrast to the $i-U$ curves on platinum, the current decreases with decreasing potential already below 0.1 V and becomes zero at about 0.042 V. This behavior reflects the influence of the cathodic formation of the β phase.

Periodic $i-U$ curves were measured [14] at 30 mV/sec in 0.5 M H_2SO_4 stirred vigorously with H_2 on the heterogeneous platinum-gold alloys (see section 8 of chapter VI). The potential sweep extended from about 0.03 V to 1.5 V. The $i-U$ curve obtained on the 30 at. % Au alloy and shown as curve a in Fig. 24 resembles the curve on platinum [15].

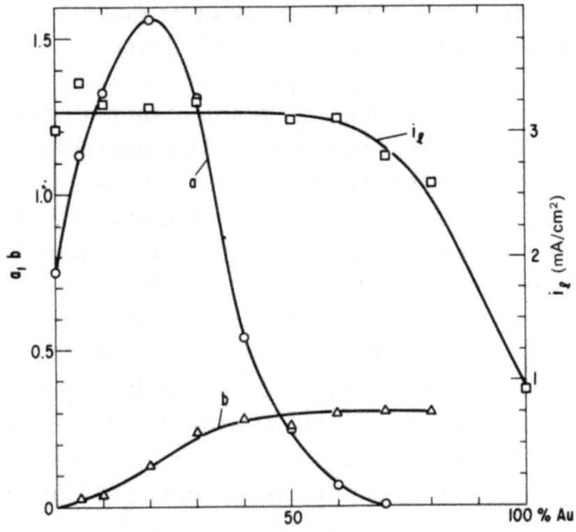

Fig. 25. Maximum rate of the H_2 oxidation as a function of the gold content of heterogeneous platinum-gold alloys. The parameter a is a relative measure of the platinum-rich phase on the surface, the parameter b for the gold-rich phase

The magnitude of the current fluctuations due to the gas stirring is indicated. The limiting current density i_{d,H_2} is reached between about 0.05 V and 0.7 V during the anodic and cathodic sweep. The decrease of i with U above 0.8 V during the anodic sweep results from the inhibiting effect of the oxygen layer formed at these potentials. This will be discussed in more detail in section 6 of chapter VIII. The shape of the $i-U$ curves on the alloys approaches that of the curve on smooth gold (curve b in Fig. 24) with increasing gold content above 50 at. %. The width of the potential range in which a limiting current is found becomess narrower until only a current peak is observed as on gold. The shape of the $i-U$ curve on smooth platinum measured [16] at large stirring rates in electrolytes with specifically adsorbable anions is similar to that of curve b. The maximum value i_l of the current density on gold is about four times smaller than the value on the alloys with gold contents below 60 at. %.

The maximum current density i_l obtained at potentials of the hydrogen region is plotted as a function of the gold content of the bulk in Fig. 25. Simultaneously the surface composition is characterized by the parameters a and b which are relative measures for the amount of the platinum-rich and gold-rich phase respectively on the surface of the alloys. The maximum rate of reaction 1 is identical with i_{d,H_2} up to 50 at. % Au. Such an independence of the composition can only result in the case of mass transport control. If i_l values were determined by kinetic control, for instance by the Volmer reaction, an influence of the composition should exist. The current density i_l decreases slowly with gold content above 60 at. % to the value on gold. Predominant mass transport control is present up to 80 at. % Au. The decrease of i_l is paralleled by the decrease of the platinum-rich phase. There is no correlation with the gold-rich phase since the amount of this phase is constant above 60 at. % Au on the surface. It was concluded [14] that hydrogen oxidation occurs predominantly on the platinum-rich phase of the alloys. Small amounts which can only be detected by impedance measurements [14] cause a considerable increase of i_l above the value of smooth gold. Combined mass transport and kinetic control exists on high purity gold as demonstrated by a small increase of i_l with the stirring intensity [14]. The purity of gold has a large influence on the rate-determining step of the H_2 oxidation. Previous measurements [10] on gold of lower purity did not reveal any influence of mass transport processes.

The large effect of small amounts of the platinum-rich phase on the rate of reaction 1 at gold contents above 60 at. % indicates that the H_2 oxidation does not occur uniformly on the whole surface. A similar conclusion on the involvement of active sites was already drawn for smooth platinum by Frumkin and Aikasyan [17]. The rate [3, 4, 16] of the Volmer reaction is much higher (at least by a factor 10) on activated smooth platinum than the limiting rate obtained [16—18] for reaction 1 on the rotating disc at high angular velocities at potentials of the hydrogen region. This may imply that different parts of the surface participate in the reaction. The small value [19] of the limiting current density of H_2 oxidation on platinum electrodes of small reactivity is probably due to the small number of sites that are not blocked by other species so that the H_2 oxidation can occur there. Adsorbed impurities are considered the cause for the relatively large anodic overvoltage of reaction 1 reported from early measurements [20] on platinum metals.

The dependence of i_{d,H_2} upon the stirring intensity was studied [10] at high stirring rates on smooth wires of Pt, Pd, Ir, and Rh and on platinized platinum. The electrolyte passed the electrodes which were mounted in the axis of a narrow glass tube at different flow rates. The electrolyte was saturated with H_2 ($p_{H_2}=1$ atm). The current density was found to increase at 0.1 V with the flow rate v on all the electrodes in 1 M H_2SO_4 at room temperature and on Pt, Pd, and Rh in 1 M NaOH.

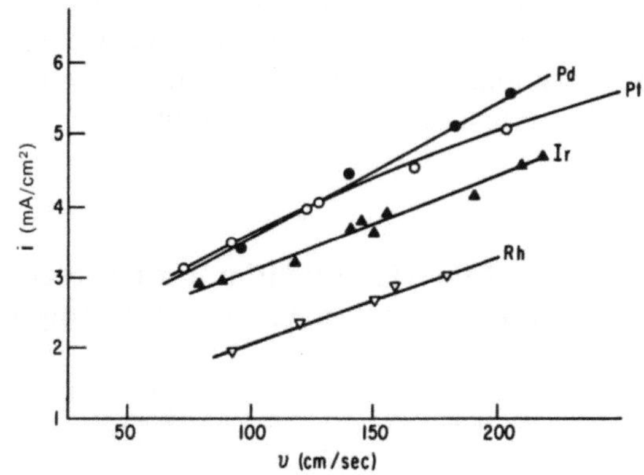

Fig. 26. Dependence of the rate of H_2 oxidation on different smooth electrodes at 0.1 V upon the flow rate of the streaming electrolyte (1 M H_2SO_4)

The results are shown in Fig. 26. A tendency of the current to reach a limiting value with increasing flow rate was not detected. Extensive measurements [21] of the anodic current at 0.1 V on the rotating platinum disc electrode as a function of the hydrogen partial pressure between 1 and 60 atm and angular velocities up to 20000 rpm in 0.5 M H_2SO_4, 1 M $HClO_4$, and 1 M HBr did not reveal that a limiting rate of reaction 1 is approached with increasing angular velocity of the disc. The current depended linearly upon ${}_{b}p_{H_2}$. In contrast a limiting rate, which is evidence for the achievement of kinetic control of reaction 1, was found [16—18] at constant potential in 0.5 M H_2SO_4, 1 M HCl, and 1 M HBr at sufficiently large revolution of the Pt disc electrode. Reaction 1 was a zero-order reaction with respect to ${}_{b}p_{H_2}$ in this case. The difference between the results reported in references 16 and 21 was explained [18] recently. Hydrogen oxidation can proceed either by a first-order or by a zero-order reaction. The pretreatment of the electrode surface is the decisive factor.

If the concentration of hydrogen ions is small, Eq. IV, 38 has to be applied in the unabridged form. Computed and experimental $i-U$ curves agree well on platinized platinum [22] and smooth electrodes of Pt, Ir, and Rh [23] in acid electrolytes. Mass transport of OH^- ions has to be taken into account [22—24] besides that of H_2 molecules at low concentrations of OH^-. Two limiting anodic currents are observed [22, 24, 25] in alkaline electrolytes if $i_{d,OH^-} < i_{d,H_2}$. Transition [22—25] from hydrogen evolution in acid media to that in alkaline media occurs at $-i > -i_{d,H^+}$ on the other hand. Theoretical treatments of the two possible transitions were given [22—24]. They are based on the assumption that reactions 1 and 2 contribute simultaneously while the supply of H^+ or OH^- from the dissociation of water molecules, considered as a homogeneous reaction, is negligible. Satisfactory agreement between theoretical and experimental $i-U$ curves is found [22—25] on platinum metals.

2. Mechanism of the H_2 Oxidation for Different Types of Nickel Electrodes in Alkaline Electrolytes

Hydrogen overvoltage on bright nickel electrodes was attributed [26] to a hindrance of the Volmer reaction

$$H_2O + e^- = H_{ad} + OH^- \tag{4}$$

except possibly in the most dilute alkaline solutions. It was suggested that reaction 4 is followed by the less hindered Heyrovsky reaction:

$$H_2O + H_{ad} + e^- = H_2 + OH^- . \tag{5}$$

The anodic $i-U$ curves showed a large increase of potential with current. These curves were interpreted as resulting from the simultaneous oxidation of molecular hydrogen and of the nickel electrode. Later it was suggested [27] that reaction 4 is rate-determining on the greater part of the surface, but that hydrogen evolution is limited by the Tafel reaction on a small part.

Recently the importance of the state of the nickel surface was emphasized [28]. The Volmer reaction was considered rate-determining only on surfaces in an "initial" state with high affinity for hydrogen adsorption. Anodic-cathodic cycling produced [28] a "steady" state for which the exchange current densities of the Volmer and Tafel reaction were claimed to be about equal. The limiting anodic current was sensitive to H_2 stirring on electrodes in the initial state and was attributed [28] to mass transport control. However, the magnitude ($i < 0.1$ mA/cm^2) excludes mass transport control since i_{d,H_2} has to be about 1 mA/cm^2 in this case [24, 25]. Sensitivity with respect to stirring is not sufficient evidence for mass transport control, as was pointed out in the discussion of the studies of H_2 oxidation on smooth gold [10]. The limiting anodic current was stated [28] to be an order of magnitude smaller on nickel electrodes in the steady state. It was assigned to the Tafel reaction (see Eq. IV, 33). Simultaneous measurements of current-potential curves and of the electrode impedance on polycrystalline nickel and on the main faces of nickel single crystals in 4 M KOH confirmed [29] the Volmer reaction as the rate-determining step for hydrogen evolution on a surface sparsely covered [30] with H atoms. It was concluded [29]

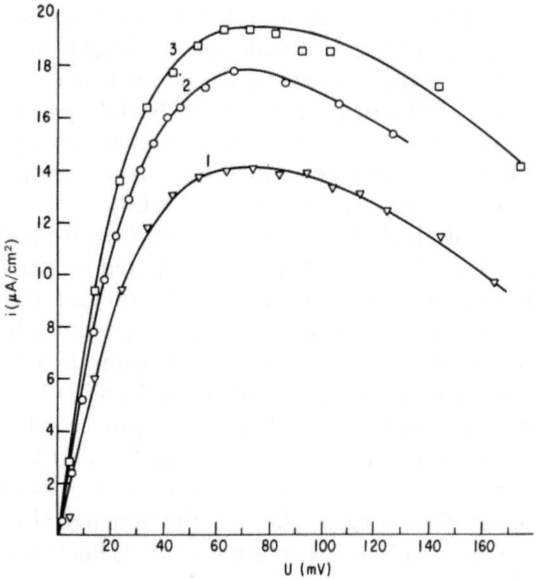

Fig. 27. Potentiostatic current-potential curves on bright nickel in alkaline electrolytes stirred with H_2.
Curve 1 : 0.1 M NaOH; curve 2 : 0.5 M NaOH; curve 3 : 1 M NaOH

that hydrogen adsorption occurs only at negative values of U. Adsorption of OH radicals was suggested [29] between 0 and 0.2 V, and the formation [31] of thin layers of $Ni(OH)_2$ between 0.1 V and 0.3 V. The partial current of H_2 oxidation was estimated smaller than 1 $\mu A/cm^2$ during anodic potential sweeps at 3 mV/sec in 4 M KOH stirred with molecular hydrogen.

Steady-state $i-U$ curves measured [32] potentiostatically in 0.1 M NaOH, 0.5 M NaOH, and 1 M NaOH in the presence of hydrogen stirring on spectrographically pure nickel rods are replotted in Fig. 27.

The current increases with potential, passes through a maximum between 0.08 V and 0.1 V and decreases subsequently. This result represents evidence that the H_2 oxidation is not limited either by the Tafel reaction or mass transport only because the current should level off in both cases. A limiting current does not exist in that sense. The decrease of i with U was attributed [32] to the inhibiting effect of nickel oxide or nickel hydroxide on the rate of the H_2 oxidation. Cathodic charging curves taken from potentials between 0 and 0.15 V displayed arrests which are characteristic for the reduction of oxides or hydroxides. The previous evidence [29] for the presence of some type of an oxygen layer between 0 and 0.2 V is corroborated. It is difficult to elucidate the rate-determining step of the H_2 oxidation on bright nickel because of the inhibiting effect of the oxygen layer. The tendency to the formation of adsorbed layers of OH or of thin films of $Ni(OH)_2$ will increase with the concentration of OH^- ions. It is conceivable that the partial current for the H_2 oxidation became very small for this reason in the measurements [29] with the sweep technique in 4 M KOH.

Reaction 2 was studied [33] on porous nickel electrodes in 1 M KOH stirred with H_2. The electrodes had been prepared by depositing nickel hydroxide electrochemically on to a sintered structure and by reducing $Ni(OH)_2$ to Ni in an atmosphere of hydrogen between 200° and 300 °C. The $i-U$ curve measured potentiostatically from $U=0$ to $U=0.8$ V looks similar to curve 3 in Fig. 27 between 0 and 0.2 V. The current decreases from a maximal value of about 0.5 mA/cm^2 between 0.08 V and 0.2 V to about 0.065 mA/cm^2 between 0.2 V and 0.3 V and remains at this value up to 0.65 V. The current decrease was attributed to the presence of chemisorbed oxygen. The current fluctuated strongly between 0.03 V and 0.2 V as on platinized platinum. Such fluctuations are characteristic for partial or complete mass transport control. The value of 0.5 mA/cm^2 for the maximum current density indicates partial control by mass transport of H_2 molecules. A considerable hysteresis is observed when measuring from 0.8 V to 0. The width of the potential region in which the maximal current is found is reduced to 0.1 V. The maximal current density is about 0.25 mA/cm^2. The inhibiting effect of the oxygen layer on the rate of the H_2 oxidation is clearly demonstrated by this result. Impedance measurements [33] at low frequencies show that the capacitance C_P has a peak between 0 and 0.25 V. A shoulder on the right side of the peak indicates that there are two overlapping peaks as on smooth nickel [29]. The peak with the shoulder cannot be attributed [29] to the adsorption of H atoms as it was suggested [33], but should be assigned to the adsorption of OH and to the formation of thin films of $Ni(OH)_2$. The small dependence of the C_P-U curves upon additions of methanol and ethanol confirms this statement. Adsorption of carbonaceous species seems to be inhibited by the oxygen layer on nickel as on platinum metals [34, 35]. The high values of C_P are evidence for a large real surface. The good performance of the porous nickel electrode [33] is probably due to the large surface.

The H_2 oxidation on electrodes with Raney nickel as electrocatalyst, especially with the double-skeleton catalyst electrode (DSK-electrode), has been studied [36—38] extensively. Raney nickel consists of a nickel lattice with a large number of defects formed during its production by dissolution of aluminum from aluminum-nickel alloys in potassium hydroxide. The residual aluminum content depends upon the conditions of dissolution and was found to vary [39] between 3 % to 25 % of the original content. The literature data on the number of H atoms bonded per Ni atom of fresh Raney nickel show a considerable spread from 0.12 in reference 40 to 2.0 in reference 41 and 42. It was realized [37, 40] that the formation of the oxygen layer occurs in the same potential region on Raney nickel and nickel DSK-electrodes in alkaline electrolytes as on smooth nickel. The potential range in which the reactivity of the Raney nickel electrode and the DSK-electrode for the H_2 oxidation is not reduced by the oxygen layer is given [37, 39] as 0 to 0.15 V, referred to a hydrogen electrode in the same electrolyte. Fine-grained Raney nickel powder which was dispersed in the electrolyte was brought into continuous or intermittent contact for charge transfer with an indifferent platinum electrode by magnetic force in a recent study [39] of the properties of Raney nickel. By eliminating the contribution of the partial currents of aluminum dissolution and of the formation of the oxygen layer to the total current, a ratio of 0.2 to 0.3 H atoms per Ni atom was established [39] for freshly prepared Raney nickel powder. The exchange current density for the H_2 oxidation was estimated from the slope of the $i-U$ curve through $U=0$ as 10 μA/cm^2, referred to the real surface area. Since the same magnitude was reported [26—28] for smooth nickel, the high reactivity of Raney nickel may be interpreted on the basis of the large real surface alone.

Nickel boride has been used [43] as electrocatalyst for the H_2 oxidation in recent years. An evaluation of this process on the basis [44] of different catalyst loadings demonstrated that the performance improved considerably with temperature between 60° and 115 °C at 4 mg nickel boride/cm^2. A linear relation between catalyst loading and performance was not found. Anodes with low loadings performed less well than those with high loadings. None of the nickel boride electrodes could sustain a current of 5 mA at room temperature. The performance of anodes with precious metal loadings of 1.25 mg/cm^2 was markedly superior to that of any of the nickel boride electrodes at a given temperature.

3. Mechanism of the H_2 Oxidation on Platinum in Contact with an Ion-Exchange Membrane

The H_2 oxidation was studied [45] on circular surfaces produced by cutting platinum wires (1 or 2 mm in diameter) perpendicularly to their axis. The wire was pressed with the circular surface against one side of an ion-exchange membrane containing SO_3^- groups linked to a polymer network. A current of 15 mA/cm² had been passed through the membranes in 0.5 M H_2SO_4 for 15 min in order to achieve a more complete conversion of the membrane into the H^+ form. Subsequent washing with double-distilled water did not remove all of the sulfuric acid from the membrane, even when the membrane stayed in the water for several months. A residual concentration of 0.01 M H_2SO_4 was estimated to remain. The thickness of the membranes was about 0.5 mm.

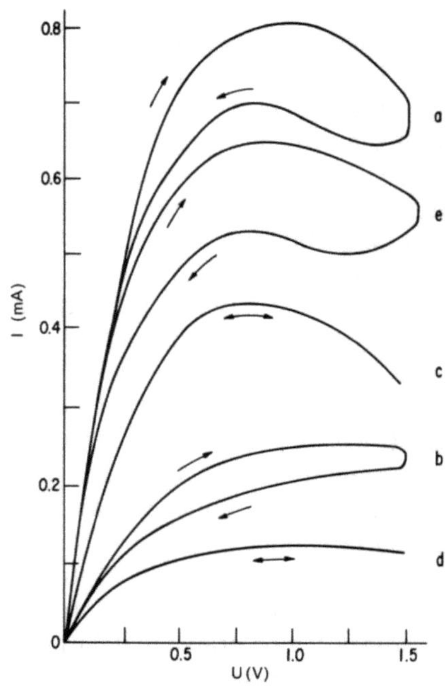

Fig. 28. Voltammetric current-potential curves of H_2 oxidation taken with different sweep rates on smooth platinum pressed against an ion exchange membrane.
Curve a: v = 34 mV/sec, diameter 2 mm
Curve b: v = 34 mV/sec, diameter 1 mm
Curve c: v = 0.86 mV/sec, diameter 2 mm
Curve d: v = 0.86 mV/sec, diameter 1 mm
Curve e: v = 0.34 mV/sec, diameter 2 mm, electrode sealed in glass

Molecular hydrogen was supplied to the other side of the membrane against which a platinum screen was pressed as a counter electrode.

Reproducible $I-U$ curves were obtained [45] by the triangular sweep technique after some initial cycles. The curves a to d in Fig. 28 were measured with sweep rates of 34 mV/sec and 0.86 mV/sec on the electrodes with 1 or 2 mm in diameter. A hysteresis was observed between the anodic and cathodic sweep at the larger sweep rates. The hysteris disappeared at sweep rates $v \leqq 0.86$ mV/sec. Curves c and d represent steady-state measurements for this reason. If the wire was completely sealed in a glass tube so that only the circular area was exposed, curve e resulted at 0.34 mV/sec.

The current becomes larger with potential, passes through a broad maximum and decreases above about 0.8 V in curves c and d. The shape of the $I-U$ curves c and d is not similar to that of curve a in Fig. 24, measured during the anodic sweep. But a strong resemblance exists between curves c and d and $I-U$ curves on semi-immersed [46] platinum wire electrodes. The shape of the latter curves could be interpreted [46] satisfactorily by assuming diffusion through a thin meniscus of electrolyte to be rate-controlling (compare chapter XVI). It was suggested [45] that the diffusion of H_2 molecules through a thin film of electrolyte which exists between certain parts of the platinum surface and the membrane leads to the shape of the curves in Fig. 28. The electrolyte is pressed nonuniformly out of the membrane. The complete sealing of the wire into glass eliminated for curve e formation of a meniscus around the platinum wire. The contribution from the occurrence of reaction 1 in the meniscus was estimated [45] to be about 20% on the basis of the different position of curves a and e in Fig. 28. Recent studies demonstrated [47] that organic species are adsorbed and block a part of the platinum surface in contact with the ion exchange membrane.

References

1. Kandler, L., C. A. Knorr, and C. Schwitzer: Z. physik. Chem. **A 180**, 281 (1937).
2. Aikasyan, E. A., and A. I. Fedorova: Doklady Akad. Nauk S.S.S.R. **86**, 1137 (1952).
3. Breiter, M., H. Kammermaier, and C. A. Knorr: Z. Elektrochem. **60**, 37 (1956).
4. — — — Z. Elektrochem. **60**, 119 (1956).
5. Kolotyrkin, Ya., and A. N. Chemodanov: Doklady Akad. Nauk S.S.S.R. **134**, 128 (1960).
6. Ludwig, F., and E. Yeager: J. Electrochem. Soc. **113**, 1109 (1966).
7. Frumkin, A. N., and L. N. Nekrassov: Doklady Akad. Nauk S.S.S.R. **126**, 115 (1959).
8. Schuldiner, S.: J. Electrochem. Soc. **99**, 488 (1952), ibid. **106**, 891 (1959).
9. Bockris, J. O'M., and S. Srinivasan: Electrochim. Acta **9**, 31 (1964).
10. Breiter, M., C. A. Knorr, and R. Meggle: Z. Elektrochem. **59**, 153 (1955).
11. —, and F. G. Will: Z. Elektrochem. **61**, 1177 (1957).

12. LINDE, J. O., and G. BORELIUS: Ann. Physik **84**, 747 (1927).
13. FRUMKIN, A., and N. ALADZHALOVA: Acta Physicochim. U.R.S.S. **19**, 1 (1944).
14. BREITER, M. W.: Ber. Bunsenges. physik. Chem. **69**, 612 (1965).
15. WILL, F. G., and C. A. KNORR: Z. Elektrochem. **64**, 258 (1960).
16. FRUMKIN, A. N.: Advances in Electrochemistry, Vol. **3**, Chap. 5. P. Delahay, ed., New York: Interscience 1963.
17. —, and E. A. AIKASYAN: Doklady Akad. Nauk S.S.S.R. **100**, 315 (1955).
18. —, V. SOKOL, and A. DMITRIEVA: J. Electroanalyt. Chem. **13**, 179 (1967).
19. VETTER, K. J., and D. OTTO: Z. Elektrochem. **60**, 1072 (1956).
20. See VETTER, K. J.: Elektrochemische Kinetik. Berlin-Göttingen-Heidelberg: Springer 1961.
21. MAKOWSKI, M. P., E. HEITZ, and E. YEAGER: J. Electrochem. Soc. **113**, 204 (1966).
22. NAGEL, K., and F. WENDLER: Z. Elektrochem. **60**, 1064 (1956).
23. BREITER, M. W.: Trans. Symposium on Electrode Processes, Chap. 17. E. Yeager, ed., New York: John Wiley and Sons 1961.
24. NAGEL, K., and F. WENDLER: Z. Elektrochem. **63**, 213 (1959).
25. BREITER, M., and K. HOFFMANN: Z. Elektrochem. **64**, 462 (1960).
26. LUKOVTSEV, P., S. LEVINA, and A. FRUMKIN: Acta Physicochim. U.S.S.R. **11**, 21 (1939).
27. FRUMKIN, A.: Discussion Faraday Soc. **1**, 57 (1947).
28. MAKRIDES, A. C.: J. Electrochem. Soc. **109**, 977 (1962).
29. WEININGER, J. L., and M. W. BREITER: J. Electrochem. Soc. **111**, 707 (1964).
30. PAST, V. E., and S. A. IOFA: Zhur. Fiz. Khim. **33**, 1230 (1959).
31. WEININGER, J. L., and M. W. BREITER: J. Electrochem. Soc. **110**, 484 (1963).
32. O'BRIEN, R. N., and K. V. N. RAO: J. Electrochem. Soc. **112**, 1245 (1965).
33. KANDLER, L., D. ROEMER, and E. HÄUSLER: Electrochim. Acta **8**, 233 (1963).
34. BREITER, M. W., and S. GILMAN: J. Electrochem. Soc. **109**, 622 (1962).
35. — Electrochim. Acta **8**, 973 (1963).
36. JUSTI, E. W., and A. W. WINSEL: Kalte Verbrennung – Fuel Cells. Wiesbaden: Franz Steiner Verlag 1962.
37. KALBERLAH, A., and A. WINSEL: Ber. Bunsenges. physik. Chem. **68**, 250 (1964).
38. BURSHTEIN, R. H., A. G. PSHENICHNIKOV, F. Z. SABIROV, and V. N. ZHURAVLEVA: Elektrokhim. **4**, 13 (1968).
39. STURM, F. v., and G. RICHTER: Electrochim. Acta **10**, 1169 (1965).
40. TUPIZIN, I. F., and I. P. TVERDOVSKY: Zhur. Fiz. Khim. **32**, 598 (1958).
41. RANEY, M.: Ind. Eng. Chem. **32**, 1199 (1940).
42. VANDAEL, C.: Ind. Chim. Belge **17**, 581 (1952).
43. JASINSKI, R. J.: Fuel Cell Systems, Advances in Chemistry Series No. **47**, Chap. 8. G. J. Young and H. R. Linden, ed., Washington, D. C.: Am. Chem. Soc., 1965
44. MEIBUHR, S. G.: Electrochim. Acta **12**, 1059 (1967).
45. HANIN, A. B., N. A. ALADZHALOVA, and N. A. FEDOTOV: Elektrokhim. **3**, 59 (1967).
46. WILL, F. G.: J. Electrochem. Soc. **110**, 145 and 152 (1963).
47. HANIN, A. B., N. A. ALADZHALOVA, and N. A. FEDOTOV: Elektrokhim. **4**, 107 (1968).

VIII. Oxygen Layers on Different Materials and Inhibition of Fuel Oxidations

The concept of the oxygen layer on platinum metals, binary alloys of platinum metals, and nickel was already mentioned in several sections of chapter VI and VII. The word "oxygen layer" is used in a general sense here to denote chemisorbed layers of OH_{ad} and O_{ad} as well as hydroxide layers and oxide layers. The potential region in which the oxygen layers are formed and reduced on platinum metals was called "oxygen region". Details about the formation, reduction and composition have yet not been described. Since the oxygen layers are important for our understanding of the mechanism of the anodic oxidation of fuels and of the oxygen electrode, a discussion is included for those metals that are of main interest in the technology of the low-temperature fuel cell.

1. Formation and Reduction of Oxygen Layers on Platinum Metals and Some Alloys

The formation and reduction of oxygen layers on platinum has been studied intensively. The scientific papers may be grouped according to their main conclusions: Immediate formation of oxides [1—10] or initial formation of chemisorbed layers [11—35] and oxide formation at larger potentials [13, 15—18, 20, 22, 25—35]. Agreement on the composition of the oxide does not exist in the papers of the first group. The opinion about the nature of the adsorbed layers and about the starting potential for the formation of some type of oxide differs greatly in the second group. Investigations of the oxygen layers were also carried out on Ir [17, 36—42], Rh [17, 36, 40, 41], Pd [19, 36, 40, 43—46], Ru [47—49], and Os [50].

The $i-U$ curves in Fig. 8 show the formation and reduction of the oxygen layer on Pt, Ir, and Rh in 2.3 M H_2SO_4 and 0.1 M NaOH above 0.4 V under the condition [21] that a periodic triangular voltage is applied between 0.05 V and 1.5 V at 0.5 V/sec. There is a similarity between the shape of the $i-U$ curves on Pt and Rh in the oxygen region. The current begins to rise at a certain potential during the anodic sweep and attains a value which becomes approximately independent of U afterwards. The reduction wave during the cathodic sweep is narrower than the oxidation wave and located at less positive potentials. The formation and reduction of the oxygen layer is an irreversible process on Pt and Rh. The irreversibility is larger for Rh than for Pt, as demonstrated by the partial overlapping of the reduction of the oxygen layer with the formation of the hydrogen layer. In contrast, the $i-U$ curves on Ir are practically symmetrical with respect to the U axis between about 0.4 V and 1.4 V during both sweeps. The reactions the steps of which are outlined in section 2 of this chapter are reversible on iridium. Palladium displays [40] a behavior similar to that of platinum. Ruthenium [47—49] and osmium [50] are similar to rhodium. The oxygen layer begins to form at more positive potentials in the sequence Os, Ru, Rh, Pt. The affinity for the formation decreases in the same order. The information which is easily derived from the voltammetric curves agrees well with that from charging curves or from studies of the oxidation and reduction at constant potential.

The curves in Fig. 8 were recorded after some time of cycling. The shape of the initial $i-U$ curves and the cause for the transition from the initial shape to the final shape under voltammetric conditions are discussed in section 7 of chapter VI.

The $i-U$ curves measured on the platinum-gold alloys, which consist of a platinum-rich phase α_1 and a gold-rich phase α_2, demonstrate [51] that the formation and reduction of the oxygen layers on patches of these two phases on the surface occur independently of each other (compare section 8 in chapter VI). On the platinum-chromium alloys the formation and reduction of oxygen layers on patches of chromium overshadows that on platinum patches.

If the capacitive component $1/\omega C_S$ of the electrode impedance in a series analog circuit of capacitance C_S and resistance R_S is measured [53—55] at constant frequency during the sweep by voltammetry with superimposed ac current, an increase of the capacitance with potential is observed between about 0.6 V and 0.9 V on platinum in 0.5 M H_2SO_4. The capacitance starts [18, 53—56] to increase at a less positive potential than the current due to the formation of the oxygen layer (see curve b in Fig. 20). The capacitance measured between 100 to 10000 Hz decreases gradually with U above 1.0 V when the charge Q_0 of the formation of the oxygen layer becomes larger. It remains nearly constant during the cathodic sweep between 1.5 V and 0.8 V (see dashed part of curve b in Fig. 20) where the oxygen layer remains unchanged. The capacitance was attributed [53, 54] as a faradaic capacitance to the reaction 1 in acid solutions.

$$(1) \qquad H_2O = OH_{ad} + H^+ + e^-$$

The shape of the $1/\omega C_S$ curve depends [54, 55] upon the type and concentration of the anion. The described dependence of C_S upon U

and Q_0 suggests that a large part of the oxygen layer consists of chemisorbed oxygen atoms or platinum oxides. Reaction 1 which requires free sites is affected by O_{ad} and oxide. As would be expected on the basis of

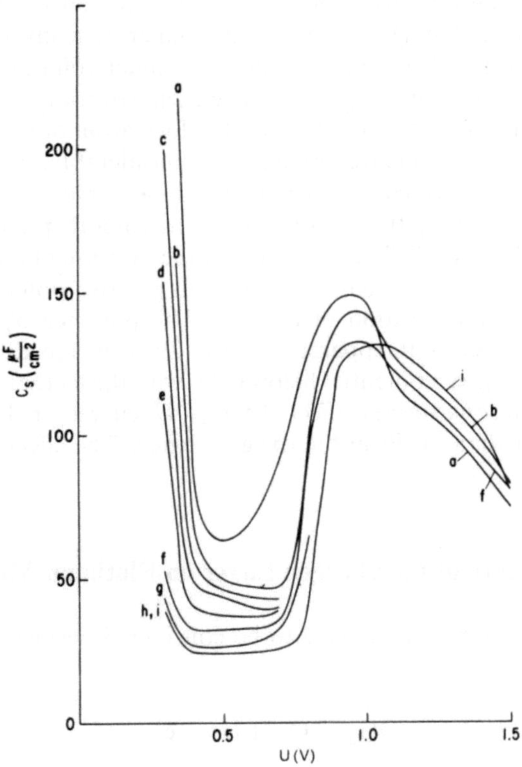

Fig. 29. Capacitance of the electrode impedance of platinum at 300 Hz in an analog series circuit during the anodic sweep with 30 mV/sec in 1 M $HClO_4$+X M CH_3OH.

a: X=0 d: X=0.02 g: X=0.2
b: X=0.005 e: X=0.04 h: X=0.4
c: X=0.01 f: X=0.08 i: X=0.8

the $i-U$ curves, the capacitance on Ir [39, 42] is much larger at a given potential than that on Pt and Rh because the formation and reduction of the oxygen layer occur reversibly.

If specifically adsorbable anions are present, the beginning of the layer formation during the anodic sweep is shifted [55] towards more positive potentials with increasing concentration of the anions (compare

Fig. 13). Simultaneously the rise of the capacitance, due to reaction 1 appears at more positive potentials. Both effects result [3, 13, 16, 57—59] from the competitive adsorption of anions.

The study of the influence of methanol [60, 61] and formic acid [62—64] on the formation of the oxygen layer on platinum in 1 M $HClO_4$ revealed that Q_0 is somewhat smaller at a given potential in the presence of 1 M fuel than in its absence under voltammetric [60, 62] and steady-state conditions [61, 64]. The chemisorption of carbonaceous species does not have a large effect on the formation of the total oxygen layer. However, it influences reaction 1 considerably. This is demonstrated by the $C_S - U$ curves in Fig. 29. These curves were determined [65] at 300 Hz during the anodic sweep on smooth platinum in 1 M $HClO_4 + X$ M CH_3OH. The beginning of the increase of the capacitance with potential is shifted [65—67] to more positive potentials with increasing bulk concentration of the fuel. The presence of chemisorbed carbonaceous species is apparent in Fig. 29 from the decrease of the capacitance at a given potential below 0.7 V with the methanol concentration. Carbonaceous species affect the $C_S - U$ curves on Pd and Rh to a lesser extent than on Pt in the oxygen region. The effect is very small [67] on Ir.

2. Nature of the Oxygen Layer on Platinum Metals

The following reactions have to be considered besides reaction 1 in acid electrolytes:

(2) $$OH_{ad} = O_{ad} + H^+ + e^- ,$$

(3) $$M + nH_2O = M(OH)_n + nH^+ + ne^- ,$$

(4) $$M + mH_2O = MO_m + 2mH^+ + 2me^- .$$

Here reactions 3 and 4 stand for the formation of the hydroxide $M(OH)_n$ and the oxide MO_m respectively. Similar reactions apply in alkaline electrolytes. It is conceivable that the chemisorbed layers and metal hydroxide or oxide are formed simultaneously on different parts of the surface in a certain potential range. The conditions [42] under which the voltammetric $i - U$ curves give evidence for the occurrence of reaction 1 and 2 are discussed subsequently. The electrolyte concentrations are assumed such that concentration gradients of OH^- or H^+ are negligible in the diffusion layer during the formation or reduction of the oxygen layer.

The situation is simple in the case of phase hydroxide or oxide. The reversible potential U_{rev} of reactions 3 and 4 is independent of the layer thickness. If the net overvoltage of reactions 3 or 4 is small, hydroxide or oxide begin to form at potentials slightly positive to U_{rev} during anodic sweeps which were started at $U < U_{rev}$ with a surface free of $M(OH)_n$ and MO_m. If the net overvoltage is not negligible, layer formation will start at more positive potentials. When the anodic sweep is reversed the current remains anodic during the subsequent cathodic sweep until the potential U_{rev} is passed. As long as only one type of hydroxide or oxide is formed during the anodic sweep, the potential at which the current becomes cathodic during the cathodic sweep is equal to U_{rev} and, therefore, independent of the potential of sweep reversal. The $i-U$ curve measured on smooth cadmium in 1 M KOH at 3 mV/sec and shown as curve b in Fig. 2 of reference 68 is an example. The current becomes anodic during the anodic sweep at a potential close to the reversible potential of $Cd(OH)_2$ formation ($U_{rev} = 0.019$ V). The current has practically the same value at a given potential $U > U_{rev}$ during both sweeps in the potential region of $Cd(OH)_2$ formation. The reduction of $Cd(OH)_2$ takes place at $U < 0.019$ V during the cathodic sweep.

In contrast to phase hydroxides or oxides the reversible potential of layer formation becomes more positive with increasing coverage for chemisorbed layers:

$$(5) \qquad k_1(1-\theta_0-\theta_{OH}) \exp \frac{\alpha_1 UF}{RT} = k_2 c_{H^+} \theta_{OH} \exp \frac{-(1-\alpha_1)UF}{RT},$$

$$(6) \qquad k_3 \theta_{OH} \exp \frac{\alpha_2 UF}{RT} = k_4 c_{H^+} \theta_0 \exp \frac{-(1-\alpha_2)UF}{RT}.$$

Here θ_{OH} and θ_O designate the coverage with OH_{ad} and O_{ad} respectively. Langmuirian adsorption is assumed for simplicity. The influence of the potential drop in the diffuse double layer is neglected. If the formation and reduction of the chemisorbed layers require a small overvoltage, the sweep reversal leads to a rapid change from an anodic to a cathodic current. The $i-U$ curves in Fig. 30 are examples. They were measured [42] at 30 mV/sec on smooth iridium in 2.5 M H_2SO_4 at 30 °C. The curves consist of several overlapping waves. The potential of reversal was chosen close to the end of each of the three waves. If the formation and reduction of chemisorbed layers take place with considerable overvoltage, the current will remain anodic during the subsequent cathodic sweep until the U_{rev} value corresponding to the coverage at the moment of reversal is reached. An increase of the potential of sweep reversal leads to an increase of the potential where the current becomes cathodic during the cathodic sweep. The $i-U$ curves on Pt [23] and Rh [41] exhibit this behavior.

The following conclusions may be drawn for the case of simultaneous formation of chemisorbed layers and hydroxide or oxide. The sweep reversal will lead to a current reversal if the formation and reduction of one of the chemisorbed layers occur with sufficiently small overvoltage and if the rate of hydroxide or oxide formation is smaller than that of the reduction of the chemisorbed layers.

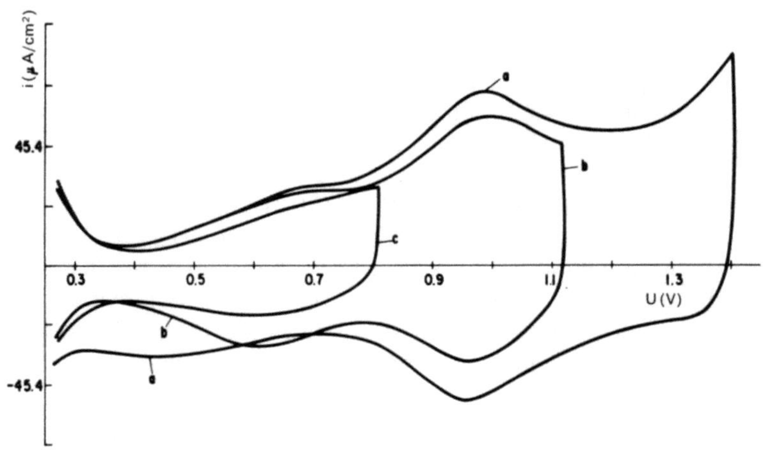

Fig. 30. Periodic i–U curves on iridium at 30 mV/sec with different potentials of reversal in 2.3 M H_2SO_4 at 30°C

Thus the results on Pt [23], Ir [41, 42], and Rh [41] are evidence for the presence of chemisorbed layers but do not exclude the possibility of simultaneous formation of phase oxide or hydroxide in patches. An increase of the sweep rate favors the appearance of chemisorbed layers.

Optical techniques have been used [29, 69] in recent years for the study of the formation of oxygen layers on platinum. The results of these investigations were interpreted in terms of oxide formation starting at certain positive potentials. However, since the optical effects may be due to patches of oxide, the simultaneous formation of chemisorbed layers cannot be ruled out.

Estimates of the ratio $Q_{\mathrm{oxide}}/Q_{\mathrm{O,cath}}$ between the charge for oxide formation and the charge for the reduction of the total oxygen layer yielded [42] about 0.12 for Pt at 1.5 V and 0.20 for Ir at 1.4 V. The oxygen layers in these studies [18, 23, 41, 42] consisted mainly of chemisorbed layers. The ratio $Q_{\mathrm{oxide}}/Q_{\mathrm{O,cath}}$ may be different for other studies in which larger $Q_{O,cath}$ values or Q_O values were obtained [21, 27—29] on Pt than those in reference 18.

It follows from Eq. 6 that θ_O/θ_{OH} increases with U if equilibrium is established. Since the formation of the chemisorbed layers on Ir fulfills this condition, the coverage with O_{ad} becomes predominant with increasing potential in each of the three potential regions. The preceding conclusion does not depend upon the type of isotherm since more complex isotherms are similar to the Langmuir isotherm when saturation coverage is approached. The extent to which the coverage consists of OH_{ad} and O_{ad} depends upon the ratio $(k_4/k_3)/(k_2/k_1)$. Theoretical curves were computed for one region with the G. E. 235 computer to find an estimate for this ratio which would lead to a width comparable to that of the experimental waves. The results suggest that $(k_4/k_3)/(k_2/k_1)$ is between 5 and 10. Adsorbed OH radicals are formed initially in each region and converted to O_{ad} afterwards. The same computation cannot be made for Pt and Rh. However the $i-U$ curves on Pt and Rh in the oxygen region appear [23, 24] to consist of at least two overlapping waves. The widths of the waves at $U>0.8$ V are comparable. This indicates a similar situation with respect to the formation of OH_{ad} as on iridium. A kinetic expression for the rate of formation of the chemisorbed layer was formulated for platinum [23, 26] and rhodium [41] by assuming a dependence of the Temkin type [70] for O_{ad} and of the Langmuir type for OH_{ad}. The expression holds approximately between 1.0 V and 1.5 V.

The preceding considerations may also be applied to the discussion of the formation and reduction of oxygen layers on Pt, Rh, and Ir in alkaline electrolytes since the respective $i-U$ curves display similar properties in acid and alkaline solutions. The extent of phase hydroxides and oxides on platinum metals is small in the potential region which is of interest for the anodic oxidation of fuels and for the oxygen electrode in fuel cells. The formation of thicker oxide layers at more positive potentials will not be considered here.

3. Oxygen Layers on Nickel in Alkaline Electrolytes

The formation of oxygen layers on nickel electrodes in alkaline solutions has been investigated [71—90] with different techniques. Anodic charging curves taken with small currents on electrodeposited [75] or smooth [74, 85] nickel from a potential of hydrogen evolution display two well defined arrests before the oxygen evolution starts. Fine details are also observable in the $i-U$ curves measured during anodic potential sweeps. The curves in Fig. 31 resulted [83] during the first sweep with 110 mV/sec on different planes of nickel single crystals and on polycrystalline nickel in 0.2 M KOH. The sweeps started at −0.2 V. The curves may be divided into three parts: an area between 0 and 0.5 V, an intermediate region between 0.5 V and 1.2 V, and one at large potentials between about 1.2 V and 1.5 V. The latter wave was attributed [83, 84] to the formation of NiOOH in agreement with results [76—81] on the charged state of nickel plates in batteries. More recent work [86, 88, 90] indicates that the charged state is even more complex than NiOOH.

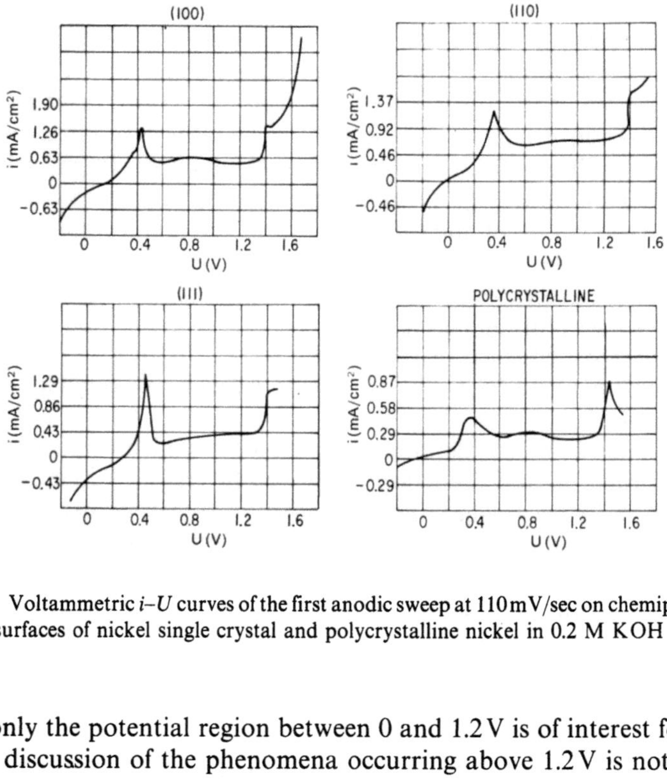

Fig. 31. Voltammetric *i–U* curves of the first anodic sweep at 110 mV/sec on chemipolished surfaces of nickel single crystal and polycrystalline nickel in 0.2 M KOH

Since only the potential region between 0 and 1.2 V is of interest for fuel cells, a discussion of the phenomena occurring above 1.2 V is not given here.

The following reactions have to be considered at $U < 1.2\,V$:

(7) $$OH^- = OH_{ad} + e^-,$$

(8) $$OH_{ad} + OH^- = O_{ad} + H_2O,$$

(9) $$Ni + 2\,OH^- = Ni(OH)_2 + 2e^-,$$

(10) $$Ni(OH)_2 + nOH^- = NiO_{2+m}H_{2+2m-n} + (n-m)H_2O + ne^-.$$

The stoichiometric coefficients n and m in net reaction 10 need not be integers.

The capacitance C_P in an analog parallel circuit at 100 Hz during the first anodic sweep is shown [84] as a function of U for a polycrystalline electrode in Fig. 32. The sweep started at -0.3 V. The sweep rate was 3 mV/sec. The C_P-U curve consists of at least two overlapping waves between -0.3 V and 0.4 V. It shows a minimum of 22 μF/cm² between

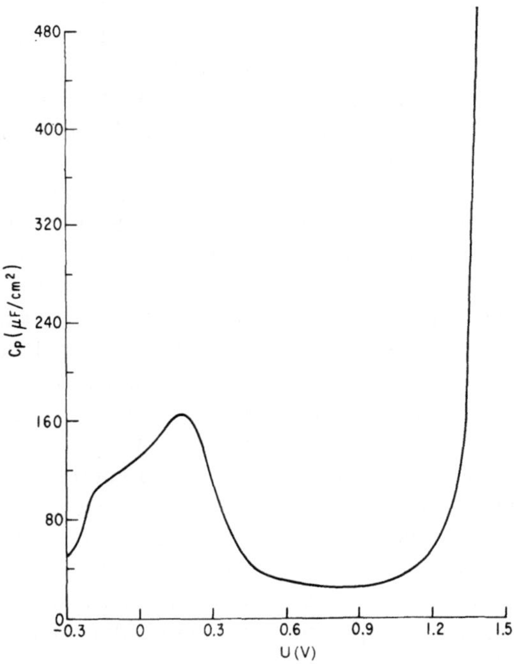

Fig. 32. Capacitance of the impedance of a polycrystalline nickel electrode in an analog parallel circuit during the first anodic sweep in 4 M KOH

0.6 V and 0.8 V. This C_P value was taken [84] as the double layer capacitance. Above 0.6 V the capacitance rises again with U. A similar shape of the C_P-U curve had already been found [73] by measurements made point by point from -0.2 V or 0 to positive potentials on polished nickel electrodes. The capacitance at 100 Hz decreased [73] from 40 μF/cm² at 0.2 V to 7 μF/cm² at 0.7 V and achieved a value of 45 μF/cm² at 1.2 V in this case. The first wave of the C_P-U curve between -0.3 V and 0 was attributed [84] as a faradaic capacitance to the Volmer reaction. The second wave was ascribed to reaction 7. However, it is conceivable that both waves are due to reaction 7 occurring on sites with different energy of adsorption of OH_{ad}. The decrease of C_P between 0.3 V and 0.5 V indicates that free sites are blocked by a layer the formation of which is irreversible. Layer formation according to reaction 9 was suggested [83, 84] since the reversible potential of reaction 9 is [91] about 0.15 V. The formation of thin layers of $Ni(OH)_2$ in this potential region had already been proposed [74, 75, 85] on the basis of charging curves. The intermediate region begins after about two layers of $Ni(OH)_2$ are formed (see Fig. 31). The simultaneous occurence of reaction 9 and reaction 10 (possibly with $n=1$, $m=0$) was assumed [83] between 0.6 V and 1.2 V. The beginning of the formation of NiOOH or of a more complex structure is marked by the rise of current and capacitance with potential at $U > 1.2$ V during the initial sweep. The increase of C_P implies that one of the first steps of net reaction 10 is relatively reversible.

The $i-U$ curves in Fig. 31 are slightly different on the various planes during the first anodic sweep. The first peak, whose location is a rough measure for the hindrance of reaction 9, appears at 0.48 V, 0.45 V, and 0.37 V for the planes (111), (100), and (110). It is shifted to less positive potentials in the order of closest to least-packed planes. When the oxygen layer formed during the first sweep had been reduced at -0.2 V, the shape of the $i-U$ curves during a second sweep was the same on all three crystal planes and similar to that of the first sweep on the polycrystalline electrode. The anodic sweep went up to 1.6 V and encompassed the formation of 8 to 10 layers of NiOOH, assuming NiOOH as the composition of the oxide for simplicity. The nickel planes are not regenerated on subsequent reduction, but a disordered surface is formed. A series of measurements with increasing amplitude of the sweep indicated that a change in crystal structure occurred already between 0.6 V and 0.8 V when 3 to 4 monolayers are present. The later result lead to the conclusion [83] that $Ni(OH)_2$ is formed and partially converted to NiOOH above 0.6 V. A single crystal surface of fcc nickel appears to support, for reasonably short times, a few $Ni(OH)_2$ layers, ordered pseudomorphically on nickel, but not ordered according to the hexagonal structure of $Ni(OH)_2$. On further oxidation of $Ni(OH)_2$ a drastic irreversible process takes place. Details about this process are not known.

If the amplitude of a triangular voltage sweep which starts at -0.2 V is gradually increased [89] on a smooth nickel electrode in 0.1 M KOH, the potential at which the current becomes cathodic during the cathodic sweep becomes more positive. This result indicates the presence of chemisorbed layers according to the discussion in section 2 of this chapter. Chemisorption of oxygen was suggested [89] between -0.15 V and 0.15 V. The interpretation [89] differs in this point from the one in reference 84. The irreversibility of the formation and reduction of $Ni(OH)_2$

is demonstrated by the appearance [89] of the peak of the reduction wave at smaller values of U than that of the oxidation wave at 0.1 V/sec.

The $Ni(OH)_2$ layers consist of a few molecular layers only. Thin layers do not yet possess [92] the properties of the bulk phase. The free energy of formation approaches [92] that of the bulk phase with increasing layer thickness. Thus the formation of the initial patches of hydroxide or oxide may be expected during the anodic sweep at potentials which are smaller than the reversible potential of the corresponding bulk phase if a kinetic hindrance is absent. The presence of thin layers cannot be predicted on thermodynamic grounds but has to established experimentally. Few experimental results on the equilibrium properties of thin films have been published [92], but thin films may be classified as taking an intermediate position between chemisorbed layers and phase hydroxides and oxides.

4. Oxygen Layers on Silver in Alkaline Electrolytes

The formation of the lowest silver oxide, Ag_2O, according to reaction 11

$$2Ag + 2OH^- = Ag_2O + H_2O + 2e^- \tag{11}$$

occurs [7, 93, 94] at about $U_{rev} = 1.17$ V in alkaline electrolytes. Since working potentials below 0.9 V are encountered at reasonable reduction rates of molecular oxygen on silver electrodes, the participation of silver oxides can be disregarded in fuel cells.

The $i-U$ curves measured [95] with 0.1 V/sec on silver between 0.1 V and 1.1 V in 1 M KOH, saturated with argon, show waves with peaks at about 0.35 V and 0.75 V during both the anodic and cathodic sweep. The waves were attributed to the formation or reduction respectively of two "adsorption compounds" of oxygen. The silver electrodes had been polished mechanically, washed in hot 1 M KOH, and reduced with a current of -1.5 mA. Anodic pretreatment at $U > 1.5$ V, carried out before the reduction, increased the roughness of the silver surface. The current at a given potential had a larger value than without anodic pretreatment.

A different interpretation was advanced by Loobmaa, Past, and Khaga [96] for the waves observed between 0 and 0.5 V during the anodic or cathodic sweep on silver in 0.5 M KOH. The triangular voltage sweep covered the potential range 0 to 0.5 V. The waves were assigned [96] to the removal or to the formation of a layer of H_{ad}. The charge of 0.45 $mcoul/cm^2$ corresponding to the anodic wave was found equal to that of the cathodic wave and independent of the sweep rate between 0.36 V/sec and 3.6 V/sec. The formation of oxygen layers at $U < 0.5$ V was ruled out.

The $i - U$ curve in Fig. 33 was measured by ZHUTAEVA and SHUMILOVA [97] in 1 M KOH with 0.1 V/sec. The silver electrode was kept at -0.4 V for 10 min before applying the sweep once. The curve displays five waves during the anodic sweep. There are three waves during the cathodic sweep. By choosing the limits of the potential sweep appropriately,

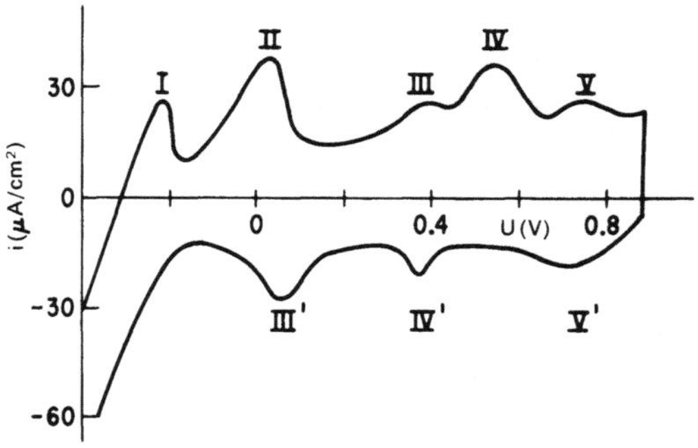

Fig. 33. Voltammetric current-potential curve on smooth silver in 1 M KOH measured with a sweep rate of 0.1 V/sec

the cathodic waves III', IV', and V' were shown to result from the reduction of the compounds formed under the anodic waves III, IV, and V respectively. Waves III, IV, and V were attributed [97] to the production of different forms of chemisorbed oxygen. Wave I was assigned to the dissolution of intermetallic compounds between K and Ag that are formed cathodically [98]. Removal of H_{ad} was suggested as the cause for wave II.

Voltammetric $i - U$ curves, started [99] in the anodic direction from about 0.25 V on a degassed silver electrode in 1 M KOH do not display the fine structure of the curve in Fig. 33 between 0.25 V and 1.0 V during the first sweep. Both charging curves and $i - U$ curves reflect only that the potential increases linearly with the anodic charge. The anodic charge computed from charging curves with 1 $\mu A/cm^2$ was found to be equivalent [99] to about 5 monolayers of O_{ad}. In contrast, the anodic charge from an $i - U$ curve measured with 1 V/sec represented only about a tenth of the charge of a monolayer of O_{ad}. A considerable dissolution of oxygen atoms in the silver was supposed [99] to occur at small charging rates on degassed silver electrodes.

5. Oxygen Layers on Carbon

It was demonstrated by FRUMKIN and coworkers [100—105] that the adsorption of acids and bases depends strongly upon the amount of gases adsorbed on activated carbon and platinized carbon. Two different mechanisms are effective for the adsorption of anions in the presence of oxygen:

$$C_qO + 2H^+ + 2A^- = C_q + 2A^-_{ad} + H_2O + 2p^+ , \tag{12}$$

$$C_qO_2 + 2H^+ + 2A^- = C_q + 2A^-_{ad} + H_2O_2 + 2p^+ . \tag{13}$$

Here C_qO and C_qO_2 denote the surface complexes due to the adsorption of oxygen, and p^+ is a hole. If the oxygen coverage is small, the amount of acid adsorbed is equivalent to the amount of oxygen adsorbed in dilute solutions. The amount of acid adsorbed increases at larger oxygen coverages to a lesser extent than expected according to Eq. 12. The formation of H_2O or H_2O_2 during the acid adsorption was considered as evidence that two different forms of adsorbed oxygen exist. It was suggested [105] that oxygen molecules are dissociated during their adsorption on sites with large heat of adsorption. The acid adsorption follows reaction 12 on these sites. A dissociation is not involved [105] for sites that are occupied at higher oxygen coverage. There the acid adsorption is accompanied by the production of H_2O_2. A similar composition of the oxygen layer may be expected when it is formed electrochemically on carbon electrodes. If the carbon electrode is held at 1.4 V for 1 min in 1 M H_2SO_4 and a cathodic current is applied subsequently, the potential decreases [106] nearly linearly with the cathodic charge. The cathodic charge was assigned in part to the charging of the double layer and, probably, in a greater measure to the reduction of adsorbed oxygen.

The anodic oxidation of graphitized carbon, charcoal, soot, and graphite was studied in concentrated acid and alkaline solutions between 50° and 100 °C by BINDER, KÖHLING, RICHTER, and SANDSTEDE [107]. The investigations included the measurements of $i-U$ curves and of galvanostatic $U-t$ curves, the cathodic reduction of oxygen layers, volumetric determinations of carbon dioxide, chemical analysis, X ray analysis, and measurements of the electric resistance of the electrode under anodic polarization. About 80 % of the coulombs were used for the production of CO_2, and 20 % for the formation of an oxygen-containing compound at potentials below oxygen evolution in acid electrolytes. The compound may contain up to 20 weight percent of oxygen. The formation of an oxygen layer precedes the CO_2 production. While a part of the oxygen layer is reducible by cathodic polarization, a large portion remains unaffected.

Voltammetric $i-U$ curves were measured on different carbon electrodes in alkaline electrolytes by MRHA, VIELSTICH, and VOGEL [108]. If the carbon electrode was kept for 3 hr at 850 °C in an atmosphere of NH_3, the curve in Fig. 34 was obtained at 29.3 mV/sec in 1 M KOH stirred with nitrogen. Carbon electrodes pretreated in this way are characterized by a relatively small overvoltage for hydrogen evolution as well as oxygen evolution. They possess [108] a high reactivity for the reduction of molecular oxygen too. In contrast, pretreatment of carbon at 1000 °C in an atmosphere of hydrogen for 5 hr leads to electrodes having a larger overvoltage for H_2 evolution, O_2 evolution, and O_2 reduction. Thus the reactivity of carbon electrodes for the O_2 reduction can be easily compared on the basis of $i-U$ curves like that in Fig. 34. The large reactivity of carbon activated in NH_3 was attributed [108] to the presence of unsaturated carbon compounds on the surface.

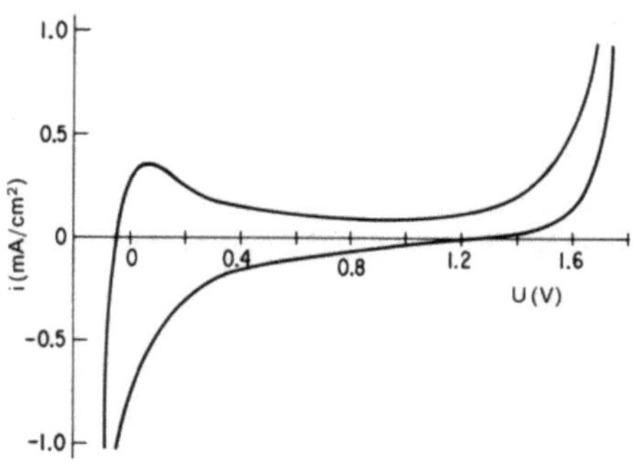

Fig. 34. Voltammetric current-potential curve measured with 29.3 mV/sec in 1 M KOH on carbon pretreated in an atmosphere of NH_3

Although MOMOT, BONNEMAY, DONIAT, and LEVART [109] demonstrated by gas adsorption (BET technique) and capacity measurements on porous carbon in alkaline electrolyte that the total surface area is much larger than the area accessible to electrolyte, they found a proportionality between the total area and the area in contact with electrolyte. If the proportionality constant is known, only one type of measurement yields the desired information on both areas. The increase of the capacitance with the concentration of KOH was interpreted as resulting from the adsorption of OH^- ions.

6. Inhibition of Fuel Oxidations by Oxygen Layers

The decrease in the rate of the anodic oxidation of molecular hydrogen on platinum with increasing potential above a certain value in the oxygen region (compare Fig. 24a) was already found and discussed by NERNST and MERRIAM [110] and SACKUR [111]. SACKUR stated that the effect is related to a decrease of the reactivity of the surface caused by the increasing oxygen coverage. THALINGER and VOLMER [112] concluded on the basis of steady-state current-potential curves, charging curves and studies with polarized light that the inhibition is not caused by the formation of oxide layers with a thickness of several Å but by chemisorbed oxygen. Iridium [38—40, 113], rhodium [40, 113, 114], palladium [40, 114], and platinum-gold alloys [115] display a similar behavior as platinum.

The three net reactions

(14) $$H_2 = 2H^+ + 2e^-,$$

(15) $$H_2 + OH_{ad} = H_2O + H^+ + e^-,$$

(16) $$H_2 + O_{ad} = H_2O$$

have to be considered besides reactions 1 and 2 in acid electrolytes. It was demonstrated [53] that the oxygen coverage on platinum is practically the same at a given potential under voltammetric conditions ($v = 30$ mV/sec) in solutions stirred with hydrogen or an inert gas. This implies that $\frac{dQ_{OH}(U)}{dt} + \frac{dQ_O(U)}{dt}$ has the same value independent of the type of gas. Let us designate by i_{15} and i_{16} the partial current densities due to reactions 15 and 16 respectively. Since the relation

(17) $$\frac{dQ_{OH}}{dt} + \frac{dQ_O}{dt} = i_1 + i_2 - (i_{15} + i_{16})$$

holds under the given conditions, it follows that $i_1 + i_2 \gg i_{15} + i_{16}$. The sum $i_1 + i_2$ can be obtained directly from the $i-U$ curve taken with argon stirring. It was verified [53] that

(18) $$i_{14} = i - (i_1 + i_2) = i_{d,H_2}(1 - p_{H_2}/{}_b p_{H_2})$$

under voltammetric conditions.

In order to see the true inhibition in the potential region where the current decreases with potential, a correction has to be made for the gradient of the concentration of H_2 in the diffusion layer. The partial current density i_{14} is proportional to the hydrogen pressure p_{H_2} adjacent to the surface. It is according to Eq. 18:

$$i_{14}\frac{{}_{b}p_{H_2}}{p_{H_2}} = i_{14}\left(1 - \frac{i - i_1 - i_2}{i_{d,H_2}}\right). \tag{19}$$

The quantity $i_{14} \cdot {}_{b}p_{H_2}/p_{H_2}$ is plotted in Fig. 35 as a function of U. At the same time the oxygen coverage expressed by $Q_O/2{}_{S}Q_H$ is shown. The results in Fig. 35 demonstrate that the inhibition is large at small oxygen

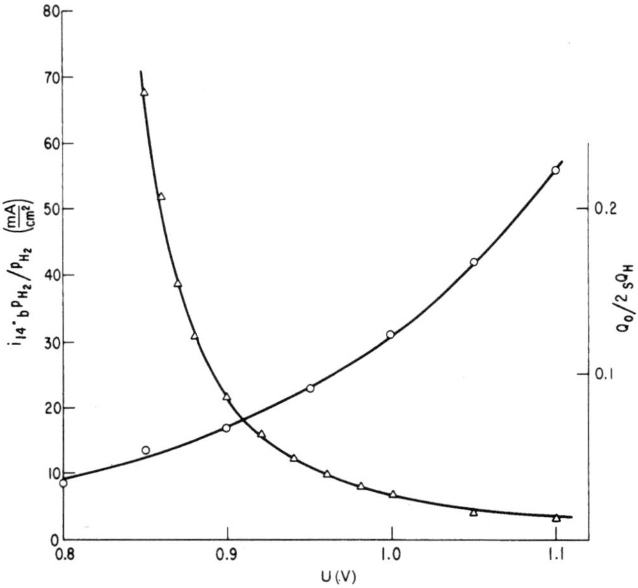

Fig. 35. Corrected rate of H_2 oxidation on platinum in 1 M $HClO_4$ and oxygen coverage as functions of potential during the anodic sweep under voltammetric conditions

coverage and levels off above 1.0 V. Adsorption of small amounts of specifically adsorbable anions caused a considerable decrease of the rate of reaction 14, too [116]. It was suggested [53] that reaction 14 occurs at a fast rate on active sites which constitute about 10 % of the surface area. After these sites are covered by OH_{ad} or O_{ad}, reaction 14 takes place on sites with lower reactivity. The inhibiting effect of the oxygen layer is not so large on these sites. The decrease of the rate with increasing coverage seems to be compensated by the increase of the rate of reaction 14 with potential so that i_{14} becomes nearly independent of U between about 1.2 V and 1.5 V.

The inhibition of the oxidation of organic fuels by the oxygen layer is similar to that of the H_2 oxidation on platinum. The methanol oxidation wave obtained [117] during the anodic sweep with 30 mV/sec on smooth platinum in 1 M $HClO_4$ + 1 M CH_3OH at 30 °C is shown as an example in Fig. 36a. The current rises with potential between 0.7 V and 0.9 V.

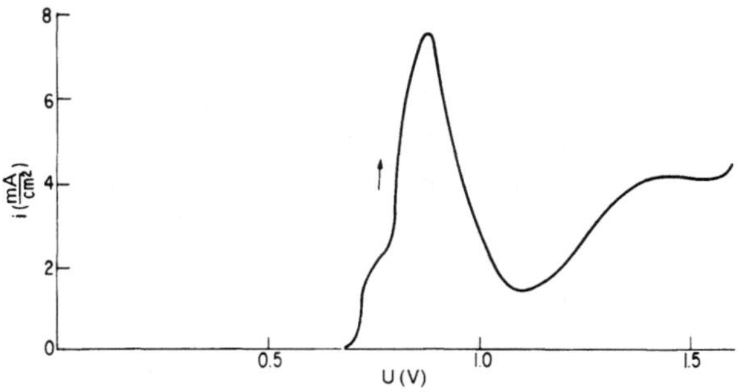

Fig. 36a. Voltammetric $i-U$ curve of the anodic sweep at 30 mV/sec on platinum in unstirred 1 M $HClO_4$ + 1 M CH_3OH

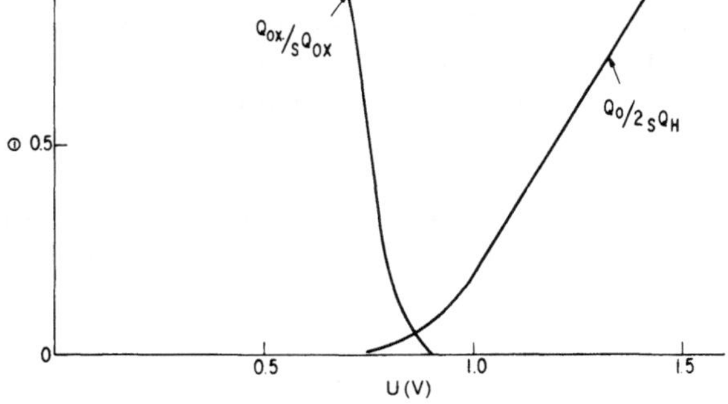

Fig. 36b. Coverage $Q_{Ox}/_sQ_{Ox}$ with carbonaceous species and oxygen coverage $Q_O/2_sQ_H$ under the same conditions

Subsequently it decreases with U between 0.9 V and 1.1 V. The coverage $Q_{Ox}/_sQ_{Ox}$ with adsorbed carbonaceous species (see section 4 of chapter IX) and the oxygen coverage were also determined [60] under the same voltammetric conditions. Both coverages are plotted as a function of potential in Fig. 36b. The coverage with carbonaceous species decreases when the current increases. Since it is small at $U > 0.9$ V, the decrease of current with U between 0.9 V and 1.1 V cannot be assigned to a change of this coverage. The decrease of current was attributed [117] to the blocking of active sites by the oxygen layer. The inhibition is greatest at $Q_O/2_s Q_H < 0.1$ as in the case of the H_2 oxidation. A similar interpretation with respect to the participation of active centers holds for methanol oxidation above 0.8 V. It is conceivable [117] that the second oxidation wave of the $i-U$ curve in Fig. 36a between 1.1 V and 1.5 V results because the increase of the oxidation rate with potential overcompensates the inhibiting effect until the surface is completely covered.

The influence of the oxygen layer has to be taken into account when the rate of oxidation of the same fuel is compared on electrodes of different platinum metals. This was demonstrated [67] in a comparative voltammetric study of methanol oxidation and adsorption on noble metal electrodes in perchloric acid solution at 30 °C. The $i-U$ curves measured at 30 mV/sec look similar on Pt and Pd on the one hand, and Ir and Rh on the other hand. There is no noticeable methanol oxidation on gold. The peak current of the first wave during the anodic sweep (compare Fig. 36a) decreases strongly in the order Pt > Pd > Rh > Ir. This sequence is not related [67] to the coverage with adsorbed carbonaceous species but to the potentials at which the oxygen layer begins to form on the respective metal at 30 °C. Recently the oxidation of ethylene was studied by DAHMS and BOCKRIS [118] in the oxygen region on Ir, Pd, Pt and Rh in 1 M H_2SO_4 at 80 °C. The strong inhibiting effect of the oxygen layer on the oxidation rate at $Q_O/2_s Q_H < 0.2$ was disregarded in the discussion of the relative electrocatalytic activity of these metals. The location of the Tafel lines of ethylene oxidation in Fig. 12 of reference 118 suggests that the order of the experimental reactivity (Pt > Ir > Rh) in this group of metals is governed by the tendency to form the oxygen layer at 80 °C. It should be pointed out that the oxygen coverage at 80 °C is larger at a given potential on rhodium than on iridium [41, 118].

References

1. BOWDEN, F. P.: Proc. Roy. Soc. London **A 125,** 446 (1926).
2. HOAR, T. P.: Proc. Roy. Soc. London **A 142,** 628 (1933).
3. HICKLING, A.: Trans. Faraday Soc. **41,** 333 (1945).
4. EL WAKKAD, S. E. S., and S. H. EMARA: J. Chem. Soc. 461 (1952).
5. KOLTHOFF, I. M., and N. TANAKA: Analyt. Chem. **26,** 632 (1954).
6. ANSON, F. C., and J. J. LINGANE: J. Am. Chem. Soc. **79,** 4901 (1957).
7. RÄDLEIN, G.: Z. Elektrochem. **61,** 727 (1957).
8. NAGEL, K., and H. DIETZ: Electrochim. Acta **4,** 141 (1961).
9. MAYELL, J. S., and S. H. LANGER: J. Electrochem. Soc. **111,** 438 (1964).

10. Fleischmann, M., J. R. Mansfield, and Lord Wynne-Jones: J. Electroanalyt. Chem. **10**, 511 (1965).
11. Butler, J. A. V., and G. Armstrong: Proc. Roy. Soc. London **A 137**, 604 (1932).
12. Armstrong, G., F. R. Himsworth, and J. A. Butler: Proc. Roy. Soc. London **A 143**, 89 (1934).
13. Ershler, B. V.: Discussions Faraday Soc. **1**, 269 (1947).
14. Salkind, Z. I., and B. V. Ershler: Zhur. Fiz. Khim. **25**, 565 (1951).
15. Nestorova, V. I., and A. N. Frumkin: Zhur. Fiz. Khim. **26**, 1178 (1952).
16. Obrucheva, A. D.: Zhur. Fiz. Khim. **26**, 1448 (1952).
17. Breiter, M., C. A. Knorr, and W. Völkl: Z. Elektrochem. **59**, 681 (1955).
18. Becker, M., and M. Breiter: Z. Elektrochem. **60**, 1080 (1956).
19. Vetter, K. J., and D. Berndt: Z. Elektrochem. **62**, 378 (1958).
20. Giner, J.: Z. Elektrochem. **63**, 386 (1959).
21. Will, F. G., and C. A. Knorr: Z. Elektrochem. **64**, 258 (1960).
22. Laitinen, H. A., and C. G. Enke: J. Elektrochem. Soc. **107**, 773 (1960).
23. Breiter, M., and W. Böld: Electrochim Acta **5**, 145 (1961).
24. —, and J. L. Weininger: J. Electrochem. Soc. **109**, 1135 (1962).
25. Dietz, H., and H. Göhr: Electrochim. Acta **8**, 343 (1963).
26. Feldberg, S. W., C. G. Enke, and C. E. Bricker: J. Electrochem. Soc. **110**, 826 (1963).
27. Visscher, W., and M. A. V. Devanathan: J. Electroanalyt. Chem. **8**, 127 (1964).
28. Gilman, S.: Electroanalytical Chemistry, Vol. 2, Chap. 3. A. Bard, ed., New York: M. Dekker 1967.
29. Reddy, A. K. N., M. Genshaw, and J. O'M. Bockris: J. Electroanalyt. Chem. **8**, 406 (1964).
30. Hoare, J. P.: J. Electrochem. Soc. **111**, 989 (1964).
31. Tikhomirova, V. I., A. I. Oshe, V. S. Bagotzky, and V. I. Luk'yanycheva: Doklady Akad. Nauk S. S. S. R. **159**, 644 (1964).
32. Luk'yanycheva, V. I., and V. S. Bagotzky: Doklady Akad. Nauk S. S. S. R. **155**, 160 (1964).
33. Schuldiner, S., and T. B. Warner: J. Electrochem. Soc. **112**, 212 (1965).
34. Luk'yanycheva, V. I., V. I. Tikhomirova, and V. S. Bagotzky: Elektrokhim. **1**, 262 (1965).
35. Hoare, J. P.: Electrochim. Acta **11**, 203 (1966).
36. Butler, J. A. V., and G. Drever: Trans. Faraday Soc. **32**, 427 (1936).
37. Obrucheva, A., and I. Burinstein: Doklady Akad. Nauk S. S. S. R. **63**, 403 (1948).
38. Breiter, M., K. Hoffmann, and C. A. Knorr: Z. Elektrochem. **61**, 1168 (1957).
39. Franke, K., C. A. Knorr, and M. Breiter: Z. Elektrochem. **63**, 226 (1959).
40. Will, F. G., and C. A. Knorr: Z. Elektrochem. **64**, 270 (1960).
41. Böld, W., and M. Breiter: Electrochim. Acta **5**, 169 (1961).
42. Breiter, M. W.: Z. Physik. Chem. N. F. **52**, 73 (1967).
43. El Wakkad, S. E. S., and A. M. S. El Din: J. Chem. Soc. 3094 (1954).
44. Hickling, A., and G. G. Vrjosek: Trans. Faraday Soc. **57**, 123 (1961).
45. Blackburn, T. R., and J. J. Lingane: J. Electroanalyt. Chem. **5**, 216 (1963).
46. Hoare, J. P.: J. Electrochem. Soc. **112**, 1129 (1965).
47. Stoyanovskaya, T. N., G. P. Khomchenko, and G. D. Vovchenko: Vestnik Moscow Univ., Ser. Khim. No. **5**, 30 (1962) and No. **2**, 20 (1963).
48. Llopis, J., I. M. Tordesillas, and J. M. Alfayate: Electrochim. Acta **11**, 623 (1966).
49. —, and M. Vazquez: Electrochim. Acta **11**, 633 (1966).
50. Khomchenko, G. P., N. G. Ul'ko, and G. D. Vovchenko: Elektrokhim. **1**, 659 (1965).
51. Breiter, M. W.: J. Phys. Chem. **69**, 901 (1965).
52. — J. Electroanalyt. Chem. **10**, 191 (1965).
53. — Electrochim. Acta **7**, 601 (1962).

54. — J. Electroanalyt. Chem. **7,** 38 (1964).
55. — Electrochim. Acta **8,** 925 (1963).
56. GILMAN, S.: Electrochim. Acta **9,** 1025 (1964).
57. KASARINOV, V. E., and N. A. BALASHOVA: Doklady Akad. Nauk S.S.S.R. **139,** 641 (1961).
58. LLOPIS, J., and A. SANCHO: J. Electrochem. Soc. **108,** 720 (1961).
59. SCHWABE, K.: Electrochim. Acta **6,** 223 (1962).
60. BREITER, M. W., and S. GILMAN: J. Electrochem. Soc. **109,** 622 (1962).
61. BAGOTZKY, V. S., and YU. B. VASILIEV: Electrochim. Acta **12,** 1323 (1967).
62. BREITER, M. W.: Electrochim. Acta **8,** 447 (1963).
63. — Electrochim. Acta **8,** 457 (1963).
64. BRUMMER, S. B., and A. C. MAKRIDES: J. Phys. Chem. **68,** 1448 (1964).
65. BREITER, M. W.: Electrochim. Acta **7,** 533 (1962).
66. — J. Electrochem. Soc. **109,** 42 (1962).
67. — Electrochim. Acta **8,** 973 (1963).
68. —, and J. L. WEININGER: J. Electrochem. Soc. **113,** 651 (1966).
69. KOCH, D. F. A., and D. E. SCAIFE: J. Electrochem. Soc. **113,** 302 (1966).
70. TEMKIN, M.: Zhur. Fiz. Khim. **15,** 296 (1941).
71. KUCHINSKI, E. M., and B. V. ERSHLER: Zhur. Fiz. Khim. **20,** 539 (1946).
72. HICKLING, A., and J. E. SPICE: Trans. Faraday Soc. **43,** 762 (1947).
73. RAKOV, A. A., T. I. BORISOVA, and B. V. ERSHLER: Zhur. Fiz. Khim. **22,** 1390 (1948).
74. VOLCHKOVA, L. M., L. G. ANTONOVA, and I. KRASIL'SHIKOV: Zhur. Fiz. Khim. **23,** 714 (1949).
75. EL WAKKAD, S. E. S., and S. H. EMARA: J. Chem. Soc. 3504 (1953).
76. JONES, E., and W. F. K. WYNNE-JONES: Trans. Faraday Soc. **52,** 1260 (1956).
77. LUKOVTSEV, P. D.: Proc. 4th Conf. Electrochem. 1956, Acad. Sci. Moscow (published 1959), p. 773.
78. BRIGGS, G. W. D., and W. F. K. WYNNE-JONES: Trans. Faraday Soc. **52,** 1272 (1956).
79. FALK, S. U.: J. Electrochem. Soc. **107,** 661 (1960).
80. BRIGGS, G. W. D., and W. F. K. WYNNE-JONES: Electrochim. Acta **7,** 241 (1962).
81. SALKIND, A. J., and P. F. BRUINS: J. Electrochem. Soc. **109,** 356 (1962).
82. CONWAY, B. E., and P. L. BOURGAULT: Trans. Faraday Soc. **58,** 593 (1962).
83. WEININGER, J. L., and M. W. BREITER: J. Electrochem. Soc. **110,** 484 (1963).
84. — — J. Electrochem. Soc. **111,** 707 (1964).
85. DAVIES, D. E., and W. BARKER: Corrosion **20,** 47 (1964).
86. KOBER, F. P.: J. Electrochem. Soc. **112,** 1064 (1965).
87. O'BRIEN, R.N., and K. V. N. RAO: J. Electrochem. Soc. **112,** 1245 (1965).
88. BODE, H., K. DEHMELT, and J. WITTE: Electrochim. Acta **11,** 1079 (1966).
89. LININ, G., N. A. SHUMILOVA, and V. S. BAGOTZKY: Elektrokhim. **3,** 460 (1967).
90. AIA, M. A.: J. Electrochem. Soc. **114,** 418 (1967).
91. POURBAIX, M., N. DE ZOUBOV, and E. DELTOMBE: CITCE VII, Lindau **193** (1955).
92. WEIL, K. G.: J. Electrochem. Soc. **110,** 640 (1963).
93. HAMER, W. J., and D. N. CRAIG: J. Electrochem. Soc. **104,** 206 (1957).
94. DIRKSE, T. P.: J. Electrochem. Soc. **106,** 453 (1959).
95. SHUMILOVA, N. A., G. V. ZHUTAEVA, and M. P. TARASEVICH: Electrochim. Acta **11,** 967 (1966).
96. LOOBMAA, V. P., V. E. PAST, and M. E. KHAGA: Elektrokhim. **2,** 927 (1966).
97. ZHUTAEVA, G. V., and N. A. SHUMILOVA: Elektrokhim. **4,** 99 (1968).
98. KABANOV, B. N., I. I. ASTAKHOV, and I. G. KISELEVA: Uspekhi Khim. **34,** 1813 (1965).
99. ZHUTAEVA, G. V., N. A. SHUMILOVA, and V. I. LUK'YANYCHEVA: Elektrokhim. **4,** 196 (1968).
100. BRUNS, B., and A. FRUMKIN: Z. Physik. Chem. **141,** 141 (1929).

101. Burstein, R., and A. Frumkin: Z. Physik. Chem. **141,** 158 (1929).
102. Frumkin, A., R. Burstein, and P. Levin: Z. Physik. Chem. **157,** 442 (1932).
103. Kuchinski, E., R. Burshtein, and A. Frumkin: Zhur. Fiz. Khim. **14,** 441 (1940).
104. Burshtein, R. H., and A. N. Frumkin: Doklady Akad. Nauk S. S. S. R. **32,** 327 (1941).
105. —, and N. B. Miller: Zhur. Fiz. Khim. **23,** 43 (1949).
106. Connolly, J. F., R. J. Flannery, and G. Aronowitz: J. Electrochem. Soc. **113,** 577 (1966).
107. Binder, H., A. Köhling, K. Richter, and G. Sandstede: Electrochim. Acta **9**, 255 (1964).
108. Mrha, J., W. Vielstich, and U. Vogel: Z. Physik. Chem. N. F. **52,** 215 (1967).
109. Momot, E., M. Bonnemay, G. Bronoël, D. Doniat, and E. Levart: Compt. rend. **265,** 444 (1967).
110. Nernst, W., and E. S. Merriam: Z. Physik. Chem. **53,** 235 (1905).
111. Sackur, O.: Z. Physik. Chem. **54,** 641 (1906).
112. Tahlinger, M., and W. Volmer: Z. Physik. Chem. **A 150,** 401 (1930).
113. Breiter, M., C. A. Knorr, and R. Meggle: Z. Elektrochem. **59,** 153 (1955).
114. —, and F. G. Will: Z. Elektrochem. **61,** 1177 (1957).
115. — Ber. Bunsenges. Physik. Chem. **69,** 612 (1965).
116. Frumkin, A. N., and E. A. Aikasyan: Doklady Akad. Nauk S. S. S. R. **100,** 315 (1955).
117. Gilman, S., and M. W. Breiter: J. Electrochem. Soc. **109,** 1099 (1962).
118. Dahms, H., and J. O'M. Bockris: J. Electrochem. Soc. **111,** 728 (1964).

IX. Adsorption of Carbonaceous Species on Platinum Metals

It is generally recognized [1—16] now that chemisorbed carbonaceous species are formed during the anodic oxidation of many organic fuels. These species play an important role in our understanding of electrocatalysis. The techniques for their study, their properties and coverage of the surface under different conditions are discussed in this chapter. The role of adsorbed species in the oxidation of various fuels is dealt with in chapter X.

1. Non-Equilibrium Aspects of the Chemisorption of Strongly Bonded Carbonaceous Species

Recent studies [15, 16] of the chemisorbed species formed during the anodic oxidation of CH_3OH, CH_2O, and HCOOH suggest that more than one type of strongly bonded species is produced even in the case of these simple fuels. The formation [16] of $COOH_{ad}$ and HCO_{ad} from methanol may be expressed by two net reactions in acid electrolytes:

(1) $$CH_3OH = CHO_{ad} + 3H^+ + 3e^- ,$$

(2) $$CH_3OH + H_2O = COOH_{ad} + 5H^+ + 5e^- .$$

Here, $COOH_{ad}$ and HCO_{ad} stand for the net composition of the chemisorbed species. All the configurations having the same net composition as COOH or HCO respectively are permissible. If both reactions occur to the same extent, the net composition of the chemisorbed species is stoichiometrically equivalent [12] to the molecule $H_2C_2O_3$. However, one of the reactions may be favored over the other one because of the different reactivity of the sites on a heterogeneous surface. The stoichiometric coefficients s, p, q in the net composition $C_sH_pO_q$ of the chemisorbed species need not be integers. Equations similar to Eq. 1 and Eq. 2 apply to the formation of $C_sH_pO_q$ from fuels containing more carbon atoms. In alkaline electrolytes, the participation of OH^- has to be taken into account.

The experimental results reported so far in the literature indicate that, independent of the fuel, the formation of $C_sH_pO_q$ is irreversible. Reversibility of the adsorption of ethylene [17, 18] and benzene [19, 20] on platinized platinum in 0.5 M H_2SO_4 has been claimed by BOCKRIS and coworkers with respect to a change in U between 0.1 V and 0.7 V.

However, as discussed in section 6 of this chapter, radiometric measurements alone do not allow to decide whether a dehydrogenation occurs during the adsorption process. Reversibility of this type does not necessarily imply an adsorption equilibrium in the thermodynamic sense but may represent [21] a balance between the rates of formation I_F and hydrogenation I_{Hy} of $C_sH_pO_q$ at potentials in the hydrogen region

$$(3) \qquad I_F(\theta_{C_sH_pO_q}, c, U) = I_{Hy}(\theta_{C_sH_pO_q}, \theta_H, U)$$

or between the rates of formation and further oxidation I_{Ox} at larger potentials.

$$(4) \qquad I_F(\theta_{C_sH_pO_q}, c, U) = I_{Ox}(\theta_{C_sH_pO_q}, \theta_{OH}, U).$$

The rates I_F, I_{Hy}, and I_{Ox} at constant pH are functions of the different coverages, the concentration of fuel adjacent to the electrode, and the potential.

The steady-state coverage with carbonaceous species may be determined rather by the balance of the rates of different electrochemical reactions than by an adsorption equilibrium. The treatment [13] of the dependence of the coverage with $C_sH_pO_q$ upon potential and methanol bulk concentration ${}_bc_M$ is an example based on Eq. 4. This treatment by Biegler and Koch predicts variations of coverage and current with ${}_bc_M$ which are similar to those obtained [11] by Bagotzky and Vasiliev under the assumption that the adsorption equilibrium is established and follows a Temkin isotherm. It is questionable for this reason whether the recent determination [22] of the heat of chemisorption of $C_sH_pO_q$, formed under anodic polarization from methanol on platinum in sulfuric acid solution, is meaningful.

The maximum coverage with $C_sH_pO_q$, designated as saturation coverage in this book, is reached when the rate of formation from bulk species at high fuel concentration becomes equal to the rate of hydrogenation or of oxidation. The rate of formation is a function of the nature of both the initial reactant and $C_sH_pO_q$. Even if a group of fuels gives rise to identical chemisorbed species, the steady-state and saturation coverage will depend upon the initial species.

2. Adsorption of Weakly Bonded Carbonaceous Species

Adsorbed species that desorb when the pressure of the bulk species is greatly reduced by pumping, in gas phase studies, are designated weakly bonded. The analogous procedure in electrochemical studies consists in replacing the mixture of fuel and electrolyte by pure electrolyte in the case of liquid fuels or by reducing the pressure of gaseous fuels by prolonged stirring with an inert gas. A shift of the potential of the test electrode during the latter procedures has to be avoided to eliminate undesirable hydrogenation or oxidation processes. Adsorbed carbonaceous species that disappear from the surface during the washing procedure or during the stirring with inert gas are called "weakly bonded" in this chapter. In contrast to the weakly bonded species, the removal of strongly bonded carbonaceous species from the surface is negligible [1, 3, 10, 12, 15] during the washing procedure.

Experimental evidence for the presence of weakly bonded species was reported recently by PODLOVCHENKO, FRUMKIN and STENIN [23]. The adsorption of carbonaceous species on a platinized platinum electrode with large real surface at open circuit was determined from the decrease of the methanol bulk concentration in 0.5 M H_2SO_4 at room temperature. Methanol (${}_bc_M = 4.5 \cdot 10^{-3}$ M) was added at $U = 0.5$ V. While the potential approached values between 0.156 V and 0.165 V, the bulk concentration dropped to about $2.15 \cdot 10^{-3}$ M. Let us suppose [3] that only strongly bonded species are formed by dehydrogenation:

(5) $$\mathrm{s\,CH_3OH + (q-s)\,H_2O = C_sH_pO_q + (2s+2p-q)\,H_{ad}}.$$

Under this assumption the sum

(6) $$Q_{El} = Q_{Ox} + Q_H$$

of the charges Q_{Ox} for the oxidation of $C_sH_pO_q$ to CO_2 and of Q_H for the oxidation of H_{ad} to H^+ should be equal to Q_{Anal}, required for the oxidation to CO_2 of the number of CH_3OH molecules which disappeared from the solution.

The charges Q_{Ox} and Q_H were obtained [23] from anodic charging curves after an appropriate washing procedure. The evaluation of similar charges was already outlined in section 6 of chapter VI. The relative ratio $(Q_{Anal} - Q_{El})/Q_{El}$ had values between 0.49 and 0.58 in three runs. Since it was established that CH_2O [23], HCOOH [23], and CO_2 [16] are not formed at open circuit after methanol addition, the above result represents evidence for the adsorption of weakly bonded species. The conclusion was confirmed [23] by another experiment. Methanol (${}_bc_M = 3.5 \cdot 10^{-3}$ M) was added once more at the end of the washing procedure. Although the potential changed only slightly, the bulk concentration decreased within 13 minutes by an amount corresponding to about $0.5\,Q_{El}$. The latter value is in agreement with the result $(Q_{Anal} - Q_{El})/Q_{El} \approx 0.5$. Physical adsorption of CH_3OH molecules was suggested [23] as the probable origin of weakly bonded species.

After the surface had been covered [10—15] largely with $C_sH_pO_q$ by anodic polarization in $0.5\,M\,H_2SO_4 + 0.15\,M\,CH_3OH$, the washing procedure was carried [23] out. Then methanol ($1.6 \cdot 10^{-3}$ to $3.5 \cdot 10^{-3}$ M) was added to the solution. The change of ${}_bc_M$ was not larger than 20 μcoul/cm² (referred to real surface). This represents about 10 % of Q_{Anal}. Adsorption of weakly bonded species occurs also in the presence of $C_sH_pO_q$. However, it is small at large values of $\theta_{C_sH_pO_q}$. Adsorption of weakly bonded species was not detected [23], if methanol was injected at potentials close to zero where the surface is nearly covered with a monolayer of H atoms. It is not known if a positive potential exists above which the adsorption of weakly bonded species is negligible because of the charge of the double layer.

3. Rate of Formation of Strongly Bonded Species at Constant Potential

The rate of formation of $C_sH_pO_q$ is controlled by mass transport as long as the rate of the slowest step of the formation is faster than that of convective diffusion. Diffusion control has been reported for ethylene [18, 21, 24] and acetylene [21, 24] during the initial time of interaction between fuel and surface. When the coverage increases with time, both the kinetic step and mass transport and finally the kinetic step alone become rate-controlling.

The nature of the rate-determining step may vary with the fuel. For methanol the rate of formation of $C_sH_pO_q$ follows [11, 13] Eq. 7:

$$i_1 = k_1\,{}_bc_M \exp(-m_1\theta - \alpha_1 n_1 FU/RT)\,. \tag{7}$$

Here θ designates the coverage with $C_sH_pO_q$, and ${}_sQ_F$ is the charge corresponding to the formation of the saturation coverage. The integration of Eq. 8

$$ {}_sQ_F \frac{d\theta}{dt} = i_1 \tag{8}$$

at constant U and a simple transformation give:

$$\frac{1}{i_1} = \frac{1}{k_1\,{}_bc_M \exp \alpha_1 \dfrac{UF}{RT}} + \frac{m_1 t}{{}_sQ_F} = \frac{1}{i_{in}} + \frac{m_1 t}{{}_sQ_F}\,. \tag{9}$$

The reciprocal value of i_1 increases linearly with time t. The analysis of current-time curves on smooth platinum at 25 °C in 0.5 M H_2SO_4 containing CH_3OH is shown in Fig. 37. The agreement between experiment and theory is good in an intermediate time interval. This result suggests that the weakly bonded species are adsorbed methanol mo-

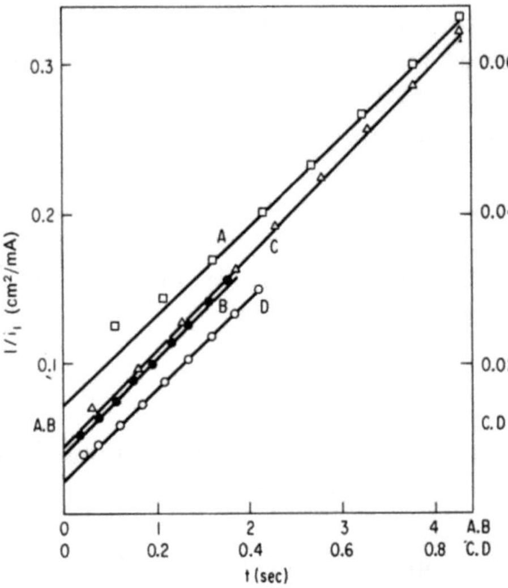

Fig. 37. Analyses of current-time curves during the initial interaction of methanol with platinum in $0.5\,M\,H_2SO_4 + X\,M\,CH_3OH$ at different potentials.

A: X=0.05,	U=0.42 V	C: X=1,	U=0.37 V
B: X=0.5,	U=0.32 V	D: X=2,	U=0.37 V

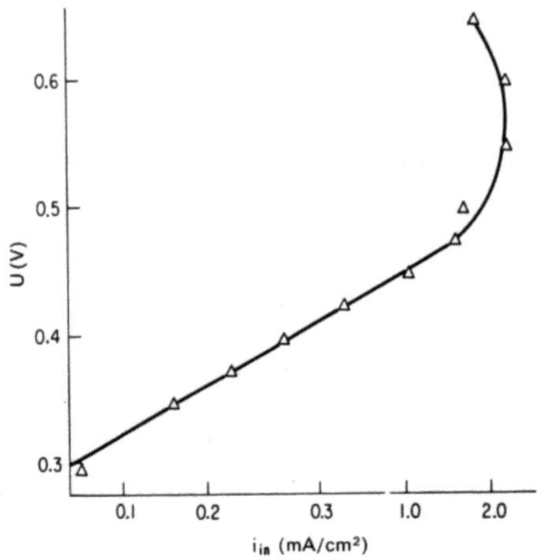

Fig. 38. Tafel plot of the initial current of the formation of $C_sH_pO_q$ in $0.5\,M\,H_2SO_4 + 0.2\,M\,CH_3OH$

lecules. Since no oxidation is involved, these molecules do not contribute to the anodic current.

The plots of $1/i_1$ versus t revealed [13] that the curves are parallel to each other for $0.3\,\mathrm{V} \leq U < 0.45\,\mathrm{V}$ at ${}_{\mathrm{b}}c_{\mathrm{M}} \leq 0.5\,\mathrm{M}$. The Tafel plot [13] of i_{in} versus U in Fig. 38 yields a straight line with a slope of 125 mV per decade of current at $U < 0.45\,\mathrm{V}$. Above 0.45 V the slope of the $1/i_1 - t$ curves changes with U. The rate of formation of $C_sH_pO_q$ depends [13] only slightly upon potential between 0.45 V and 0.65 V.

Although the applicability of Eq. 7 to the formation of $C_sH_pO_q$ from methanol has been established, the rate-determining step of the reaction and the cause for the small dependence of i_{in} upon U between 0.45 V and 0.65 V are not known with certainty. This statement is made in view of the result [15, 16] that more than one type of species seem to be formed.

The rate of formation of $C_sH_pO_q$ from ethane at 60 °C in 1 M $HClO_4$ was found [8, 21] to follow the law

$$ {}_{\mathrm{S}}Q_{\mathrm{F}} \frac{d\theta}{dt} = k_1 \, {}_{\mathrm{b}}c_{\mathrm{E}} (1-\theta)^2 \tag{10} $$

up to $\theta = 0.7$ at 0.3 V and 0.4 V in 1 M $HClO_4$.

The rate-determining step is second order in unoccupied sites and first order in bulk concentration. The following mechanism was suggested as consistent with Eq. 10:

$$ C_2H_6 = (C_2H_5)_{ad} + H_{ad}, \tag{11} $$

$$ H_{ad} = H^+ + e^- . \tag{12} $$

The charge Q_{F} due to the formation of $C_sH_pO_q$ is obtainable by the integration of the current-time traces under the assumption that the contribution of other electrochemical processes is negligible. Other processes may occur simultaneously with the reaction during the initial time, and during the whole process. The rapid reduction of the oxygen layer produced during the activation procedure on smooth platinum is an example of the first of these cases, the hydrogenation or further oxidation of $C_sH_pO_q$ of the second case. Integration of current-time traces corrected according to Eq. 9 was suggested [13] in the case of methanol to eliminate the contribution of other processes. The steady-state coverage can be computed from Q_{F} values. However, other methods which do not require the knowledge of kinetic expressions equivalent to Eq. 7 appear to be more attractive.

4. Coverage from Anodic Pulses

Two different approaches have been developed. If the oxidation of $C_sH_pO_q$ is to be carried out at a slow rate, the concentration of the fuel dissolved in the electrolyte has to be decreased to an acceptable level by washing procedures [1, 3, 10, 12, 23] in the case of liquid fuels and by intensive stirring with an inert gas in the case of gaseous fuels. These procedures make the contribution of the oxidation of fuel from the solution negligible. The method which was first applied to the oxidation of adsorbed carbonaceous species by PAVELA [25] is useful if the species are strongly bonded so that losses during the washing or stirring are small. The technique of determining the coverage with adsorbed species by anodic pulses of large amplitude was developed [26—28] by BREITER and GILMAN. The oxidation occurs at such a rate that the simultaneous oxidation of fuel, diffusing to the surface within the duration of the pulse, is negligible. The fact that oxygen layers on platinum metals inhibit [27] the oxidation of fuels is of advantage for smooth electrodes on which the oxidation of $C_sH_pO_q$ and the formation of the oxygen layer occur simultaneously at large pulses.

The net reaction for the oxidation of $C_sH_pO_q$ is given by Eq. 13 in acid electrolytes

$$(13)\quad C_sH_pO_q + (2s-q)H_2O = sCO_2 + (4s+p-2q)H^+ + (4s+p-2q)e^- .$$

The 100% efficiency of CO_2 production [23] is a consequence of the relatively large potentials of oxidation and the high oxidation state of $C_sH_pO_q$ itself. In alkaline solutions, the net reaction is:

$$(14)\qquad C_sH_pO_q + (6s+p-2q)OH^- = sCO_3^{--} + (3s+p-q)e^- .$$

The application of the technique involving a washing procedure is illustrated in Fig. 39. A platinized platinum electrode with 50 cm^2 of apparent surface was polarized anodically with 50 mA in 0.5 M H_2SO_4 +0.1 M CH_3OH for 40 minutes at 0°, 20°, 40°, and 60°C. Afterwards, the electrode and the compartments of the vessel were washed five times with 0.5 M H_2SO_4, saturated previously with helium. The curves a′, b′, c′, and d′ were taken at the above temperatures. They exhibit a short hydrogen branch and an arrest at potentials of the double layer region. The arrest is due to reaction 13. As example of the adsorption of a gaseous fuel, the charging curves a, b, c, and d obtained with 50 mA in the presence of CO_{ad}, are shown in Fig. 39 for the given temperatures. The experimental procedure was already described in section 6 of chapter VI. The charging curves in the absence of chemisorbed carbonaceous species are not added. But each of the curves, a, b, c, d and a′, b′, c′, and d′ coincides at potentials in the oxygen region with the corresponding

$U-t$ curve without adsorbed species. The electrode is free of $C_sH_pO_q$ or CO_{ad} at $U > 0.75$ V. The transition times for the oxidation of $C_sH_pO_q$ or CO_{ad} can be determined as illustrated by curve b in Fig. 15. The charge Q_{Ox} for the oxidation of $C_sH_pO_q$ or CO_{ad} is the product of current density and transition time. Instead of oxidizing the chemisorbed species at constant current, a voltage that increases linearly with potential at

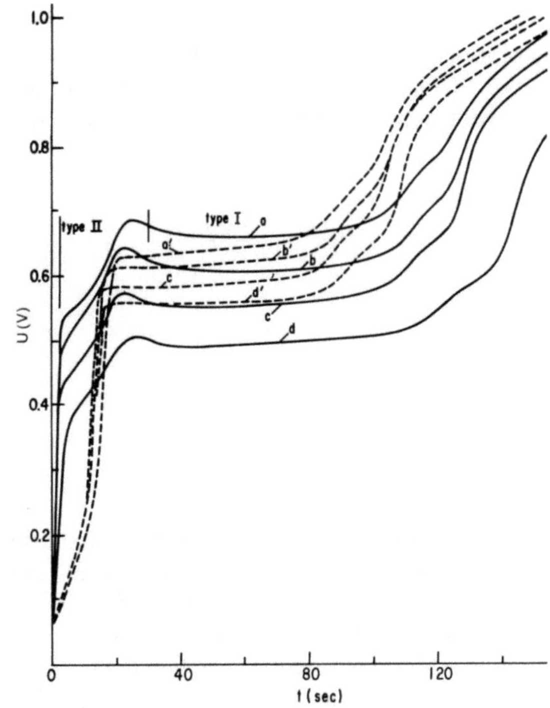

Fig. 39. Anodic charging curves on platinized platinum with 50 mA in 0.5 M H_2SO_4 at different temperatures. Chemisorbed species formed previously by bubbling with CO for curves a, b, c, d and by anodic polarization in 0.5 M H_2SO_4 + 0.1 M CH_3OH for curves a′, b′, c′, d′. a, a′: 0 °C, b, b′: 20 °C, c, c′: 40 °C, d, d′: 60 °C

a slow rate ($v \leq 1$ mV/sec) can be applied [2] potentiostatically. The integration of the wave due to the oxidation of the chemisorbed species yields Q_{Ox}. As for the adsorption of H atoms or CO molecules the coverage θ with $C_sH_pO_q$ may be defined:

$$\theta = \frac{Q_{Ox}}{{}_sQ_{Ox}}. \tag{15}$$

Again the subscripts in front of a symbol denote saturation coverage for the respective species. This definition is identical with the regular definition of coverage as ratio of surface excesses

$$\theta_r = \frac{\Gamma}{{}_s\Gamma} \tag{16}$$

if there is only one type of adsorbed species since

$$Q_{Ox,j} = n_j F \Gamma_j . \tag{17}$$

Here n_j is the number of electrons required for the oxidation of a single species of type j to CO_2, Γ_j is the adsorbed amount in M/cm^2 (surface excess). When several types of species are present, the coverage θ

$$\theta = \frac{Q_{Ox}}{{}_sQ_{Ox}} = \frac{\sum_j Q_{Ox,j}}{\sum_j {}_sQ_{Ox,j}} \tag{18}$$

is only identical with θ_r

$$\theta_r = \frac{\sum_j \Gamma_j}{\sum_j {}_s\Gamma_j} \tag{19}$$

if

$$\Gamma_m = \Gamma_n \qquad n, m = 1, \ldots, j . \tag{20}$$

The curves in Fig. 40 and Fig. 41 are examples of the use of anodic pulses with large amplitudes. Anodic charging curves taken [28] with 91 mA/cm^2 from 0.4 V on smooth platinum in 1 M $HClO_4$ (curve a) and in 1 M $HClO_4$ + 0.1 M HCOOH (curve b) are replotted in Fig. 40. The current was applied when the potential reached 0.4 V during the anodic sweep of periodic voltammetric $I-U$ curves with $v = 30$ mV/sec. The graphical determination of the transition time τ_O for the completion of about a monolayer of adsorbed oxygen atoms is demonstrated for curve a. It is $Q_O = \tau_O \cdot i$. Curve b exhibits an arrest at about 1.0 V, followed by a gradual rise of potential which leads to the coincidence of curves a and b in the region of oxygen evolution. The gradual potential rise makes a graphical determination of the transition time of the arrest at 1.0 V inaccurate. The procedure of determining τ (see curve b) takes into account the possibility of a simultaneous oxidation of $C_sH_pO_q$ and of the formation of the oxygen layer. If $i \cdot \tau$ is designated as Q, the coverage defined on the basis of charges is:

$$\theta = \frac{Q - Q_O}{{}_sQ - Q_O} . \tag{21}$$

The same information may be obtained [26] by applying potentiostatically a rapid voltage sweep to positive potentials. Trace A in Fig. 41 is the $i-t$ curve measured from 0.4 V at 800 V/sec in 1 M $HClO_4$. Trace B was measured under the same conditions in 1 M $HClO_4$ + 1 M CH_3OH after the formation of $C_sH_pO_q$ had occurred for some time at open circuit. Shaded area 2 less shaded area 1 is proportional to the difference $(Q-Q_O)$.

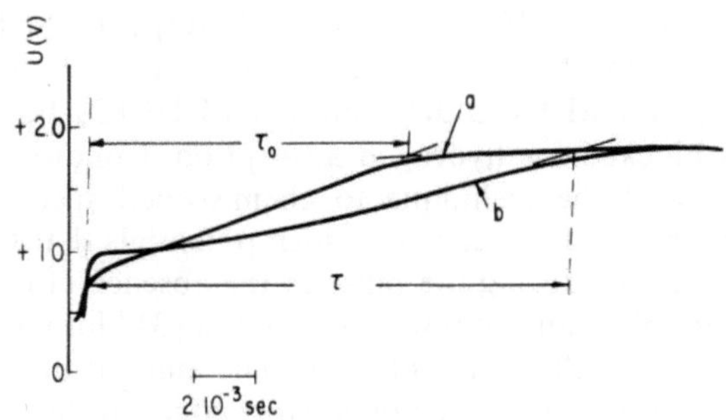

Fig. 40. Anodic charging curves on platinum from 0.4 V during the anodic potential sweep in 1 M $HClO_4$ (curve a) and in 1 M $HClO_4$ + 0.1 M HCOOH (curve b) at 91 mA/cm²

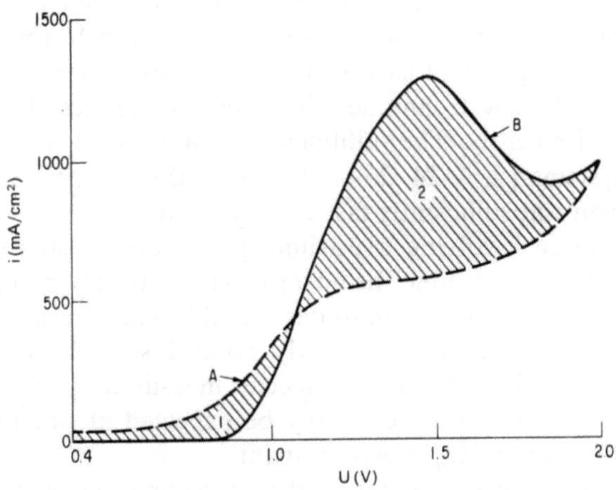

Fig. 41. Voltammetric $i-U$ curves on platinum with 800 V/sec in 1 M $HClO_4$ (curve A) and in 1 M $HClO_4$ + 1 M CH_3OH (curve B) starting from about 0.4 V

The above technique gives correct results if the following conditions [26, 28] are fulfilled:

a) The chemisorbed material is completely oxidized by the time that traces a and b in Fig. 40, or traces A and B in Fig. 41, merge.

b) Identical charges Q_O were involved in the formation of the oxygen layer when traces a and b or traces A and B coincide.

c) Capacitive corrections are negligible.

d) Negligible oxidation of fuel from the electrolyte occurs during the pulse.

Evidence for the validity of condition a was established by applying [29, 30] a cathodic current of large amplitude at the potentials where traces a and b merge. The cathodic charging curve exhibits the same hydrogen arrest in 1 M $HClO_4$ and 1 M $HClO_4$+0.1 M HCOOH. If adsorbed species were still present, the hydrogen arrest should be shorter in 1 M $HClO_4$+0.1 M HCOOH than in 1 M $HClO_4$ because part of the surface is blocked for hydrogen adsorption. Condition b restricts the applicability of the technique to chemisorbed species for which traces a and b or A and B coincide at such potentials that O_2 evolution is not yet appreciable during the pulse in the absence of fuel. Evidence for the fulfilment of condition b was established [31] for CO. The initial adsorption of CO is diffusion-controlled. The amount of CO adsorbed in a certain time was found to be practically equal to the amount computed from $(Q-Q_O)$. It may be concluded that condition b holds for those types of chemisorbed species which are oxidized in the same potential region as CO_{ad}. A small error is introduced in the determination of $(Q-Q_O)$ because the charge for the double layer charging is slightly smaller in the presence of adsorbed species than in their absence (condition c). The adsorption of carbonaceous species decreases the double layer capacity. A new technique which aims to reduce this error was described [32] recently. The fulfilment of condition d was verified for numerous systems [2, 4, 24, 26—28, 31] by the indepence of $(Q-Q_O)$ upon i or upon v at sufficiently fast charging rates.

The technique involving a washing procedure or stirring with an inert gas allows to determine the adsorption of strongly bonded species. Anodic pulses with large amplitudes should yield the charge due to the oxidation of strongly and weakly bonded species. An agreement of the coverage with carbonaceous species measured by the two techniques on the same electrode can only be expected at potentials where the coverage with $C_sH_pO_q$ is predominant.

The maximum relative error in the determination of θ by the two techniques is given by Eq. 22 or Eq. 23 respectively:

$$\frac{\Delta\theta}{\theta} \leq \left|\frac{\Delta Q_{Ox}}{Q_{Ox}}\right| + \left|\frac{\Delta_s Q_{Ox}}{{}_sQ_{Ox}}\right|, \tag{22}$$

$$\frac{\Delta\theta}{\theta} \leq \left|\frac{\Delta Q}{{}_sQ-Q_O}\right| + \frac{\Delta Q_O}{{}_sQ-Q_O} + \left(\left|\frac{\Delta_s Q_O}{{}_sQ-Q_O}\right| + \left|\frac{\Delta Q_O}{{}_sQ-Q_O}\right|\right)\frac{Q-Q_O}{{}_sQ-Q_O}. \tag{23}$$

If it is assumed that the different Q's can be obtained with an accuracy of 3%, the error for the first technique will be smaller or equal to 6%. It is more difficult to estimate the error limit for the second technique, but instead of 6% it will be at least 10%. The author is not sure if sufficient attention has been paid to the error limits in recent studies of chemisorbed species on smooth electrodes.

5. Coverage from Hydrogen Deposition

The effect of adsorbed carbonaceous species on hydrogen adsorption was already discussed in section 6 of chapter VI. The quantity

$$\theta' = 1 - \frac{{}_{S}Q'_{H}}{{}_{S}Q_{H}} \tag{24}$$

or

$$\theta' = \frac{{}_{S}Q_{H} - {}_{S}Q'_{H}}{{}_{S}Q_{H} - {}_{R}Q'_{H}} \tag{25}$$

is a relative measure for the number of sites on which hydrogen adsorption does not occur. The definition in Eq. 24 refers [26, 28] θ' to the total part of the surface which is available for hydrogen adsorption. In contrast, the portion of the hydrogen adsorbing area on which carbonaceous species are not adsorbed under saturation conditions is excluded [11] in Eq. 25. This portion is measured by the charge ${}_{R}Q'_{H}$ obtained from the hydrogen arrest at maximum coverage.

Fig. 42 shows a series of cathodic charging curves which were taken [28] from 0.4 V on smooth platinum with 91 mA/cm² in 1 M $HClO_4$ containing different additions of HCOOH. The cathodic pulse was applied when the potential reached 0.4 V during the anodic sweep of periodic voltammetric $I-U$ curves. The determination of the transition

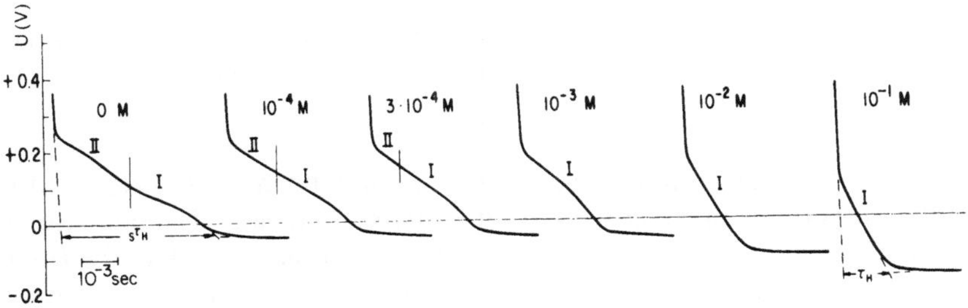

Fig. 42. Cathodic charging curves on platinum with 91 mA/cm² from 0.4 V during the anodic sweep at different additions of HCOOH to 1 M $HClO_4$

time ${}_{S}\tau_{H}$ for the formation of a monolayer of H atoms is shown in the curve on the left side. The transition time τ_{H} on the right side of Fig. 42 corresponds to ${}_{R}Q'_{H}$ since saturation coverage with $C_sH_pO_q$ is present. The region of the strongly bonded hydrogen is marked as II, the region of the weakly bonded hydrogen as I. The increase in the bulk concentration ${}_{b}c_{F}$ leads to a decrease of ${}_{S}Q'_{H}$ because more $C_sH_pO_q$ is chemisorbed. It appears from Fig. 42 that region II disappears more rapidly than region I with ${}_{b}c_{F}$. A similar situation was found in the case of CO adsorption (compare Fig. 15). One cathodic charging curve is sufficient for the determination of ${}_{S}Q'_{H}$ as long as $U \geq 0.4\,V$ in acid electrolytes. At potentials in the hydrogen region, both a cathodic and an anodic charging curve have [28, 30] to be taken. If the same current density is used, the sum of the anodic ($\tau_{H,an}$) and cathodic ($\tau_{H,cath}$) transition time is equal to τ_{H}.

The quantity θ' is representive of the coverage with carbonaceous species if a desorption for electrostatic reasons or a reduction does not occur. These two conditions are fulfilled for many strongly bonded

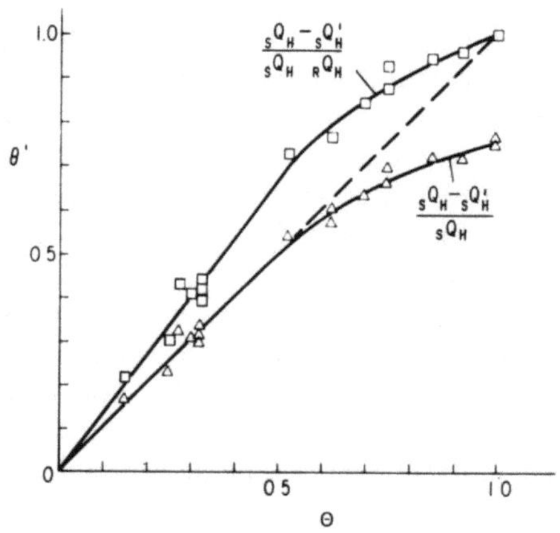

Fig. 43. Plots of the coverage θ' according to Eq. 24 and 25 as a function of θ for $C_sH_pO_q$ formed on platinum during the methanol oxidation

species. The same cathodic charging curve may be measured repeatedly if the potential is brought back to the starting potential. The increase of the hydrogen overvoltage by the chemisorbed species (compare the $U-t$ curves on the right side of Fig. 42) favors the deposition of H_{ad} to the production of H_2. It is not known if a desorption is absent for weakly adsorbed molecules.

The quantity θ' as defined by Eq. 24 and Eq. 25 is plotted in Fig. 43 as a function of θ from the data [26] in methanolic solutions. A linear relation exists between θ' and θ for $\theta<0.5$. Above $\theta=0.5$ the $\theta'-\theta$ curves show a curvature. The lower curve tends towards $\theta'=0.75$ for $\theta=1$. It appears that only 75% of the sites accessible to hydrogen adsorption are available for the adsorption of carbonaceous species (largely $C_sH_pO_q$). The curvature which is also present in the upper curve may be due to a steric effect resulting from the adsorption of one or more types of species. Similar $\theta'-\theta$ curves were obtained [28] for species formed from HCOOH on smooth platinum. Although the interpretation of the shape of the curves in Fig. 43 is uncertain, the curves are useful for conversion purposes.

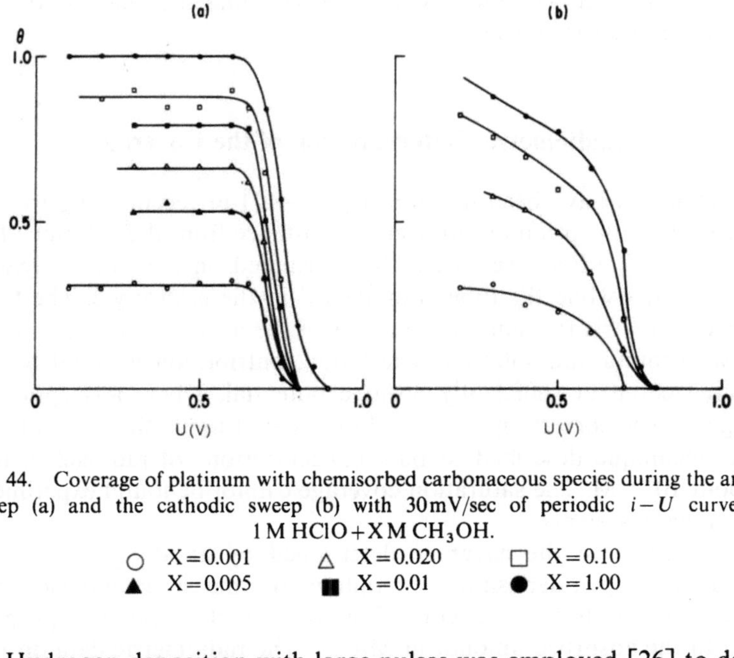

Fig. 44. Coverage of platinum with chemisorbed carbonaceous species during the anodic sweep (a) and the cathodic sweep (b) with 30 mV/sec of periodic $i-U$ curves in 1 M HClO + X M CH_3OH.

○ X = 0.001 △ X = 0.020 □ X = 0.10
▲ X = 0.005 ■ X = 0.01 ● X = 1.00

Hydrogen deposition with large pulses was employed [26] to determine the coverage of smooth platinum with carbonaceous species in 1 M $HClO_4$ containing different amounts of CH_3OH. A short rectangular current was passed by switching rapidly from a potentiostatic to a galvanostatic circuit whenever the potential reached the desired value during the anodic or cathodic sweep. The periodic voltage of triangular shape applied at 30 mV/sec between 0.1 V and 1.5 V served to achieve a reproducible surface state for the start of the pulse. The coverage θ during the anodic and cathodic sweep is plotted as a function of potential at different methanol bulk concentration, in Fig. 44a and Fig. 44b respectively. It increases during the cathodic sweep after the oxygen layer formed during the last part of the anodic sweep has been largely reduced. The increase of θ with U is rapid between about 0.8 V and 0.6 V. A constant value, which is maintained during a large part of the subsequent anodic sweep, is reached at 0.1 V. The coverage decreases rapidly with U above 0.6 V during the anodic sweep. The decrease results [26, 27] because the rate of formation of carbonaceous species does not increase with potential as does the rate of oxidation of $C_sH_pO_q$.

The coverage was attributed to adsorbed methanol molecules in the original [26, 27] and subsequent papers [33, 37]. This view cannot be maintained any more. A large part of the species formed at ${}_bc_M \geq 0.005$ M consists of $C_sH_pO_q$ at $U \leq 0.6$ V during both the cathodic and anodic sweep. It is not known to what extent weakly bonded species contribute to the coverage above 0.6 V.

6. Radiometric Determination of the Coverage

Techniques have been developed [38—42] in recent years for the study of the adsorption of radioactive substance from the solution. The adsorption of radioactive species is determined on metallized Geiger-tube windows while the tube is immersed in the electrolyte. The total recorded count is the sum of the count of the adsorbed species and of the background and solution. The latter contribution is equal to the observed count at sufficiently positive potentials where adsorption is negligible. The correction for the solution count limits the applicability of the technique described to bulk concentrations of radioactive fuel of about 10^{-2} M. The saturation coverage cannot be found experimentally in many systems.

In contrast to the coverage determined by anodic pulses or by cathodic hydrogen deposition, the radiometric measurements yield the regular coverage θ_r since the count from the adsorbed species is proportional to the adsorbed amount in M/cm². The radiometric data do not allow a direct decision if a dehydrogenation occurs during the adsorption process, since only $\sum_j \Gamma_j$ is obtainable if j types of carbonaceous species exist. The same limitations apply here as for the coverage determined from anodic charges or from hydrogen deposition.

Curve c in Fig. 45 represents the coverage determined [42] radiometrically on platinum at ${}_bc_M = 10^{-2}$ M as a function of U. The window of the Geiger tube consisted of a polyester film with a thickness of 10^{-3} cm, a vapor deposited gold film of about 1000 Å on top of the polyester, and an electroplated platinum film on top of the gold. The electrode was stepped to 1.8 V and held there for 12 sec in order to remove species adsorbed during the preceding operations. Then the potential was switched to the potential of interest. The steady-state value for the adsorbed species was assumed to be established when the count no longer changed with time. The steady-state coverage developed within

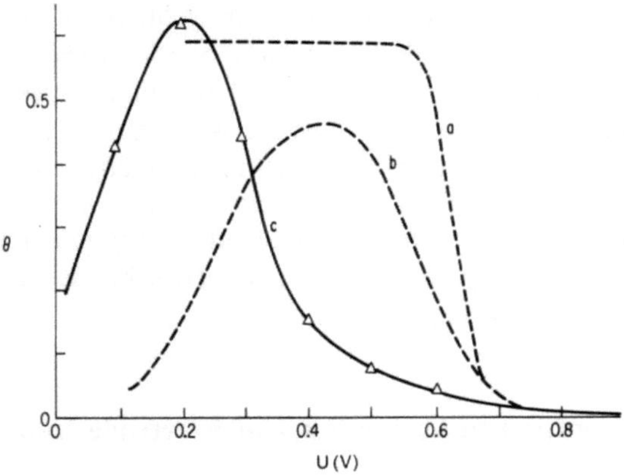

Fig. 45. Coverage of $C_sH_pO_q$ on platinum as a function of U under different conditions.
a: determined by hydrogen deposition[24] during the anodic sweep in $1\,M\,HClO_4 + 10^{-2}\,M\,CH_3OH$
b: determined by hydrogen deposition[43] after 2 minutes of formation at a given potential in $0.5\,M\,H_2SO_4 + 10^{-2}\,M\,CH_3OH$
c: determined by radiometric measurements[42] after a steady-state count had been achieved in $0.5\,M\,H_2SO_4 + 10^{-2}\,M\,CH_3OH$

less than 3 minutes at $U > 0.4$ V. The time increased with decreasing potential and exceeded 40 minutes at 0.1 V. The saturation coverage was estimated to be $4.5 \cdot 10^{14}$ methanol molecules per real square centimeter of electrode area. Curve a in Fig. 45 is the $\theta - U$ curve at ${}_bc_M = 10^{-2}$ M from Fig. 44a. The shape of curve c is similar to that of curve b which was determined [43] by hydrogen deposition according to Eq. 25. In contrast to curve c, the $C_sH_pO_q$ formation which was preceded by an anodic pretreatment was allowed [43] for 2 minutes only.

Curves c and b differ considerably in the location of the maximum. The difference was attributed [42] to insufficient time of adsorption. Although the insufficient time of adsorption is a valid objection [13] against the acceptance of curve b as representative for steady-state coverage, the marked difference between curve c on the one hand and curves b and a on the other hand cannot be resolved in this way. An insufficient time of formation implies that the θ' values of curve b are too small at $U < 0.3$ V and $U > 0.45$ V. Recent steady-state determinations of θ' according to Eq. 24 of the author [44] on platinized platinum in 0.5 M H_2SO_4 + 0.01 M CH_3OH yielded a $\theta - U$ curve close to curve a for $U > 0.4$ V. The steady-state measurements had been made by adding methanol at 0.05 V, polarizing with 1 mA for 2 hr and increasing the current subsequently every 2000 sec.

The reason for the different results of electrochemical and radiometric measurements is not known at present. Simultaneous electrochemical and radiometric determination of the coverage with carbonaceous species might give the answer. The rough agreement between the maximum of the θ value of curve c with that of curve a is just a consequence of the estimate of the saturation coverage. There is no experimental evidence that the maximum possible coverage with methanol molecules (not with COOH and HCO) as derived by visual examination of closely packed Fischer-Hirschfelder-Taylor atomic models is achieved in practice.

7. Coverage and Capacitance of Electrode Impedance

Capacity measurements are very useful [45, 46] for the study of the adsorption of organic species on liquid electrodes (mercury, amalgams, gallium). The work on solid electrodes [33, 35—37, 47—51] is not so comprehensive as that on mercury. The reason for this situation lies in the difficulty of achieving a satisfactory reproducibility and of interpreting the data. On platinum metals, the double layer region is bordered on the left side by the hydrogen region and on the right side by the oxygen region. The existence of these two regions on noble metals and their alloys has a profound effect [49, 52—54] on the adsorption of organic species. The implications of this effect were discussed [52, 54] under the simplifying assumption of adsorption equilibrium of the hydrogen atoms and of the undissociated organic molecules. Some general predictions were made with respect to the influence of the irreversible adsorption of oxygen. These effects have been disregarded in recent discussions [18, 20, 55—57] of radiometric studies of the adsorption of organic species on noble metal electrodes.

Voltammetry [50] with superimposed ac current was used [33, 36—37] by the author to measure the impedance of platinum metals at 1000 Hz in acid solutions containing methanol. The periodic $i - U$ curves a and a′ in Fig. 46 were measured at 30 mV/sec and 30 °C in 1 M $HClO_4$ and 1 M $HClO_4$ + 1 M CH_3OH respectively. The shape of the curves is easily reproduced, and this was one of the main reasons for the voltammetric measurements. Curve a and curve a′ during the anodic sweep were discussed in section 1 of chapter VI and in section 6 of chapter VIII respectively. An anodic wave is observed during the cathodic sweep in a potential range where the oxygen layer has been largely reduced in the case of curve a′.

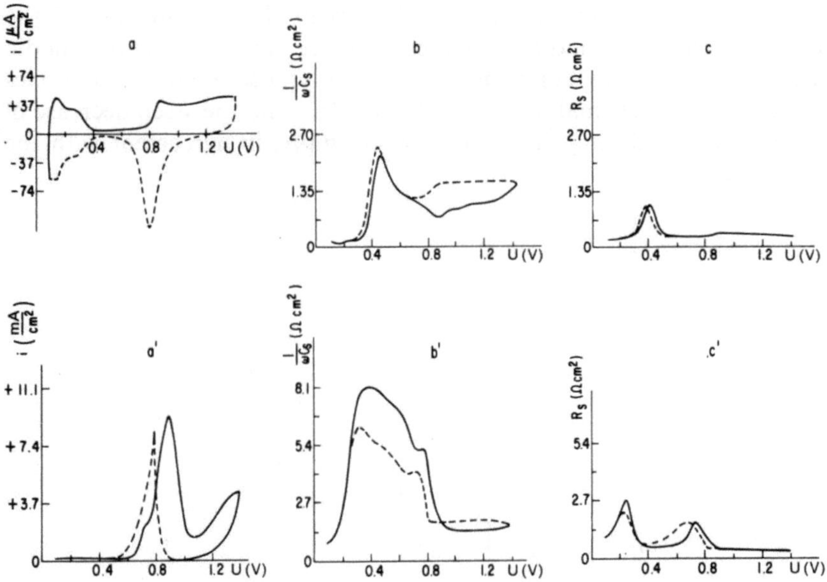

Fig. 46. Current density, capacitive component and ohmic component of the electrode impedance as a function of potential on platinum under voltammetric conditions. a, b, c: 1 M $HClO_4$, a′, b′, c′: 1 M $HClO_4$ + 1 M CH_3OH

Curves b and c represent the capacitive component $1/\omega C_S$ and the ohmic component R_S of the impedance in a series circuit at 1000 Hz in the absence of CH_3OH. The pseudocapacitance of the Volmer reaction makes $1/\omega C_S$ small in the hydrogen region. The capacitive component increases rapidly with U between 0.35 V and 0.45 V when the hydrogen coverage becomes small. After a maximum, $1/\omega C_S$ increases slowly with potential up to 0.85 V during the anodic sweep. A gradual decrease follows. During the cathodic sweep, $1/\omega C_S$ does not change much between 1.4 V and 0.6 V. Then the difference between the two sweeps is small. The interpretation of these results under consideration of reaction 26

$$(26) \qquad H_2O = OH_{ad} + H^+ + e^-$$

has already been given in section 1 of chapter VIII.

Curve b′ in the presence of methanol displays considerably higher values of $1/\omega C_S$ in the hydrogen region and in the double layer region than curve b during both the anodic and cathodic sweep. Hydrogen is still adsorbed, to a much smaller extent, in the hydrogen region. This suggests simultaneous adsorption of $C_sH_pO_q$ and hydrogen between 0 and 0.3 V. The carbonaceous species are formed [2] at potentials of the anodic wave during the cathodic sweep. The broadening of the double layer region and the small capacitance values during the anodic sweep (compare Fig. 29) demonstrate the presence of adsorbed species for $U<0.9$ V. The small hump at about 0.75 V and the steep decrease of $1/\omega C_S$ with U during the anodic sweep imply, in agreement with the

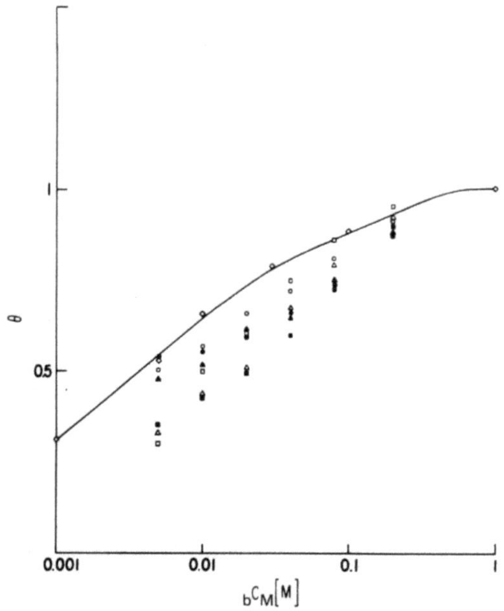

Fig. 47. Coverage of platinum with $C_sH_pO_q$ as a function of methanol bulk concentration at different potentials during the anodic sweep under voltammetric conditions.
◇ from Fig. 44a for $0.4\,V \leq U \leq 0.6\,V$
□, △, ○ from Eq. 27
■, ▲, ● from Eq. 28
□, ■ $U=0.4$ V; △, ▲ $U=0.5$ V; ○, ● $U=0.6$ V

results in Fig. 44a, that the carbonaceous species are removed when the formation of the oxygen layer starts. The removal is due [26, 28] to the oxidation of $C_sH_pO_q$. Similarly, the small hump during the cathodic sweep at about 0.70 V is indicative of the beginning of the formation of $C_sH_pO_q$. The coverage at a given potential between 0.3 V and 0.7 V is smaller during the cathodic sweep than during the anodic sweep (compare Fig. 44). The ohmic component (curve c′) also exhibits humps in the respective potential regions, but more details are displayed by the $1/\omega C_S - U$ curve.

An attempt was made [33, 49, 51] to compute the coverage with $C_sH_pO_q$ from the capacitance data as on mercury. The validity of Frumkin's model [58] of the double layer was assumed in the simplified form:

$$C_\theta = C_{\theta=O}(1-\theta) + C_{\theta=1} \cdot \theta \,. \tag{27}$$

Here the C's designate the double layer capacity for the coverage $\theta = O$, and 1. The value of $C_{\theta=O}$ at a given potential was taken [31, 46] as that in the absence of CH_3OH, the value of $C_{\theta=1}$ as that at ${}_bc_M = 1$ M. The θ values computed from Eq. 27 at three different potentials in the double layer region are represented by open symbols in Fig. 47.

On the basis of Eq. 27, $C_{\theta=O}$ has to be larger than any C_θ since θ is positive. The $1/\omega C_S - U$ curves show that this condition is not always fulfilled for the cathodic sweep. Therefore, the following relationship which is not based on the physical model was tried also (solid symbols in Fig. 47):

$$\theta = \frac{C_{\theta=1}}{C_\theta} \,. \tag{28}$$

The θ values from the capacity data [33] are plotted as a function of ${}_bc_M$ in Fig. 47 together with the corresponding θ values from Fig. 44 (solid line). Fig. 47 demonstrates that the θ values according to Eq. 27 and Eq. 28 are always lower than the respective values of the direct determinations. Very approximate values are obtained. Eq. 27 and Eq. 28 are not at all applicable at potentials outside the double layer region.

Capacity data measured [35] under the same voltammetric conditions on Pt, Pd, Ir, Rh, and Au were used for a comparison of the adsorption of carbonaceous species on these metals. In Fig. 48 the difference $C_{\theta=O} - C_\theta$ is plotted as a function of ${}_bc_M$ at the potentials given in parenthesis above the points. The potential of comparison could not be the same for all the metals because the formation of O_{ad} starts at different potentials. A fixed potential which is independent of ${}_bc_M$ and is located before the beginning of the oxygen layer formation could be chosen for Pt, Pd and Au. For Ir and Rh the potential of the highest peak of the $1/\omega C_S - U$ curve was taken. This potential shifts slightly with ${}_bc_M$.

The difference $C_{\theta=0} - C_\theta$ tends towards a limiting value with ${}_bc_M$ on Pt, Ir, and Au. About a monolayer is present on that part of the surface on which carbonaceous species are adsorbed. This part appears

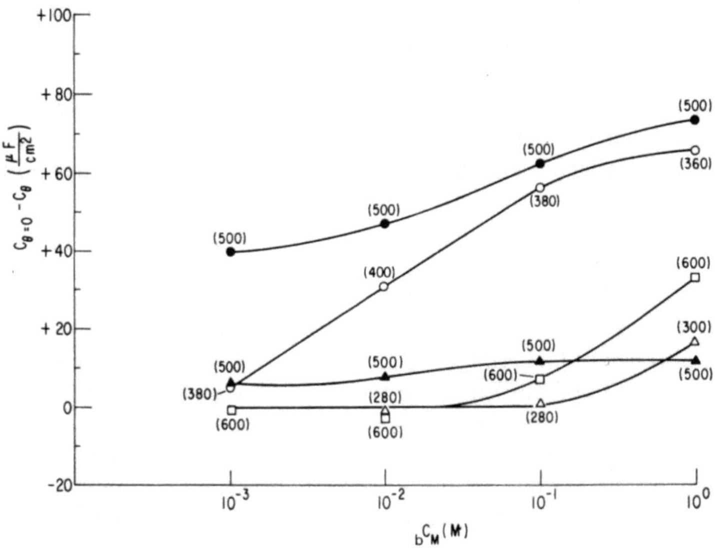

Fig. 48. Difference of capacitances $C_{\theta=0} - C_{\theta}$ as a function of methanol bulk concentration during the anodic sweep. The numbers give the potential in mV to which C_{θ} belongs
● Pt, ○ Ir, ▲ Au, □ Pd, △ Rh

to be the same on Ir and Pt (about 75 % according to Fig. 43) and much smaller on Au. The results on Rh and Pd indicate that the adsorption of carbonaceous species becomes noticeable only at ${}_{b}c_{M} > 10^{-2}$ M. A monalayer is not approached at ${}_{b}c_{M} = 1$ M under the given conditions.

8. Determination of the Number of Electrons in the Oxidation of Chemisorbed Species

A combination of charging curves and simultaneous determination of the evolved amount of CO_2 by gas chromatography was recently developed [15, 16, 59] for electrodes with large real area. The chemisorbed species are formed first. Then the concentration of fuel is reduced to an acceptable level by washing with clean electrolyte in the case of liquid fuels or by stirring with an inert gas if the fuel is gaseous. A constant flow of He is maintained at 0.5 cm^2/sec through the vessel and through the coil of a gas chromatograph. The adsorbed species are oxidized at constant current, and the $U-t$ curve is recorded by a suitable recorder. The current is turned off when the potential reaches 1.2 V. The CO_2 content of the gas is determined by gas chromatography at intervals of about 400 sec. An example is given in Fig. 49. The amount of CO_2 converted into electrical units is plotted as a function of time.

From plots of the type in Fig. 49, the charge Q_{CO_2} may be computed:

$$Q_{CO_2} = \int_0^{t_0} I(t-25)\,dt\,. \tag{29}$$

The time is counted in seconds. The time t_0 of the upper integration limit corresponds to 4 500 sec in Fig. 49. Since it took 50 sec for the gas mixture to flow through the 25 cm^3 coil of the gas chromatograph, an average value I which corresponds to $(t-25)$ was measured at the time t when the sampling coil was separated from the cell by the valve.

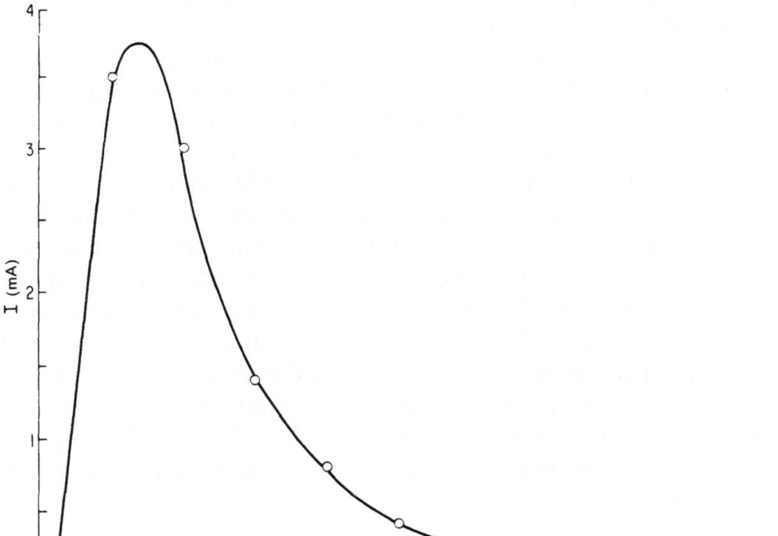

Fig. 49. Carbon dioxide content of the gas mixture coming from the cell in electrical units as a function of time after CO_2 had been produced by the oxidation of $C_sH_pO_q$

According to Eq. 13 the charge Q_{Ox} determined from the charging curve is proportional to $(4s+p-2q)$, and Q_{CO_2} to s if I_{CO_2} is obtained by calibration of the gas chromatograph with an electrochemical reaction involving one electron per CO_2 molecule. Thus, the number of electrons per CO_2 molecule

$$\frac{Q_{Ox}}{Q_{CO_2}} = 4 + \frac{p-2q}{s} \tag{30}$$

is accessible from experimental data. With an estimated uncertainty of 5 % in the determination of Q_{Ox} and Q_{CO_2}, the error limit for

$$\frac{Q_{Ox}}{Q_{CO_2}} \text{ is } \pm 20\,\% \text{ if } \frac{Q_{Ox}}{Q_{CO_2}} \approx 2\,.$$

A similar idea concerning the combination of charging curves and gas chromatography in the study of carbonaceous species on a Teflon-bonded platinum-black electrode was used in reference 60. The species were formed by adsorbing propane first and dehydrogenating the species afterwards. However, the amount of CO_2 cannot be determined quantitatively since diffusion of CO_2 into the electrolyte takes place. Only the amount of carbon dioxide that leaves the side of the electrode adjacent to the gas phase was determined.

An analytical procedure that gives the same information as the combination of charging curves and gas chromatography was employed by JOHNSON and KUHN [6] in the study of the nature of "reduced" carbon dioxide [2]. Chemisorbed species were formed at open circuit by the interaction of CO_2 with platinum in 1 M H_2SO_4 at potentials of the hydrogen region. After the washing procedure the electrode was transferred to a cell containing 2 M NaOH. The chemisorbed species were oxidized at constant current. After removal of the electrode, the amount of carbonate in the alkaline solution was determined by chemical analysis. The amount of carbonate in electrical units agreed with the charge Q_{Ox} for the oxidation of the chemisorbed species if the oxidation was assumed to be a two-electron step. Although the formation of chemisorbed species in the presence of HCOOH or CH_2O was also studied [5], the analytical technique was not applied to establish the values of the ratio Q_{Ox}/Q_{CO_2}.

9. Effect of pH and Anions on Coverage

The effect of pH on the adsorbed amount of $C_sH_pO_q$ was studied [11, 14] on smooth platinum at constant potential and methanol bulk concentration in phosphate buffer solutions. The coverage at 0.7 V decreases with increasing pH at ${}_bc_M = 0.1$ M and 0.01 M, passes through a broad minimum between pH = 4 and 7 and increases between pH = 7 and 10 again. The decrease of coverage with pH may result [14] from a change of the double layer structure. The increase starts at a pH value close to pK_{diss} of methanol. Dissociation of methanol both in the bulk and on the surface seems to occur. Two types of adsorbed particles, HCO and CO^-, were suggested [11, 14] for alkaline solutions.

Specifically adsorbable anions like Cl^- and Br^- lead [10, 36, 61, 62] to a decrease of the chemisorption of $C_sH_pO_q$. Periodic voltammetric $I-U$ curves were taken [36] at 30 mV/sec in 1 M $HClO_4$ + 0.1 M CH_3OH + Y M HCl (Y = 10^{-6} to 10^{-2} M). The impedance of the smooth Pt electrode was measured simultaneously by voltammetry with superimposed ac current. The rate of methanol oxidation at a given potential was found to decrease with ${}_bc_{Cl^-}$. The peak currents of the two waves during the anodic sweep and of the wave during the cathodic sweep

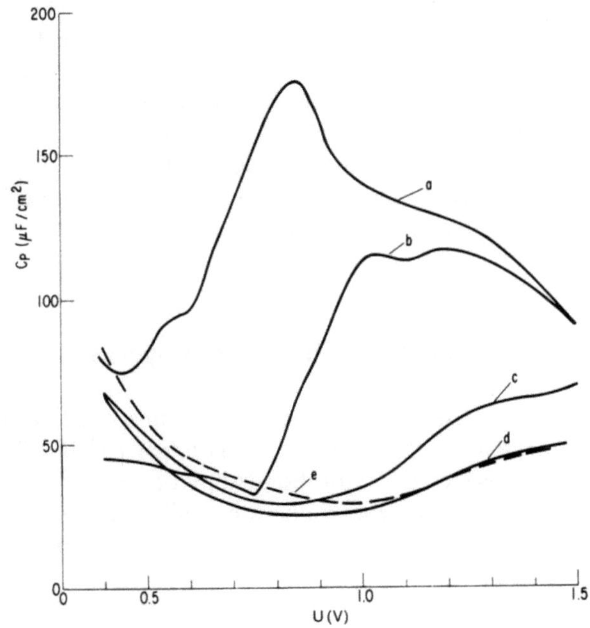

Fig. 50. Capacitance of the impedance on platinum as a function of potential during the anodic sweep in 1 M $HClO_4$ + X M CH_3OH + Y M HCl.
a: X = 0, Y = 0
b: X = 0.1, Y = 0
c: X = 0.1, Y = 3.10^{-4}
d: X = 0.1, Y = 10^{-2}
e: X = 0, Y = 10^{-2}

(compare curve a′ in Fig. 46) become smaller. The wave during the cathodic sweep disappears at ${}_bc_{Cl^-} > 10^{-3}$ M in unstirred solutions. The waves characteristic for the Volmer reaction become noticeable at $Y = 5 \cdot 10^{-5}$ and increase in height up to ${}_bc_{Cl^-} = 10^{-2}$ M. The $I-U$ curves demonstrate that the formation of $C_sH_pO_q$ is increasingly inhibited.

The capacitance C_p of the impedance in a parallel circuit was computed [36] from the data measured in a series circuit. C_p is plotted as a function of U during the anodic sweep in Fig. 50 for different additions of HCl. Curve a in 1 M $HClO_4$ displays the characteristic increase of C_p with U between 0.5 V and 0.8 V. The beginning of the increase is shifted to about 0.75 V in curve b when $C_sH_pO_q$ is present. Small amounts of HCl produce a wide minimum in the C_p-U curve. The shape of the C_p-U curves becomes increasingly similar to that for chloride ion adsorption under equivalent conditions [63]. This is clearly demonstrated by the closeness of curves d and c which were measured in the presence and absence of CH_3OH in 1 M $HClO_4$ + 0.01 M HCl. The adsorption of Cl^- prevents the formation of $C_sH_pO_q$. Competive formation of $C_sH_pO_q$ from ethane and adsorption of chloride ions were studied [62] recently. Preadsorbed chloride ions caused an initial decrease of the rate of formation of $C_sH_pO_q$ owing to a blocking of adsorption sites in 1 M $HClO_4$ saturated with ethane at 60 °C. When $C_sH_pO_q$ forms, the chloride ions desorb gradually at ${}_bc_{Cl^-}=10^{-5}$ M. The adlayers obtained in the presence and absence of Cl^- appeared similar.

10. Nature of Chemisorbed Species

The difficulty of distinguishing between chemisorbed carbonaceous species is illustrated by the results of NIEDRACH [5] in Fig. 51. The electrode which consisted of Teflon-bonded platinum black was brought to $U=0.4$ V. Different hydrocarbons were allowed to interact with the electrode at open circuit in 2.5 M H_2SO_4 at 25 °C. When equilibrium

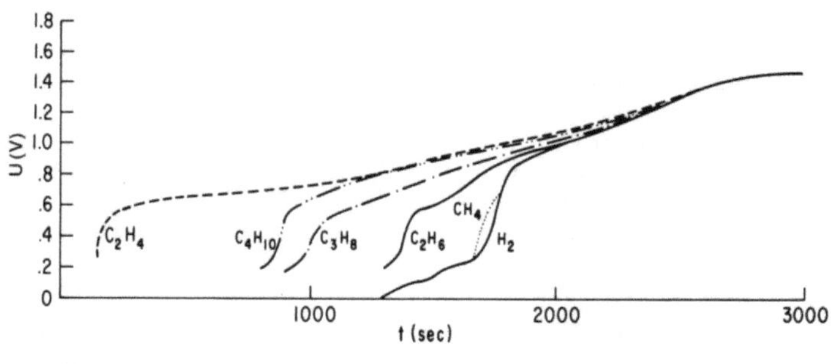

Fig. 51. Anodic charging curves for the oxidation of chemisorbed species formed at open circuit during the interaction of different hydrocarbons or hydrogen with a Teflon-bonded platinum black electrode

had been achieved, after 30 to 60 minutes, the charging curves in Fig. 51 were taken with 30 mA (=2.6 mA/cm^2). A curve for the oxidation of H_{ad} was also obtained. The $U-t$ curves are plotted in such a way that they coincide in the oxygen region. Although different hydrocarbons represent the initial species, the charging curves display arrests for the oxidation of chemisorbed carbonaceous species in the same potential region. The nature of the initial species affects the length of the arrests but not the potential region of oxidation. The results in Fig. 51 suggest that chemisorbed species of a similar net composition are formed by the interaction of the various hydrocarbons with platinum. The long arrests resulting from C_2H_4 and C_4H_{10} consist of two parts. The first part is located at potentials below 0.8 V where the oxygen layer is small. The second part is found above 0.8 V. Three different interpretations are conceivable for the fine structure of the arrests:

a) The same chemisorbed species are present on two types of sites differing in the free energy of adsorption. Oxidation to CO_2 occurs in two potential ranges.

b) Species of one type exist originally. These species are oxidized to a second type of a higher oxidation state below 0.8 V. The oxidation of the species of the second type to CO_2 takes place above 0.8 V.

c) Two different types of species are adsorbed. The species of the first type are oxidized below 0.8 V to CO_2, the species of the second type above 0.8 V.

Interpretation a is not considered likely by the author because the width of the upper potential region with about 0.6 V is too large compared to a width of about 0.25 V for the lower region. Interpretation b is in disagreement with the results on hydrocarbons discussed later in this section. The oxidation state of the species that are oxidized at more positive potentials is lower than that of the species oxidized at less anodic potentials. Interpretation c is favored.

The adsorption of carbonaceous species on platinum from methanolic solutions was erroneously attributed solely to the adsorption of methanol molecules in the first papers [25, 26] dealing with the electrochemical determinations of the coverage. Other interpretations, namely the formation of carbonaceous species by dehydrogenation of the fuel, were advanced [1—6], subsequently. Although the latter interpretations turned out to be correct [11, 13, 15, 16], the original evidence [1—6] on which they were based was not sufficient. This holds especially for reference 3. Methanol was added [3] at certain potentials. The formation of $C_sH_pO_q$ and of H_{ad} was allowed for a certain time at open circuit on platinized platinum in 0.05 M H_2SO_4 + X M CH_3OH. The charge Q_H of the hydrogen arrest and the charge Q_{Ox} due to the oxidation of $C_sH_pO_q$ were determined from charging curves taken after intensive washing. It was assumed that solely the formation of $C_sH_pO_q$ according to reaction 5 takes place:

$$\frac{Q_{Ox}}{Q_H} = \frac{4s+(p-2q)}{2s-(p-2q)}. \tag{31}$$

The absence of the formation of CH_2O [23], HCOOH [23], and CO_2 [16] was only shown recently. The net composition HCO was suggested [3] on the basis of the experimental result $Q_{Ox}/Q_H \approx 1$. Values between 1 and 2 were found [64] by experiments in 0.05 M $H_2SO_4 + 5\cdot10^{-4}$ M CH_3OH without washing procedure.

The charge Q_F for the formation of $C_sH_pO_q$ from CH_3OH was obtained [11] by integration of the $i_1 - t$ trace (see section 3), and the charge Q_{Ox} for the oxidation from an anodic pulse. The net composition HCO was postulated [11] on the basis $Q_{Ox}/Q_F = 1$. However, more recent work [13] raises doubts about the method by which Q_F was determined in reference 11. A value of about 2 was suggested [13] for Q_{Ox}/Q_F. The latter value implies that two electrons are required on the average during the oxidation of $C_sH_pO_q$. The preceding results constitute experimental evidence that methanol is dehydrogenated during the initial interaction between Pt and CH_3OH.

The technique described in section 8 supplies direct results on the nature of $C_sH_pO_q$. It was applied to the determination of the ratio Q_{Ox}/Q_{CO_2} for the conditions given in Table 2.

Table 2. *Maximum and minimum values of Q_{Ox}/Q_{CO_2} during five determinations for each condition*

Fuel		CH_3OH	CH_2O	HCOOH	CO	CO_2	Remarks
Q_{Ox}/Q_{CO_2}	Max	2.5	2.2	2.2			Species formed anodically [15]
	Min	1.7	1.8	2.0			
Q_{Ox}/Q_{CO_2}	Max	2.6	2.3	2.2	2.1	2.3	Species formed [16, 59]
	Min	1.7	1.7	1.9	2.0	1.9	at open circuit

On the average, 2.1 electrons are required for the production of one CO_2 molecule during the anodic oxidation of $C_sH_pO_q$. The chemisorbed species formed under different conditions have the same net composition. Assuming $\frac{p-2q}{s} = -2$ and considering $s>2, p>2, q>3$ as unlikely in the studied cases of simple fuels, possible net compositions are given in Table 3:

Table 3. *Possible net compositions*

	s=1		s±2	
p	0	2	0	2
q	1	2	2	3
$C_sH_pO_q$	CO	H_2CO_2	C_2O_2	$H_2C_2O_3$

A distinction between the four compositions in Table 3 cannot be made on the basis of the results in Table 1. However, the results in Fig. 39 demonstrate [65] that CO_{ad} and $C_sH_pO_q$ are not identical. There are two types [59, 66] of CO_{ad} as marked for curve a in Fig. 39. Type II species are oxidized at less anodic potentials than type I. The arrest of the oxidation of type I species is located close to the arrest of the oxidation of $C_sH_pO_q$ at a given temperature between 0° and 40°C. The difference in potential between corresponding arrests at the same temperature is not sufficient to distinguish between type I species of CO_{ad} and $C_sH_pO_q$ But such a distinction can be made on the basis of the temperature dependence. The potential region of the arrest of the oxidation of type I species of CO_{ad} decreases more rapidly with temperature than that of $C_sH_pO_q$. The determination [65] of the experimental heat of activation [67] $\Delta H^{\ddagger}_{1,\mathrm{ref}}$ at 0.6 V according to Eq. 32.

$$\Delta H^{\ddagger}_{1,\mathrm{ref}} = -R\left(\frac{\partial \ln I}{\partial\, 1/T}\right)_U \tag{32}$$

yielded values of 14±1 kcal/M for CO_{ad} and 7±1 kcal/M for $C_sH_pO_q$. The formic acid molecule may be ruled out since the oxidation of formic acid occurs [12] at considerably less positive potentials than that of $C_sH_pO_q$. The previous interpretation of the author [68, 69] that $C_sH_pO_q$ is identical with HCOOH in solutions containing formic acid cannot be maintained. The average value of Q_{ox}/Q_{CO_2} which is about half of the sum of the maximum and minimum value in Table 1 is slightly larger than 2. This suggests that more than one type of species is present. As pointed out [12], the net composition $H_2C_2O_3$ is equivalent to the simultaneous adsorption of HCO and COOH in equal amounts.

Evidence for the presence of more than one type of chemisorbed species was found in studies [7—9, 61, 62] of the species formed during the anodic oxidation or the reactions at open circuit of hydrocarbons on platinum in acid solutions. It was observed that a part of the chemisorbed species desorbs from the surface at potentials below 0.2 V. The results [21] on ethane in Fig. 52 are an example. A programmed sequence of potential steps or sweeps which has been used extensively in the work of GILMAN [8, 9, 21, 31] served to obtain reproducible results. The sequence is shown at the bottom of Fig. 52. Steps A to C constitute a pretreatment procedure. The oxygen layer is reduced during Step D. Ethane oxidation was allowed for different times t_E at 0.4 V. In step F the desorption proceeded for different times at 0.06 V. Current time curves of the subsequent linear potential sweeps with 100 V/sec

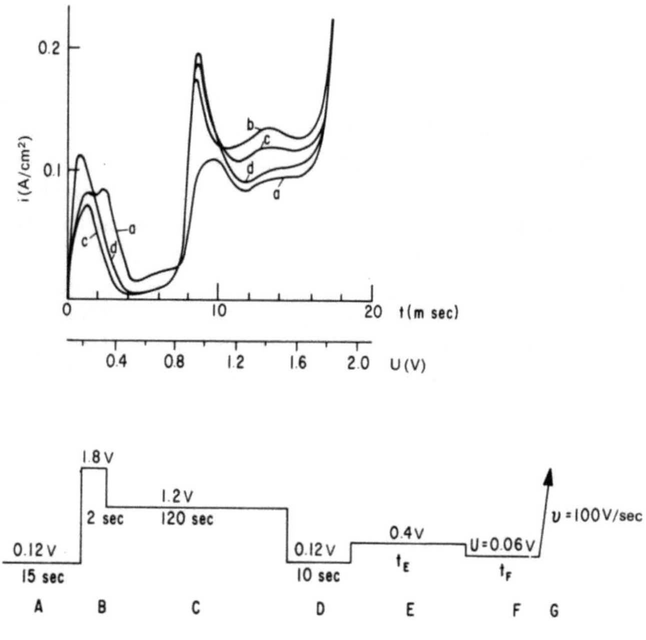

Fig. 52. Anodic $i-U$ traces obtained with 100 V/sec on platinum at 60 °C in 1 M $HClO_4$ saturated with ethane for different times of oxidation (t_E) and desorption (t_F).
a: $t_E=10$ msec, $t_F=10$ msec; c: $t_E=100$ sec, $t_F=100$ msec
b: $t_E=100$ sec, $t_F=0$; d: $t_E=100$ sec, $t_F=100$, 1000 sec

are shown in the upper part of Fig. 52. Trace a corresponds to a clean platinum surface. Trace b is obtained if only the formation of the chemisorbed species is allowed for 100 sec followed by 100 to 1000 sec of desorption respectively. The area between trace a and traces after desorption becomes smaller with t_F above 1.2 V. The species that are oxidized at relatively positive potentials desorb at 0.06 V. These species were designated as type II species by Gilman. The nondesorbable species (type I) are oxidized at potentials below 1.2 V. Anodic $I-U$ curves (see curve b in Fig. 4 of ref. 27) for the oxidation of $C_sH_pO_q$ formed in the presence of methanol display a large wave in the same potential range as the type I species if comparable sweep rates are employed. This fact, together with the nondesorbability below 0.2 V, suggests a similar net composition. A classification of the carbonaceous species into two broad classes, the O-type (designated as type I previously) and the CH type (type II), was proposed by BRUMMER [70]. The distinction is based on the judgment whether most of the C—H bonds in the original fuel have been replaced by C—O bonds.

The oxidation state of the adsorbed species is frequently characterized [4, 7, 21, 70] by the ratio $({}_{s}Q_{H}-{}_{s}Q'_{H})/Q_{Ox}$. For instance a value of about 1.3 was reported [70] for the O-type species, and of about 4 for the CH-type species. If a linear dependence is assumed between $({}_{s}Q_{H}-{}_{s}Q'_{H})$ and the number of sites blocked by chemisorbed species of j different types

(33) $$ {}_{s}Q_{H}-Q_{H}=\sum_{j} m_{j}\Gamma_{j} $$

it follows:

(34) $$ \frac{{}_{s}Q_{H}-Q_{H}}{Q_{Ox}}=\frac{\sum_{j} m_{j}\Gamma_{j}}{\sum_{j} n_{j}\Gamma_{j}}. $$

The ratio $({}_{s}Q_{H}-Q_{H})/Q_{Ox}$ depends upon Γ_{j}. It has a simple significance only for $j=1$ or for $\Gamma_{1}=\Gamma_{2}=\cdots=\Gamma_{j}$.

11. Oxidation Mechanism of Chemisorbed Carbonaceous Species

After an adequate washing procedure, the oxidation of $C_sH_pO_q$ may be studied separately if the rate of oxidation is adjusted to avoid overlapping with the formation of the oxygen layer. The latter condition is easily fulfilled on electrodes with large real surface.

It was found [1, 10, 12] that the potential of the initial part of the arrest due to the oxidation of $C_sH_pO_q$ (see Fig. 39) did not change with pH if measured at equal currents versus a hydrogen electrode in the same solution. A straight line resulted if the potential of the initial part was plotted as a function of log I. A chemical reaction between $C_sH_pO_q$ and OH_{ad} was postulated [1, 10, 12] to be rate-determining on a heterogeneous surface of the Temkin type [71]:

(35) $$ I=k'_{1}\,f(\theta)\exp\left(\frac{\mu_{OH}}{RT}\right). $$

Here f(θ) expresses the dependence upon coverage with $C_sH_pO_q$, and μ_{OH} is the chemical potential of OH_{ad}. It follows from the assumed equilibrium of reaction 26:

$$\mu_{OH} = F \cdot U + \text{const.} \tag{36}$$

The combination of Eq. 35 and Eq. 36 leads to a logarithmic dependence of I upon U. The correctness of the preceding interpretation depends

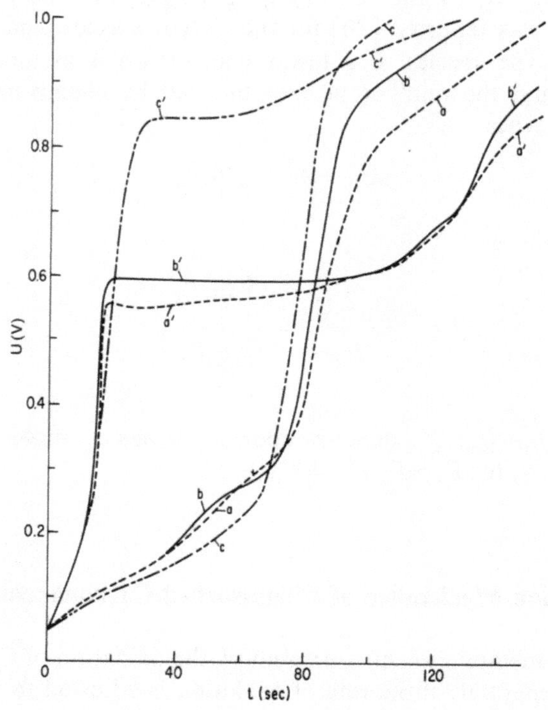

Fig. 53. Anodic charging curves taken with 50 mA in different solutions in the presence (a', b', c') and absence (a, b, c) of chemisorbed carbonaceous species. a, a': 1 M $HClO_4$, b, b': 0.5 M H_2SO_4, c, c': 1 M HCl

in a critical way upon the validity of the exponential dependence of I upon μ_{OH} in Eq. 35. Since this dependence cannot be verified in an independent way owing to the small coverage with OH_{ad}, the above oxidation mechanism is considered a tentative one by the author.

The following studies were made by the author [44] to obtain information on the participation of reaction 26 in the oxidation of $C_sH_pO_q$ from the effect of anions on it (compare section 1 in chapter VIII).

The chemisorbed species were formed on platinized platinum at 10 mA for 2000 sec in 1 M $HClO_4$ + 0.1 M CH_3OH. After the electrode potential had been adjusted to 0.05 V, the electrode and the vessel were washed four times with double distilled water and twice with the final electrolyte. The charging curves a', b', c' in Fig. 53 were subsequently recorded at 50 mA in 1 M $HClO_4$, 0.5 M H_2SO_4, and 1 M HCl respectively at 25 °C. The arrest due to the oxidation of $C_sH_pO_q$ appears at more anodic potentials in the given sequence of the solutions. The current was interrupted at $U = 1.2$ V. After the carbon dioxide had been removed by helium stirring with 0.5 cm^3/sec the curves a, b, and c were taken with 50 mA.

If two different types of species are adsorbed, a fine structure of the arrest might be expected as for the oxidation of weakly and strongly bonded hydrogen (see hydrogen region of curves a and b in Fig. 53). The existence of a fine structure and of a noticeable slope of the $U-t$ curve is demonstrated by curve a' taken in 1 M $HClO_4$. The lack of a fine structure and the remarkable flatness of the initial part of the arrest in curves b' and c' is considered an anion effect. The large influence of the anions on reaction 13 is evident from the shift of the arrest to more positive potentials with increasing adsorbability of the anions. The shift of the arrest is paralleled by the formation of the oxygen layer at more positive potentials (see curves a and b in Fig. 53). Similarly, the faradaic capacitance of reaction 26 found [50] by voltammetry with superimposed ac current at potentials below the oxygen region appears at more positive potentials during the anodic sweep in sulfuric acid than in perchloric acid solution. Chloride ions produce [63] this effect already at small concentrations (${}_bc_{Cl^-} < 10^{-4}$ M) when added to 1 M $HClO_4$.

The influence of the anions on the discharge of water molecules could be due to a decrease of the constant k_1 in Eq. 35 or to the displacement of the equilibrium of reaction 26 because of the inhibition of OH adsorption within the frame of the mechanism given [1] by FRUMKIN and PODLOVCHENKO. However, the results are also consistent with the interpretation that the equilibrium of reaction 26 is established

$$(37)\quad k_3(1-\theta-\theta_{OH})\exp\frac{\alpha_2 UF}{RT} = k_{40}\,c_{H^+}\theta_{OH}\exp\left(-\frac{(1-\alpha_2)UF}{RT}\right)$$

and that the discharge of water molecules on top of $C_sH_pO_q$ is rate-determining:

$$(38)\quad I = k_1\,f(\theta)\exp\left(\frac{\alpha_1 UF}{RT}\right).$$

Examples for the influence of the electrocatalyst on the oxidation of $C_sH_pO_q$ formed [72] at open circuit in 0.5 M H_2SO_4 + 1 M CH_3OH on platinum-black, electrolytically prepared platinum-ruthenium, and platinum-rhenium mixtures are illustrated by curves d, e, and f in Fig. 54.

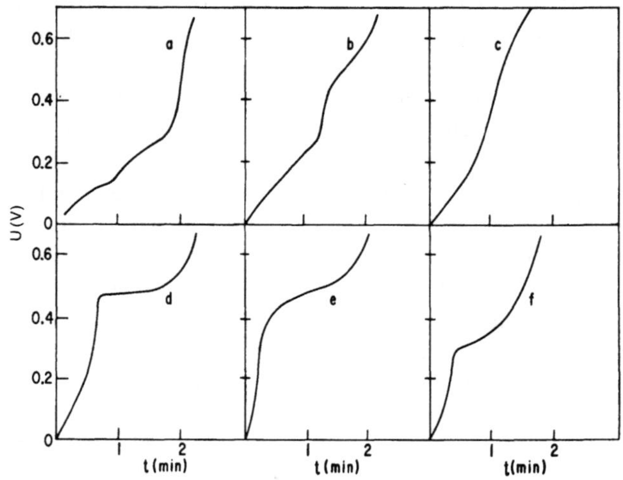

Fig. 54. Charging curves in the absence (a, b, c) and presence of chemisorbed carbonaceous species (d, e, f) on platinum (a, d), platinum-ruthenium (b, e), and platinum-rhenium (c, f) in 0.5 M H_2SO_4. Current density was 2 mA/cm² for Pt, and 5 mA/cm² for the other electrocatalysts

The arrest of the oxidation of $C_sH_pO_q$ is located at less positive potentials in the sequence platinum, platinum-ruthenium (36 % Ru) and platinum-rhenium (8 % Re). Curves a and d were taken with 2 mA/cm², the other curves with 5 mA/cm². The upper curves were obtained in the absence of $C_sH_pO_q$. The appearance of the oxidation region on platinum-rhenium at less positive potentials was attributed [72] to the oxidation of the chemisorbed species by a rhenium oxide and the electrochemical reoxidation of rhenium.

References

1. Frumkin, A. N., and B. I. Podlovchenko: Doklady Akad. Nauk S.S.S.R. **150,** 349 (1963).
2. Giner, J.: Electrochim. Acta **9,** 63 (1964).
3. Podlovchenko, B. I., and E. P. Gorgonova: Doklady Akad. Nauk S.S.S.R. **156,** 673 (1964).
4. Brummer, S. B., and A. C. Makrides: J. Phys. Chem. **68,** 1448 (1964).

5. Niedrach, L. W.: J. Electrochem. Soc. **111**, 1309 (1964).
6. Johnson, P. R., and A. T. Kuhn: J. Electrochem. Soc. **112**, 599 (1965).
7. Brummer, S. B., J. I. Ford, and M. J. Turner: J. Phys. Chem. **69**, 3424 (1965).
8. Gilman, S.: Trans. Faraday Soc. **61**, 2546 (1965).
9. Niedrach, L. W., S. Gilman, and I. Weinstock: J. Electrochem. Soc. **112**, 1161 (1965).
10. Petry, O. A., B. I. Podlovchenko, A. N. Frumkin, and Hira Lal: J. Electroanalyt. Chem. **10**, 253 (1965).
11. Bagotzky, V. S., and Yu. B. Vasiliev: Electrochim. Acta **11**, 1439 (1966).
12. Podlovchenko, B. I., O. A. Petry, A. N. Frumkin, and Hira Lal: J. Electroanalyt. Chem. **11**, 12 (1966).
13. Biegler, T., and D. F. A. Koch: J. Electrochem. Soc. **114**, 904 (1967).
14. Bagotzky, V. S., and Yu. B. Vasiliev: Electrochim. Acta **12**, 1323 (1967).
15. Breiter, M. W.: J. Electroanalyt. Chem. **14**, 407 (1967).
16. — J. Electroanalyt. Chem. **15**, 221 (1967).
17. Wroblowa, H., B. J. Piersma, and J. O'M. Bockris: J. Electroanalyt. Chem. **6**, 401 (1963).
18. Gileadi, E., B. T. Rubin, and J. O'M. Bockris: J. Phys. Chem. **69**, 3335 (1965).
19. Bockris, J. O'M., H. Wroblowa, E. Gileadi, and B. J. Piersma: Trans. Faraday Soc. **61**, 2531 (1965).
20. Heiland, W., E. Gileadi, and J. O'M. Bockris: J. Phys. Chem. **70**, 1207 (1966).
21. Gilman, S.: J. Electrochem. Soc. **113**, 1036 (1966).
22. Gonz, I., S. S. Beskorovainaya, Yu. B. Vasiliev, and V. S. Bagotzky: Elektrokhim. **3**, 315 (1968).
23. Podlovchenko, B. I., A. N. Frumkin, and V. F. Stenin: Elektrokhim. **3**, 339 (1968).
24. Gilman, S.: Trans. Faraday Soc. **62**, 466 (1966).
25. Pavela, T. O.: Ann. Acad. Sci. Fennicae, Series A, II. Chemica **59** (1954).
26. Breiter, M. W., and S. Gilman: J. Electrochem. Soc. **109**, 622 (1962).
27. Gilman, S., and M. W. Breiter: J. Electrochem. Soc. **109**, 1099 (1962).
28. Breiter, M. W.: Electrochim. Acta **8**, 457 (1963).
29. Brummer, S. B.: J. Phys. Chem. **69**, 562 (1965).
30. Breiter, M. W.: J. Electrochem. Soc. **112**, 1244 (1965).
31. Gilman, S.: J. Phys. Chem. **67**, 78 (1963).
32. Brummer, S. B.: J. Phys. Chem. **71**, 2838 (1967).
33. Breiter, M. W.: Electrochim. Acta **7**, 533 (1962).
34. — J. Electrochem. Soc. **110**, 449 (1963).
35. — Electrochim. Acta **8**, 973 (1963).
36. — Electrochim. Acta **9**, 827 (1964).
37. — J. Phys. Chem. **69**, 3377 (1965).
38. Power, W. H., and J. W. Heyd: Analyt. Chem. **28**, 523 (1956).
39. Cook, H. D.: Rev. Sci. Instr. **27**, 1081 (1956).
40. Kafalas, J. A., and H. C. Gatos: Rev. Sci. Instr. **29**, 47 (1958).
41. Blomgren, E. A., and J. O'M. Bockris: Nature **186**, 305 (1960).
42. Smith, R. E., H. B. Urbach, J. H. Harrison, and N. L. Hatfield: J. Phys. Chem. **71**, 1250 (1967).
43. Khazova, O. A., Yu. B. Vasiliev, and V. S. Bagotzky: Elektrokhim. **1**, 84 (1965).
44. Breiter, M. W.: Discussions Faraday Soc. **45**. 79 (1968).
45. Parsons, R.: Advances in Electrochemistry, Vol. **1**, Chap. 1. P. Delahay, ed., New York: Interscience Publishers 1961.
46. Frumkin, A. N., and B. B. Damaskin: Modern Aspects of Electrochemistry, Vol. **3**, Chap. 3. J. O'M. Bockris and B. E. Conway, ed., Butterworths 1964.
47. Borisova, T., B. Ershler, and A. Frumkin: Zhur. Fiz. Khim. **22**, 925 (1948).
48. Kheifets, V. L., and B. S. Krasikov: Doklady Akad. Nauk S.S.S.R. **94**, 101 (1954).

49. BREITER, M. W.: J. Electrochem. Soc. **109,** 42 (1962).
50. — J. Electroanalyt. Chem. **7,** 38 (1964).
51. RIGHTMIRE, R. A., R. L. ROWLAND, D. L. BOOS, and D. L. BEALS: J. Electrochem. Soc. **111,** 242 (1964).
52. FRUMKIN, A.: Doklady Akad. Nauk S.S.S.R. **154,** 1432 (1964).
53. —, N. A. BALASHOVA, and V. E. KAZARINOV: J. Electrochem. Soc. **113,** 1011 (1966).
54. — J. Electrochem. Soc. **113,** 1022 (1966).
55. DAHMS, H., and M. GREEN: J. Electrochem. Soc. **110,** 1075 (1963).
56. BOCKRIS, J. O'M., and D. A. J. SWINKELS: J. Electrochem. Soc. **111,** 736 (1964).
57. —, M. GREEN, and D. A. J. SWINKELS: J. Electrochem. Soc. **111,** 743 (1964).
58. FRUMKIN, A. N.: Z. Phys. **35,** 792 (1926).
59. BREITER, M. W.: Electrochim. Acta **12,** 1213 (1967).
60. GRUBB, W. T., and M. E. LAZARUS: J. Electrochem. Soc. **114,** 360 (1967).
61. BRUMMER, S. B., and M. J. TURNER: Hydrocarbon Fuel Cell Technology, p. 409. B. Baker, ed., New York: Acedemic Press Inc. New York 1965.
62. GILMAN, S.: J. Phys. Chem. **71,** 2424 (1967).
63. BREITER, M. W.: Electrochim. Acta **8,** 925 (1963).
64. — Abhandlungen Sächsische Akad. Wiss., **49**, 221 (1968).
65. — J. Electroanalyt. Chem., **19**, 131 (1968).
66. — J. Phys. Chem. **72,** 1305 (1968).
67. TEMKIN, M.: Zhur. Fiz. Khim. **22,** 1081 (1948).
68. BREITER, M. W.: Electrochim. Acta **8,** 447 (1963).
69. — Electrochim. Acta **10,** 503 (1965).
70. BRUMMER, S. B.: J. Electrochem. Soc. **113,** 1041 (1966).
71. TEMKIN, M.: Zhur. Fiz. Khim. **15,** 296 (1941).
72. CATHRO, K. J.: Electrochem. Techn. **5,** 441 (1967).

X. Anodic Oxidation of Fuels at Low Temperatures

The anodic oxidation of fuels in low temperature cells, mainly on platinum metals, platinum metal alloys and alloys of platinum metals with other metals, is the subject of this chapter. Most oxidation studies were made on these metals because the efficiency of other electrocatalysts is too low. The mechanism for the oxidation of carbon monoxide, mixtures of hydrogen and carbon monoxide, formic acid, methanol, higher alcohols, hydrocarbons, and hydrazine is discussed in separate sections.

1. Classification of the Oxidation Mechanisms

The oxidation mechanisms may be divided in two groups according to the nature of the rate-determining step as

a) a chemical reaction without charge transfer or

b) an electrochemical reaction with charge transfer.

Group a includes dehydrogenation reactions [1—4], reactions between adsorbed carbonaceous species and OH_{ad} [4—6] or O_{ad} [7—10], preelectrochemical steps [11, 12], and desorption steps [13, 14] as rate-determining steps. Examples are given below:

$$CH_3OH = CH_3O_{ad} + H_{ad}, \tag{1}$$

$$HCO_{ad} + OH_{ad} = CO_2 + 2H_{ad}, \tag{2}$$

$$HCOOH = HCOO^- + H^+, \tag{3}$$

$$H_{2,ad} = H_2. \tag{4}$$

Mechanisms with rate-determining discharge steps [15—19] involving adsorbed carbonaceous species, or with a rate-determining water discharge [20, 21], belong to group b. Examples for this type of rate-determining step are:

$$CO_{ad} + H_2O_{ad} = CO_2 + 2H^+ + 2e^-, \tag{5}$$

$$H_2O = OH_{ad} + H^+ + e^-, \tag{6}$$

$$H_2O = O_{ad} + 2H^+ + 2e^-. \tag{7}$$

It was already pointed out in section 4 of chapter IV that the definition of one rate-determining step is meaningful for a sequence of reactions leading to one final product under steady-state conditions. The above classification is restricted to net reactions of this type. The oxidation of carbon monoxide to carbon dioxide is an example.

2. Oxidation of Carbon Monoxide

The simplest organic fuel is carbon monoxide. The adsorption of CO and the subsequent oxidation of CO_{ad} to CO_2 were studied [19, 22—27] on smooth platinum electrodes in acid electrolytes. Voltammetric [19, 22—24, 27] and galvanostatic techniques [25, 26] were used to determine Q_{CO} due to the oxidation of CO_{ad} (compare Fig. 15). A value of 1.8 was reported in ref. 23 for $2\,{}_sQ_H/{}_sQ_{CO}$ and of 1.08 in ref. 26. GILMAN [23] attributed this ratio to the presence of two forms of CO_{ad} in analogy to the interpretation [28] of infrared studies of the adsorption of carbon monoxide on platinum in the gas phase. One molecule of the bridged or linear form occupies two or one platinum atom respectively. Physical adsorption of CO on top of a monolayer of chemisorbed carbon monoxide was suggested [25] as interpretation of $2\,{}_sQ_H/{}_sQ_{CO}=0.94$. Codeposition of H atoms on a platinum surface having the saturation coverage of CO_{ad} lead to the conclusion [22] that carbon monoxide is only adsorbed on about 80% of the Pt atoms. In contrast, a percentage of 98% was given in ref. 26. GILMAN [19], postulated the "reactant-pair" mechanism in which an adsorbed carbon monoxide molecule reacts with a water molecule adsorbed at an adjacent site (compare reaction 5). The transfer of the first electron from the adsorbed complex is supposed rate-determining. WARNER and SCHULDINER [25] proposed a reaction between CO_{ad} and O_{ad}. Reaction 7 is assumed rate-determining.

The investigations of carbon monoxide adsorption on platinized platinum [29, 30] were recently supplemented [31] to allow the establishment of similar relations as on smooth platinum. SOKOLSKY and coworkers [29, 30] concluded that physical adsorption of carbon monoxide occurs at about 10 °C in 0.5 M H_2SO_4 on platinized platinum. Conversion to CO_2 and H_2 was postulated for temperatures between 20 °C to 30 °C. Initial conversion and subsequent chemisorption were suggested between 50 °C and 70 °C. BINDER, KÖHLING, and SANDSTEDE [32] concluded from measurements on Raney platinum electrodes in phosphoric acid and sulfuric acid solutions that chemical conversion is the first step in the anodic oxidation of CO at temperatures between 90 °C and 150 °C.

It had been noticed by the author [33] that a certain portion of the adsorbed carbon monoxide is oxidized at less positive potentials than the rest (see the short arrest between about 0.3 V and 0.55 V of curve c in Fig. 15) The transition from the lower to the upper arrest is sometimes accompanied [31] by an overshoot (see curves a, b, c, d in Fig. 39). Adsorbed carbon monoxide that is oxidized in the upper arrest was designated [31] as type I, and the corresponding charge as Q_1. The adsorbed species oxidized in the short arrest were denoted [31] as type II, and the charge as Q_2. It was demonstrated [31] by gas chromatography that the net reaction in acid solutions

$$CO_{ad} + H_2O = CO_2 + 2H^+ + 2e^- \tag{8}$$

is the same for type I and type II species.

The adsorption of carbon monoxide at open circuit was studied on a platinized platinum cylinder of 50 cm² with equal adsorption times of 2000 sec in 0.5 M H_2SO_4, 0.1 M Na_2SO_4 + 0.05 M H_2SO_4, and 0.1 M Na_2SO_4 + 0.005 M H_2SO_4. The anodic $U-t$ curves with 50 mA in Fig. 55 coincide nearly in 0.5 M H_2SO_4 and 0.1 M Na_2SO_4 + 0.05 M

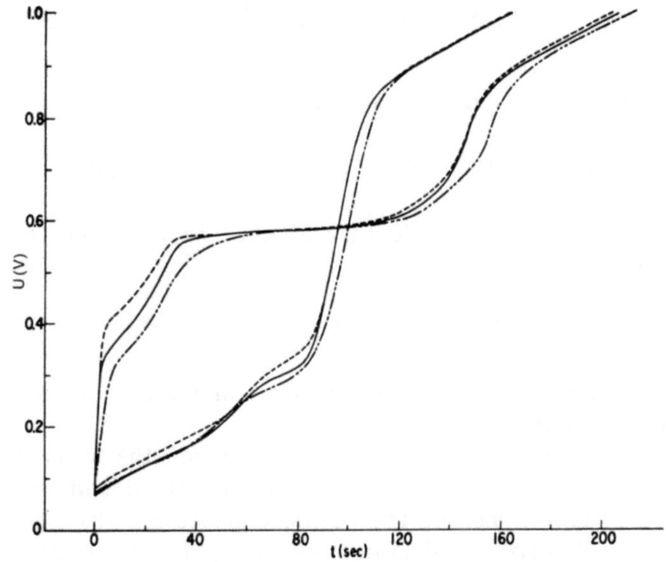

Fig. 55. Charging curves with 50 mA on platinized platinum in the absence (lower curves) and presence (upper curves) of CO_{ad} at 25 °C.
- - - - - 0.5 M H_2SO_4,
——— 0.1 M Na_2SO_4 + 0.05 M H_2SO_4,
- - - - - - 0.1 M Na_2SO_4 + 0.005 M H_2SO_4

H_2SO_4 at potentials of the hydrogen region if CO_{ad} is absent. The curve in 0.1 M Na_2SO_4+0.005 M H_2SO_4 is located between 10 to 20 mV above the curves in the other two solutions. The anodic removal of H atoms is nearly independent of pH, as would be expected, since a hydrogen electrode in the same electrolyte as the test electrode served as reference electrode. A large difference exists for the pH dependence of the processes occurring in the upper and lower arrest of the oxidation of type I and type II species. The $U-t$ curves are shifted by 40 to 50 mV against each other at a given time between 10 to 30 sec in the region of the lower arrest. In contrast, they coincide nearly in the upper arrest. The different pH dependence is evidence of different oxidation mechanisms of type I and type II species. A reaction mechanism in which the electrode potential is not determined by an equilibrium between H_{ad} and H^+ or between OH_{ad} and H_2O is consistent with the results on the type II species. In contrast, the pH dependence during the oxidation of type II species requires [4] the assumption of such an equilibrium.

The rate of oxidation was found [31] to depend upon U and the coverage $Q_2/{}_sQ_2$ of type II species in the following way:

$$I = k_2 \exp\left(\beta_2 \cdot \frac{Q_2}{{}_sQ_2} + \alpha_2 n_2 \frac{UF}{RT}\right). \tag{9}$$

The parameters in Eq. 9 have the numerical values:

$$k_2 = 1.3 \cdot 10^{-4}\,\text{mA}, \quad \beta_2 = 6.1, \quad \alpha_2 n_2 = 0.58.$$

Eq. 9 with an $\alpha_2 n_2$ value smaller than 1 is consistant with the assumption of reaction 10

$$CO_{ad} + H_2O = COOH_{ad} + H^+ + e^- \tag{10}$$

as rate-determining step. The dependence of the rate upon coverage is of the Temkin type [34].

Neither the reactant-pair mechanism [19] nor the water discharge mechanism [25] are in agreement with the pH dependence for the oxidation of type I species. The pH dependence may be explained as suggested by FRUMKIN and PODLOVCHENKO [4] in their discussion of the mechanism of ethanol oxidation. A chemical reaction between adsorbed carbonaceous species and OH_{ad} is supposed rate-determining on a heterogeneous surface of the Temkin type [34]. The discussion is the same as that for the oxidation mechanism of chemisorbed carbonaceous species in section 11 of chapter IX.

Ratios of characteristic charges for platinized platinum are given in Table 4 for saturation coverage of the surface with CO_{ad}:

Table 4. *Ratios of characteristic charges at 25 °C.*

	$2_sQ_H/_sQ_{CO}$	$_sQ_1/_sQ_{CO}$
0.05 M H_2SO_4	1.44±0.14	0.82±0.08
0.5 M H_2SO_4	1.27±0.13	0.76±0.08
0.1 M Na_2SO_4 + 0.05 M H_2SO_4	1.37±0.14	0.78±0.08
0.1 M Na_2SO_4 + 0.005 M H_2SO_4	1.40±0.14	0.80±0.08
1 M KOH	1.16±0.12	0.77±0.08

Since $_sQ_H$ is representative for the number of exposed Pt atoms, the ratio $2_sQ_H/_sQ_{CO}$ corresponds to the number of Pt atoms per adsorbed molecule in acid solutions. The average value of $2_sQ_H/_sQ_{CO}$ is 1.37 in these solutions. It lies between the two values reported for smooth platinum in ref. 22 and ref. 26. The average value of 0.79 for $_sQ_1/_sQ_{CO}$ demonstrates that about 21 % of the adsorbed molecules is of type II. The latter value on platinized platinum is nearly the same as the value of 0.25 found [35] for the ratio of the volumes of weakly and strongly adsorbed carbon monoxide on platinum powder in the gas phase at room temperature. In this case the amount of weakly bonded CO_{ad} desorbs by pumping alone. The value of $2_sQ_H/_sQ_{CO}$ is somewhat lower in 1 M KOH than the corresponding values in the acid solutions. The ratio $_sQ_1/_sQ_{CO}$ is about the same.

The adsorption of carbon monoxide was studied [36], on different samples of platinum/silica catalysts in the gas phase. The ratio of exposed Pt atoms to adsorbed molecules varied from about 1 to 2 with increasing platinum content. It was suggested [36] that the coverage with CO_{ad} depends upon the crystallite size. Linear bonding occures predominantly on small, highly dispersed crystallites. Bridge bonding is favored when the uninterrupted area of platinum crystallites has attained a reasonable size. A distinction between weakly and strongly bonded CO_{ad} was not made. It is demonstrated by the upper curve in Fig. 17 that the parameter $\left(1 - \frac{_sQ'_H}{_sQ_H}\right)$ is equal to $Q_{CO}/_sQ_{CO}$. The same number $2_sQ_H/_sQ_{CO}$ of Pt atoms is covered on the average by one molecule of both types of CO_{ad} on platinized platinum for this reason. The result that the average number of exposed Pt atoms per adsorbed molecule is larger than 1 may imply the presence of linear and bridge bonded CO_{ad}. A quantitative determination of the amount of these two forms is difficult because of the influence [36] of the crystallite size and because "packing" rules [37] apply to the linearly bonded CO_{ad} but not to the bridged CO_{ad} on crystallites of sufficient size.

The values of $2_sQ_H/_sQ_{CO}$ and $_sQ_1/_sQ_{CO}$ obtained in 1 M KOH indicate that the oxidation of CO_{ad} is a two-electron process, as in acid solutions. The following scheme was suggested [31]:

(11) $$CO_{ad} + OH^- = COOH_{ad} + e^-,$$

(12) $$COOH_{ad} + 3\,OH^- = CO_3^{--} + 2H_2O + e^-.$$

The scheme is supported by the result [31] that the upper arrest of the oxidation of type I species consists of two parts of about equal lengths, independently of Q_1. The first part may be attributed to reaction 11, the second part to reaction 12. A separation of the two one-electron steps may be observable in alkaline solutions since three OH^- ions are required in the second step. In contrast, a hydrogen atom has only to be split off from $COOH_{ad}$ in acid solutions, making the second step a rapid process.

The two arrests during the oxidation of $C_sH_pO_q$ are also of about equal length in 0.1 M KOH and 1 M KOH (see Fig. 7b in ref. 38 and Fig. 8 in ref. 39). A two-electron process that is similar to the mechanism of the oxidation of type I species of CO_{ad} is considered likely, in agreement with results [40] on the oxidation of $C_sH_pO_q$ in acid solutions.

The charging curves a, b, c, d in Fig. 39 represent evidence that carbon monoxide is chemisorbed between 0° and 60 °C. The reaction

(13) $$CO + H_2O = CO_2 + H_2$$

is considered a side reaction at 25 °C in agreement with the gas analysis [30] that only about 3% of CO is converted to CO_2 and H_2 at $U=0$ and $U=0.5$ V in 0.5 M H_2SO_4 at 20 °C and 30 °C.

Current and coverage determined as $\left(1-\frac{_sQ'_H}{_sQ_H}\right)$ under galvanostatic conditions are plotted as function of U in Fig. 56. The steady-state measurements were made [31] under CO stirring on the platinized Pt cylinder with 50 cm² in 0.5 M H_2SO_4 at 25 °C. The coverage is equal to 1 between 0.2 V and 0.4 V. A rapid decrease of coverage with potential is observed between 0.5 V and 0.6 V. Although the general shape of the $I-U$ curve was reproducible, the potential may vary by as much as 0.1 V at $I>10$ mA in different runs. The drift of potential with time to more positive values is about 10 mV/hr at $I<10$ mA. A steady-state potential could no longer be achieved at 30 mA. A current of 30 mA (0.6 mA/cm²) represents the limiting diffusion current of CO. A potentiostatic $I-U$ curve was obtained point by point by starting the CO stirring at a given potential and by measuring the current when the change of I with time had become small after 500 sec. The potentiostatic $I-U$ curve has a shape similar to the $I-U$ curve in Fig. 56. The current is small initially and increases rapidly with potential in a narrow region. However, the rapid increase of I with U occurs in the potentiostatic case at potentials about 0.2 V more positive than in the galvanostatic case.

The potentiostatic and galvanostatic measurements indicate independently of the experimental conditions that the decrease of $Q_{CO}/_{S}Q_{CO}$ starts when the current becomes larger than 10 mA. Tafel plots [31] for the oxidation of type II species of CO_{ad} showed rates which are larger

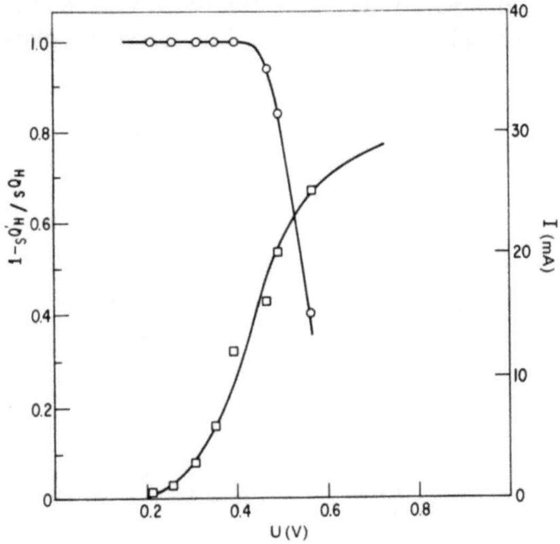

Fig. 56. Current (□) and coverage (○) with CO_{ad} as function of potential for steady-state measurements under galvanostatic conditions on platinized platinum in 0.5 M H_2SO_4 at 25 °C and CO stirring

or comparable to the oxidation rates of disolved CO under steady-state conditions. It was suggested [31] that the oxidation of CO below 0.4 V involves the type II species as intermediates. The decrease of $Q_2/_{S}Q_2$ between 0.1 V and 0.4 V implies that the rate of adsorption of type II species is insufficient to maintain the full coverage with increasing oxidation rates. Adsorption of the type II species is considered the rate-determining step at $U < 0.4$ V in acid electrolytes.

The type I species may be the intermediates during the oxidation of dissolved carbon monoxide at $I > 10$ mA. The oxidation rates of type I species were found [31] to be comparable to the rates of CO oxidation under steady-state conditions. The coverage $Q_1/_sQ_1$ decreases with potential at currents which are approaching the limit imposed by mass transport. Insufficient supply, which leads to a smaller adsorption rate, is the reason for the decrease of $Q_1/_sQ_1$. Partial mass transport control was also concluded [22, 41] for the CO oxidation on smooth platinum free of CO_{ad}. In contrast to the voltammetric measurements [22, 27, 41] on smooth platinum, (compare the curve for Pt in Fig. 57), the decrease of $Q_1/_sQ_1$ with U begins on platinized platinum in a potential region in which the coverage with OH_{ad} and O_{ad} is small.

Comparative voltammetric studies of the CO oxidation on Pt, Pd, Rh and Au were carried out by STONEHART [27] on rotating wire electrodes at 25 °C in 1 M $HClO_4$ saturated with CO. The $i-U$ curves in Fig. 57 were taken [27] at 59 mV/sec. The curve on platinum displays

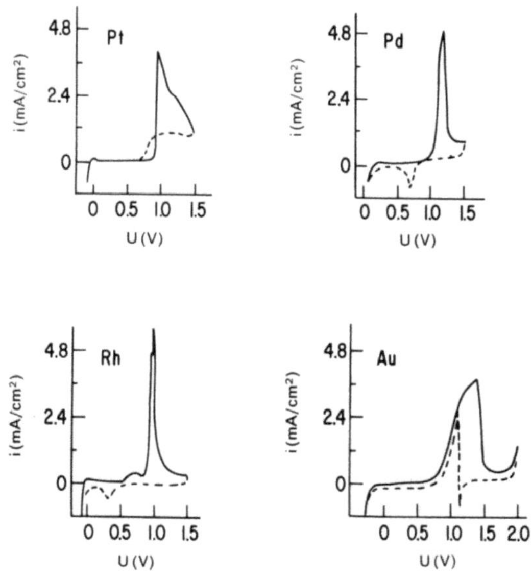

Fig. 57. Periodic current-potential curves taken with 59 mV/sec on roating wire electrodes in 1 M $HClO_4$ saturated with CO

the sudden increase [22, 27, 41] of current with potential at about 0.85 V during the anodic sweep. The coverage with CO_{ad} is zero when the peak is reached [22]. Subsequently the current decreases with U because of the increasing coverage with the oxygen layer. The gradual decrease may be [22] a combined effect of inhibition and blocking of free sites by the oxygen layer. The current maintains [22] its value at the moment of sweep reversal during the initial part of the cathodic sweep when the oxygen coverage is constant. The latter result demonstrates that CO oxidation on platinum requires sites free of the oxygen layer. The initial parts of the $i-U$ curves on Pd and Rh are similar to that on Pt during the anodic sweep. The sudden rise of current with U indicates a rapid depletion of CO_{ad}. Such a depletion was demonstrated recently by GILMAN [42] for Rh. The inhibition of the CO oxidation by the oxygen layer is greater on Pd and Rh than on Pt. The partial current of the CO oxidation is small during a large portion of the cathodic sweep. This leads to the appearance of the reduction wave of the oxygen layer on Pd and Rh during the cathodic sweep. The origin of the small oxidation wave with peak at about 0.7 V is not quite clear for Rh. In contrast to Pt, Pd, and Rh, the current rises gradually with potential above 0.7 V on Au. It decreases rapidly with U when the oxygen coverage is large at 1.4 V. After the oxygen layer has been reduced during the cathodic sweep, the current of the CO oxidation is nearly the same at a given potential as during the anodic sweep. Gold shows a larger reactivity with respect to the CO oxidation below 0.8 V than Pt, Pd, and Rh under voltammetric conditions.

The CO oxidation was studied by KATHLEEN BRUMMER [43] on a rotating disc electrode of platinum in the presence of dispersed platinum powder (about 1 g per 100 ml of 1 M $HClO_4$) or dispersed tantalum carbide powder (1 g per 200 ml of 1 M $HClO_4$) and on a gold disc electrode in the presence of tantalum carbide. Chemisorption of carbon monoxide was allowed for a sufficient time to achieve a monolayer coverage of the disc and the platinum powder. Then the dissolved carbon monoxide was removed from the electrolyte by stirring with nitrogen. As a result of the collision with powder particles, the oxidation rate of carbon monoxide reaching the platinum disc was much faster at a given potential than the steady-state rate in the same system under CO stirring. The rate increased at constant potential in a linear fashion with the disc surface free of CO_{ad}. The Tafel plot yielded [43] a slope of about 250 mV. This slope is greater than the slope of the corresponding plot [31] for the oxidation of CO_{ad} at constant coverage $\theta_{CO} > 0.25$. Measurements of the steady-state rates of the CO oxidation on the platinum or gold disc in the presence of tantalum carbide powder under CO stirring showed a better reactivity of the gold-tantalum carbide compound electrode than the platinum-tantalum carbide electrode at a given potential. An attempt [43] was made to interpret the above results by a mechanism similar to the reactant-pair mechanism, which requires the simultaneous adsorption of CO_{ad} and H_2O_{ad} on adjacent sites. However, many features, like the slope of 250 mV and the different limiting rates at 0.8 V in the presence of tantalum carbide on the platinum and gold disc, are not understandable on this basis.

3. Oxidation of Mixtures of Hydrogen and Carbon Monoxide

In the indirect approach to the hydrocarbon fuel cell, the fuel passes through a reformer before entering the cell. The effluent gas of the hydrocarbon reformer consists of hydrogen with small amounts of other gases, expecially CO_2, CO and CH_4.The performance of a platinum black anode is seriously decreased by the adsorption of CO from this gas mixture. It was reported [44, 45] recently that the problem can be alleviated by the use of composites containing ruthenium [44], iridium or rhodium [45] besides platinum. While ruthenium and iridium by themselves are poor electrocatalysts for synthetic reformer gas, pure rhodium shows a good performance.

In the platinum-ruthenium system, maximum reactivity is obtained at about 30 weight % of Ru, in the platinum-iridium system with at Pt-50% Ir electrocatalyst. Large reactivity is present from about 10% to 100% rhodium in the platinum-rhodium system. The analysis of the gas leaving the fuel cell by mass spectrometry revealed that only a small percentage of the total current results from the oxidation of CO on the composite electrocatalysts. Results [45] for the oxidation of 95% H_2 +5% CO in 2.5 M H_2SO_4 at 85°C on platinum black, Pt−30.2% Ru, Pt−50% Rh, and Pt−51.2% Ir are compared in Fig. 58. The fraction of the total current attributable to carbon monoxide oxidation is much larger on platinum than on the composite anodes. The rapid increase of the partical current of CO oxidation on platinized platinum between 20mA and 40mA is accompanied by a considerable increase of U. It was suggested [44, 45] that the improved performance results from a large number of unpoisoned sites for the H_2 oxidation on the composite anodes.

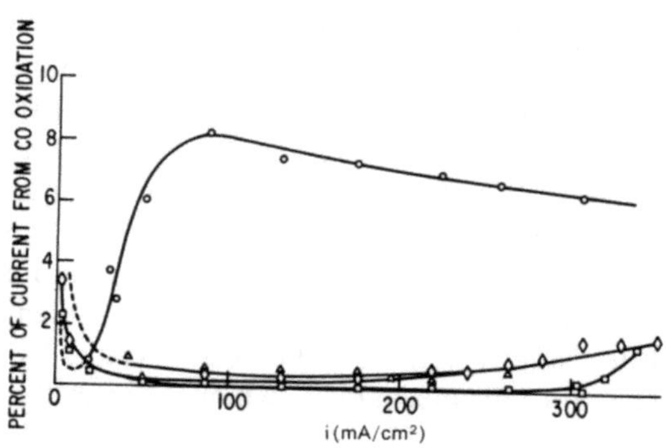

Fig. 58. Effect of current density on the percent of total current due to the CO oxidation from a fuel mixture of 95 % H_2 + 5 % CO in 2.5 M H_2SO_4 at 85 °C.
○: Pt, △: Pt − 30.2 % Ru, □: Pt − 50 % Rh, ◇: Pt − 51.2 % Ir

4. Oxidation of Formic Acid

The net reaction for the anodic oxidation of formic acid has been established [1, 46, 47] in acid electrolytes as:

$$HCOOH = CO_2 + 2H^+ + 2e^- . \tag{14}$$

This was confirmed by recent measurements [40] by the author on platinized platinum. There is some controversy in the literature as to the reaction in alkaline solutions. While a uniform two-electron process throughout the pH range was reported in some papers [1, 7, 46—48], other work [10, 11, 49] indicates that the formate ion is not easily oxidized in alkaline electrolytes. The subsequent discussion is largely based on the extensive studies of formic acid oxidation in acid electrolytes in which the formic acid molecule is the reacting species [13].

Chemisorbed carbonaceous species are formed during the anodic oxidation [9, 39, 40, 48, 50—57] of HCOOH on platinum or during the interaction [39, 52, 58] at open circuit. The net composition $COOH_{ad}$, suggested in references 9 and 56 for smooth platinum in acid electrolytes is not in agreement with results obtained [40, 58] by the combination of charging curves and gas chromatography on platinized platinum.

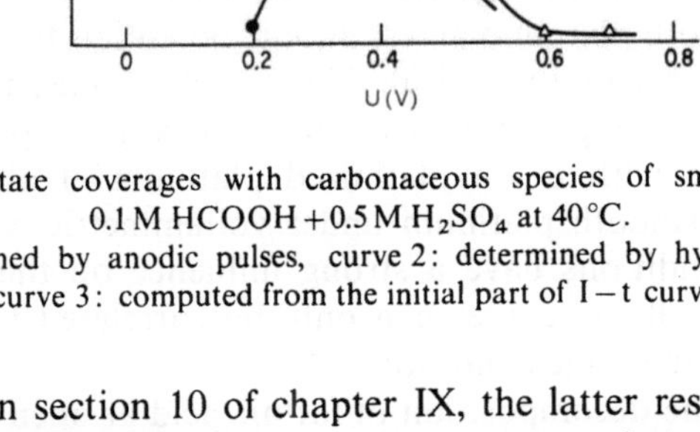

Fig. 59. Steady-state coverages with carbonaceous species of smooth platinum in 0.1 M HCOOH + 0.5 M H_2SO_4 at 40 °C.
Curve 1: determined by anodic pulses, curve 2: determined by hydrogen deposition, curve 3: computed from the initial part of I – t curves

As discussed in section 10 of chapter IX, the latter results suggest the simultaneous adsorption of more than one type of species, possibly of HCO and COOH. Slightly more than two electrons are required on the average for the anodic formation of one molecule CO_2 from one adsorbed particle $C_sH_pO_q$.

The curves [56] 1, 2 and 3 in Fig. 59 represent the coverage $Q_{Ox}/{}_{S}Q_{Ox}$, $({}_{S}Q'_{H} - {}_{R}Q'_{H})/({}_{S}Q_{H} - {}_{R}Q'_{H})$ and $Q_F/{}_{S}Q_F$ of smooth platinum in 0.1 M

$HCOOH + 0.5$ M H_2SO_4 at 40 °C as a function of U under steady-state conditions. After anodic pretreatment the electrode was switched potentiostatically to the desired potential. The current which decreases with time was recorded. The integration of the $i-t$ curves yielded Q_F. When steady-state conditions had been achieved, Q_{Ox} or $_SQ'_H$ were determined by anodic or cathodic pulses as described in section 4 and 5 of chapter IX. The curves 1 and 2 which coincide have a nearly parabolic schape. The coverages $Q_{Ox}/_SQ_{Ox}$ and $(_SQ'_H - _RQ'_H)/(_SQ_H - _RQ'_H)$ increase with U for $U < 0.1$ V. They depend only slightly upon U between 0.1 V and 0.4 V and decrease between 0.4 V and 0.6 V. The coverage is very small above 0.6 V. In contrast, the ratio $Q_F/_SQ_F$ increases rapidly with U between 0.2 V and 0.4 V and is smaller than $Q_{Ox}/_SQ_{Ox}$ at a given U. The result $Q_F < Q_{Ox}$ which had been established [54] for $U = 0.2$ V holds [56] for $U \leq 0.4$ V. The ratios $Q_F/_SQ_F$ and $Q_{Ox}/_SQ_{Ox}$ agree practically above 0.4 V in Fig. 59. A chemical reaction between H_{ad} and HCOOH was proposed [56] at potentials of the hydrogen region. The simultaneous occurrence of the reactions 15 and 16

$$HCOOH = COOH_{ad} + H_{ad}, \tag{15}$$

$$HCOOH + H_{ad} = HCO_{ad} + H_2O \tag{16}$$

would lead to the formation of HCO_{ad} and $COOH_{ad}$. Reaction 16 represents a reduction process [39].

The studies [48, 50—57] on the properties of $C_sH_pO_q$ formed during the interaction between HCOOH and platinum at open circuit or during the anodic oxidation of HCOOH support the conclusion that the net composition of $C_sH_pO_q$ remains the same between 0 and 0.6 V. The steady-state coverage reported in references [48, 53, 56] begins to decrease on smooth platinum at less anodic potentials than the coverage determined under conditions [50, 51] of cylic voltammetry. On the other hand the coverage after two minutes at a given potential begins [59] to decrease with U on platinized platinum at about the same potential as on smooth platinum under voltammetric conditions. The experimental conditions have a strong influence on the shape of the $\theta - U$ curve. The $\theta - U$ curves can only be correlated to $i - U$ curves measured under the same conditions.

The rate of the decomposition of formic acid at open circuit is [60] about 0.05 mA/cm^2 on platinized platinum in 0.5 M H_2SO_4 at room temperature. The initially large decomposition observable by the formation of bubbles was found [61] to stop after a period of a few minutes to an hour on Raney platinum in acid electrolytes. It was demonstrated [62] that a rate of about 0.05 mA/cm^2 is too large to be accounted for by a mixed potential process in which the decompositon is the anodic process and hydrogen evolution or oxygen reduction represent the cathodic process. The exchange current density of the Volmer reaction was estimated [62] to be about $1\,mA/cm^2$ on smooth platinum in 0.5 M H_2SO_4 + 0.1 M HCOOH at room temperature. Since the anodic current densities of formic acid oxidation are much smaller than $1\,mA/cm^2$ at $U < 0.4\,V$, the equilibrium of the Volmer reaction can be considered [62] established in the hydrogen region.

Steady-state $i-U$ curves of formic acid oxidation in acid media consist [48, 49, 59, 60] of two regions. There is a Tafel line in the potential region between 0.2 V and about 0.4 V where the coverage with $C_sH_pO_q$ does not change much. Slopes of about 60 mV were reported at room temperature by some investigators [48, 63], slopes of 120 mV by others [59]. A transition from the Tafel type dependence to a limiting current region occurs at potentials where the coverage decreases with U. The

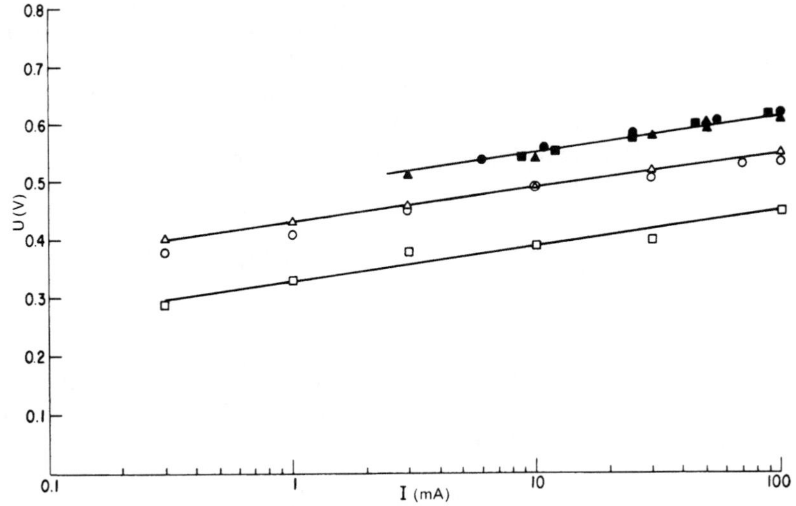

Fig. 60. Current-potential curves on platinized platinum in 0.1 M fuel + 0.5 M H_2SO_4 at 25 °C. Open symbols represent steady-state values. Solid symbols correspond to the oxidation of chemisorbed carbonaceous species at nearly constant coverage.
□, ■: HCOOH, ○, ●: CH_2O, △, ▲: CH_3OH

limiting current measured potentiostatically [48] is too small at ${}_bc_F > 0.01$ M to result from mass transport control. The voltammetric $I-U$ curves exhibit [51] a first oxidation wave during the anodic sweep at 30 mV/sec between 0.3 V and 0.6 V for ${}_bc_F \geq 0.01$ M in 1 M $HClO_4$. The coverage with $C_sH_pO_q$ is practically independent [50] of U in this region. When the coverage decreases rapidly with U within a narrow potential range of about 0.15 V, the current also rises rapidly with U. The oxygen layer which begins to form simultaneously inhibits the formic acid oxidation. A second wave is observed instead of a limiting current under voltammetric conditions. Slopes of Tafel lines determined under these conditions depend upon sweep rate and coverage.

Steady-state $I-U$ curves measured [63] in 0.1 M fuel + 0.5 M H_2SO_4 (fuel = CH_3OH, CH_2O, HCOOH) on the platinized cylinder with 50 cm^2 at 25 °C are represented in the semilogarithmic plot of Fig. 60 by open symbols. The corresponding solid symbols represent the oxidation of the chemisorbed species formed during the fuel oxidation in the three different solutions at nearly constant coverage. The curves were determined in the absence of fuel after appropriate washing of the vessel and of the electrode. The solid symbols scatter around the same Tafel line, as would be expected on the basis of the similar net composition of $C_sH_pO_q$ for the three fuels (compare Table 1). The steady-state rate of formic acid oxidation at a given potential is more than two orders of magnitude larger than the rate of the oxidation of $C_sH_pO_q$. This implies [63] in agreement with earlier conclusions [48, 53] that $C_sH_pO_q$ is not an intermediate of the oxidation of HCOOH at $U \leq 0.4$ V. The chemisorbed species are produced as a side product [48, 52].

It appears that two different paths have to be postulated for the oxidation of HCOOH.

(17) $$\mathrm{HCOOH} \longrightarrow \mathrm{HCOOH_{ad}} \begin{cases} \xrightarrow{I_1} \mathrm{C_sH_pO_q} \xrightarrow{I_2} \mathrm{CO_2} \\ \xrightarrow{I_3} \mathrm{CO_2} \end{cases}$$

The partial current I_3 is much larger than (I_1+I_2) under steady-state conditions in the Tafel region. The following mechanism is proposed for the CO_2 production by path 3:

(18) $$\mathrm{HCOOH} = \mathrm{HCOOH_{ad}},$$

(19) $$\mathrm{HCOOH_{ad}} = \mathrm{COOH_{weakly\ bonded}} + \mathrm{H_{ad}},$$

(20) $$\mathrm{HCOO_{weakly\ bonded}} = \mathrm{CO_2} + \mathrm{H^+} + e^-,$$

(21) $$\mathrm{H_{ad}} = \mathrm{H^+} + e^-.$$

The above mechanism is similar to mechanisms discussed in previous papers [48, 51]. The existence of weakly bonded species is assumed. The net composition of these species is the same as that of a portion of the strongly species, but they are more easily oxidized according to reactions 19 and 20. Reactions 18 and 19 occur only on sites which are free of $C_sH_pO_q$ and H_{ad}. The percentage of participating sites is smaller for steric reasons than the maximum of about 27% which is still available for hydrogen adsorption in the case of saturation coverage with $C_sH_pO_q$. The absence [48, 50, 53] of a separate arrest at less anodic potentials than the arrest for the oxidation of $C_sH_pO_q$ and the very small length of the hydrogen arrest in charging curves, taken on smooth platinum in 1 M $HClO_4$+X M HCOOH (X=0.01, 0.1, 1) from $U \geq 0.3$ V, is experimental evidence for a small amount of $HCOOH_{ad}$ and $HCOO_{weakly\ bonded}$. The statement about the hydrogen arrest had to be included since $HCOOH_{ad}$ and $HCOO_{ad}$ may be oxidized at the same potentials as H_{ad}. The removal of small amounts of $HCOOH_{ad}$ and $HCOO_{ad}$ by the washing procedure cannot be excluded for platinized platinum.

Recent steady-state measurements made by the author on the platinum cylinder in 0.5 M H_2SO_4+X M HCOOH (X=0.01, 0.03, 0.1, 0.3, 1) at 25 °C do not confirm a linear dependence of I upon bulk concentration at constant U (see Fig. 61). Such a dependence had been reported [48] for smooth platinum. The current at 0.38 V is only three times larger at ${}_bc_F=1$ M than at ${}_bc_F=0.01$ M. Simultaneously, the coverage with $C_sH_pO_q$ decreases [53, 56] from 1 to about 0.8. The current decreases less with ${}_bc_F$ in Fig. 61 than according to a proportionality [6] between I and ${}_bc_F{}^{1/2}$. The slope of the Tafel lines is about 60 mV. The results in Fig. 61 indicate that reaction 19 is rate-determining and that the equilibrium of reaction 18 is established in the Tafel region.

Steady-state measurements of formic acid oxidation on the cylinder in 0.5 M H_2SO_4+0.1 M HCOOH and 0.1 M Na_2SO_4+1 M HCOOH

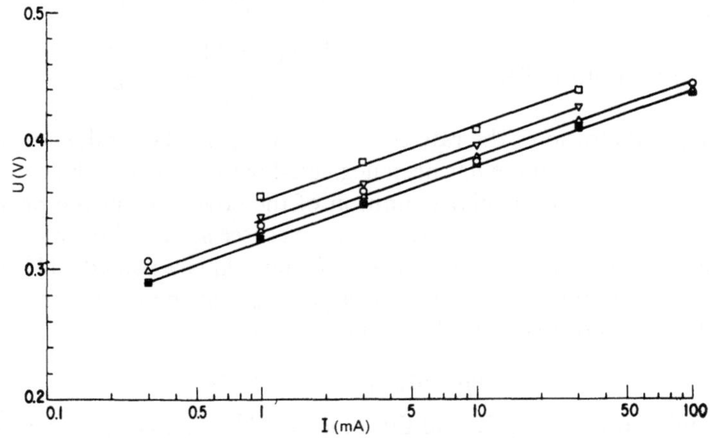

Fig. 61. Steady-state current-potential curves of formic acid oxidation on platinized platinum in X M HCOOH+0.5 M H_2SO_4 at 25 °C.
□: X=0.01, ▽: X=0.03, ○: X=0.1, △: X=0.3, ■: X=1.0

$+Y$ M H_2SO_4 (Y=0.1, 0.01) at 25 °C are shown in Fig. 62. The Tafel lines practically coincide. There is no pH dependence [48, 64] in acid electrolytes if a hydrogen electrode in the same electrolyte as the test electrode is used as reference electrode. The slope of the Tafel lines remains about 60 mV.

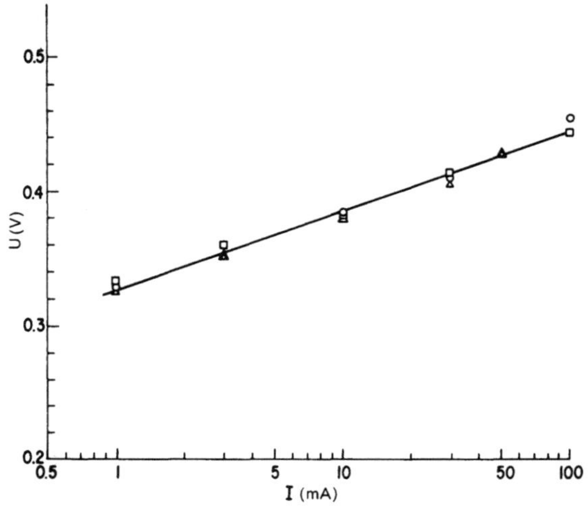

Fig. 62. Steady-state current-potential curves of formic acid oxidation on platinized platinum in 0.1 M HCOOH + X M Na_2SO_4 + Y M H_2SO_4 at 25 °C. □: X=0, Y=0.5, ○: X=0.1, Y=0.05, △: X=0.1, Y=0.005

The rate of formic acid oxidation may be expressed in the Tafel region:

$$(22) \qquad I = k_3\,\theta_{\mathrm{HCOOH}} \exp\left(\frac{-\beta\theta_{\mathrm{H}}}{RT}\right) = k_3\,\frac{\mathrm{K_b}\,c_{\mathrm{F}}(1-\theta)}{1+\mathrm{K_b}\,c_{\mathrm{F}}} \exp\left(\frac{-\beta\theta_{\mathrm{H}}}{RT}\right).$$

Here $\mathrm{K_b}$ is the ratio of the rate constants of Eq. 18. The influence of the small hydrogen coverage has been neglected with respect to the coverage θ with $C_sH_pO_q$. The regular definition of the coverages in the presence of simultaneous adsorption of several species (see section 3 of chapter IX) is used. An exponential dependence of I upon θ_H is assumed [34]. Since θ is practically independent of U at a given ${}_bc_F$ in the hydrogen region, the equilibrium of reaction 21 may be written

$$(23) \qquad -\beta' \cdot \theta_{\mathrm{H}} + \ln {}_bc_{\mathrm{H^+}} = UF + \mathrm{const}.$$

The combination of Eq. 22 and Eq. 23 leads [4] to a logarithmic dependence of I upon U. A close look at Fig. 61 and 62 reveals that the Tafel lines extend from 0.30 V to about 0.44 V. The hydrogen coverage is too small there for an experimental verification of Eq. 22 and Eq. 23. As pointed out previously in a similar context, expressions like Eq. 22 and Eq. 23 are considered tentative by the author for this reason.

Both paths are effective at potentials of the transition region and in the region of the limiting current. The existence of a limiting current suggests that the adsorption of formic acid molecules is rate-determining there. It is doubtful whether the equilibrium of the Volmer reaction is still established. A kinetic treatment of the coupling of two paths has been given by ANDREW [65] for the case of methanol oxidation. A similar treatment is applicable here.

An accelerating effect of chemisorbed layers of sulfur or selenium on the oxidation of HCOOH [61, 66] and also for CO [67] on Raney platinum and platinum black was found by BINDER, KÖHLING, RICHTER and SANDSTEDE recently. The sulfur layer was formed by cathodic polarization of platinum in hot 5 M H_2SO_4 or by treating the electrode with gaseous hydrogen sulfide. Selenium was deposited from a solution of selenium oxide in dilute sulfuric acid. The sulfur and selenium layers were resistant against removal by oxidation up to $U = 0.6$ V.

The current density of formic acid oxidation [61] at 0.2 V is shown at different temperatures in Fig. 63 as a function of the coverage θ_S with sulfur. It increases with θ_S at a given temperature, passes through a maximum at $\theta_S \approx 0.3$ and decreases afterwards. Continuous decomposition into CO_2 and H_2 occurred at open circuit in the presence of the layers even at room temperature. In the absence of adsorbed layers of S or Se, the decomposition stopped after a period of a few minutes to an hour. While the current at constant potential decreases [61] to small values with time at $U \leq 0.4$ V in the absence of S or Se, the decrease of current is small in the presence of these layers. The oxidation of $C_sH_pO_q$ takes place [68] in nearly the same potential range whether a sulfur layer is present or not. While the sulfur layer decreases the amount of $C_sH_pO_q$ by blocking of sites, it does not shift the potential range of oxidation of $C_sH_pO_q$ to more positive values as is done by specifically adsorbable anions. In contrast to acid solutions, the sulfur layer inhibits [61] the formic acid oxidation in alkaline media.

The strong increase of the reactivity of platinum with respect to net reaction 14 and the negligible influence on the oxidation of $C_sH_pO_q$ by sulfur layers confirm the previous conclusion that $C_sH_pO_q$ inhibits the net reaction by blocking of sites. It was suggested [61] that the formic acid oxidation occurs on top of $C_sH_pO_q$ in the absence of the sulfur layer and on top of a mixed layer of sulfur and $C_sH_pO_q$ in the presence of S. However, this interpretation does not explain why a limiting current is reached with increasing potential at small coverage with $C_sH_pO_q$. The acceleration of reactions 14 (or 17) in the presence of layers of sulfur or selenium is attributed by the author to a reduced inhibition of reaction 19. The available experimental data are not sufficient to give a detailed interpretation.

The kinetics of formic acid decomposition at open circuit were recently studied [69] by POLIAK, VASILIEV, BAGOTZKY, and SMIRNOVA on smooth and palladinized palladium in aqueous solutions of 0.25 M DCOOH + 0.5 M H_2SO_4 and 0.25 M DCOOH at 70 °C. The

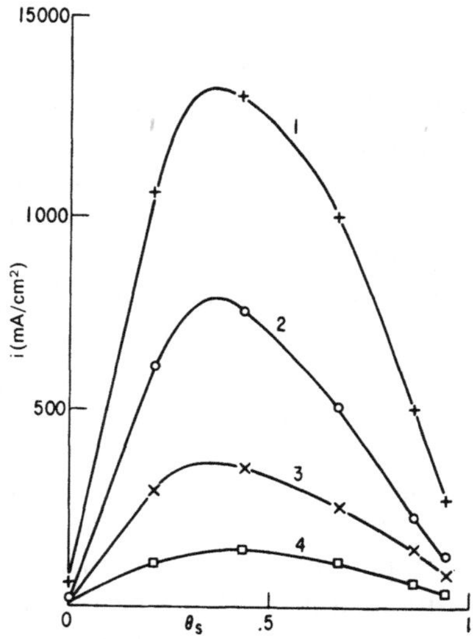

Fig. 63. Rate of formic acid oxidation at $U = 0.2$ V on Raney platinum in 2 M HCOOH + 1.5 M H_2SO_4 at different temperatures in dependence of the coverage with sulfur. Curve 1: 90 °C, curve 2: 70 °C, curve 3: 50 °C, curve 4: 30 °C

percentage of H_2 in the evolved gas was 98.4 % and 92.5 % respectively while the content of HD was 1.63 % and 7.5 %. Deuterium was not detected. The analysis was done by mass spectrometry. A direct cracking of HCOOH into H_2 and CO_2 without intermediates or the sequence of reactions 24 and 25

(24) $$2\,HCOOH = H_2 + 2\,HCOO_{ad},$$

(25) $$2\,HCOO_{ad} = H_2 + CO_2$$

were proposed [69] as possible mechanisms because HD was detected. It was argued that HD should not be detected in the case of the subsequent mechanism since the rapid exchange between D_{ad} and H^+ would replace any D_{ad} by H_{ad}.

(26) $$HCOOH = HCOO_{ad} + H_{ad},$$

(27) $$HCOO_{ad} = CO_2 + H_{ad},$$

(28) $$2H_{ad} = H_2.$$

It was found that the amount of adsorbed carbonaceous species was too small to be detected by pulse techniques.

The conclusions on the mechanism in ref. 69 are not convincing. A direct cracking of DCOOH should produce mainly HD. The sequence of reactions 24 and 25 should yield about equal amounts of D_2 and H_2. The results in acid solutions do not allow an unambiguous distinction between the two possibilities.

5. Methanol Oxidation

The general features of the oxidation of methanol, which is one [70] of the attractive fuels, were discussed in section 4 of chapter IV. The reaction appears more complex [63] in acid electrolytes than assumed in recent papers [6—8, 10, 11, 16—19, 38, 49, 52, 71—81] because CH_2O and HCOOH are formed besides CO_2 to a noticeable extent under steady-state conditions. The efficiency of CO_2 production depends strongly upon the bulk concentrations of CH_3OH, CH_2O, HCOOH and the potential. The formation of chemisorbed species during the initial oxidation of methanol on platinum, the coverage as a function of potential under different conditions, the nature of the chemisorbed species and their oxidation were described in chapter IX.

The following discussion is based on the scheme 59 in chapter IV. The maximum rate of CO_2 production by the intermediate formation and oxidation of $C_sH_pO_q$ is about $3I_2$ under steady-state conditions since about four electrons are involved in the formation of $C_sH_pO_q$ and two electrons in the oxidation. It may be estimated from the upper curve in Fig. 60 for a given potential. The influence of the coverage with $C_sH_pO_q$ is negligible for $Q_{Ox}/{}_sQ_{Ox} > 0.3$ in a first approximation since the $U-t$ curves for the oxidation of $C_sH_pO_q$ have a small slope there (see the dashed curves in Fig. 39). After a steady state had been achieved at 10 mA in 0.5 M H_2SO_4 + X M CH_3OH (X = 0.01, 0.1), the potential, the ratios I/I_{CO_2} and $(1 - {}_sQ'_H/{}_sQ_H)$ were determined [82]. Then the same quantities were obtained after the addition of small amounts of HCl. The quantities are plotted as a function of the chloride ion concentration ${}_bc_{Cl^-}$ in Fig. 64. Slow oscillations, the amplitude of which became larger with ${}_bc_{Cl^-}$ as marked, were observed at X = 0.1 for ${}_bc_{Cl^-} \geq 6.10^{-3}$ M. The addition of HCl leads to a strong decrease of the coverage with

$C_sH_pO_q$ in agreement with voltammetric results [74] and was attributed [6, 38, 80, 82] to the specific adsorption of the anion. The quantity $(1-{}_sQ'_H/{}_sQ_H)$ decreases at X=0.01 from 0.67 at ${}_bc_{Cl^-}=0$ to 0.13 at ${}_bc_{Cl^-}=3.10^{-4}$ M. Simultaneously the potential increases from 0.48 V to 0.55 V.

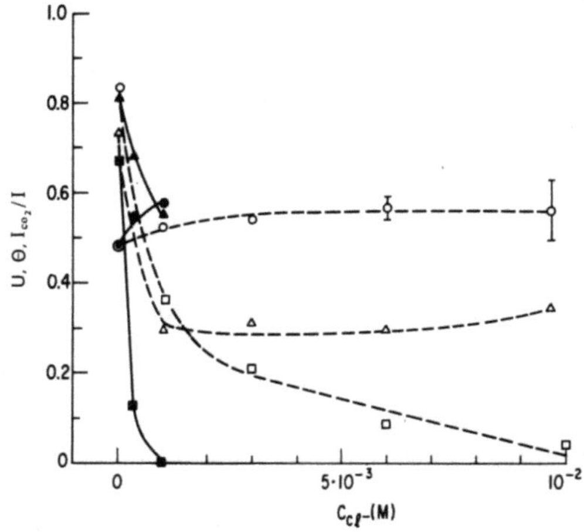

Fig. 64. Potential (●, ○), coverage (■, □) with chemisorbed carbonaceous species, and efficiency of CO_2 production (▲, △) on platinized platinum as a function of the chloride ion concentration at 10 mA in 1 M $HClO_4$+X M CH_3OH+Y M HCl.
●, ■, ▲: X=0.01, ○, □, △: X=0.1

The rate of the oxidation of $C_sH_pO_q$ at 0.55 V was estimated from the I_2-U curves in 1 M $HClO_4$ to be about 0.7 mA at $Q_{Ox}/{}_sQ_{Ox}\approx 0.15$. Thus the sum (I_1+I_2) cannot be larger than 2.1 mA. Such a rate does not account for $I_{CO_2}=8.1$ mA at ${}_bc_{Cl^-}=0$. A similar statement holds at ${}_bc_{Cl^-}=3\cdot 10^{-4}$ M. Carbon dioxide is not produced predominantly by the formation and oxidation of $C_sH_pO_q$. The small decrease of I_{CO_2}/I from 0.81 at ${}_bc_{Cl^-}=0$ to 0.68 at ${}_bc_{Cl^-}=3\cdot 10^{-4}$ confirms the conclusion that the larger amount of CO_2 is evolved by the second path in the absence and presence of Cl^- ions.

Steady-state $I-U$ curves of methanol oxidation were measured [82] on the platinized platinum cylinder of 50 cm² (${}_sQ_H=80$ mcoul/cm²) at potentials between 0.40 V and 0.55 V and on a platinized platinum foil of 8 cm² (${}_sQ_H=14$ mcoul/cm²) between 0.55 V and 0.70 V in 0.5 M H_2SO_4+X M CH_3OH (X=0.01, 0.1, 1, 5) at 25 °C. The efficiency I_{CO_2}/I and the quantity $(1-{}_sQ'_H/{}_sQ_H)$ were determined too. The data in Tables 5 and 6 were interpolated from plots of U, $(1-{}_sQ'_H/{}_sQ_H)$, and I_{CO_2} versus I and U versus I_2.

Table 5. *Values of characteristic parameters for the platinum cylinder*

U(V)	0.40				0.45				0.50				0.55			
$_bc_M$(M)	0.01	0.1	1	5	0.01	0.1	1	5	0.01	0.1	1	5	0.01	0.1	1	5
$1-{}_sQ'_H/{}_sQ_H$	0.64	0.82	0.93	0.93	0.64	0.82	0.93	0.93	0.64	0.82	0.93	0.93	0.63	0.82	0.93	0.93
I(mA)	0.23	0.34	0.11	0.34	1	1.55	1.55	0.48	4.9	8.0	8.0	2.2	22	65	65	10
$3I_2$(mA)	0.43	0.43	0.43	0.43	1.8	1.8	1.8	1.8	7.8	7.8	7.8	7.8	33	33	33	33
I_{CO_2}/I	—	—	—	—	0.81	0.76	0.68	—	0.92	0.75	0.42	0.51	0.99	0.86	0.47	0.18
$I_{CO_2}/3I_2$	—	—	—	—	0.45	0.65	0.59	—	0.58	0.77	0.43	0.14	0.63	0.17	0.09	0.05

Table 6. *Values of characteristic parameters for the platinum foil*

U(V)	0.55				0.60				0.65				0.70			
$_bc_M$(M)	0.01	0.1	1	5	0.01	0.1	1	5	0.01	0.1	1	5	0.01	0.1	1	5
$1-{}_sQ'_H/{}_sQ_H$	0.62	0.82	0.93	0.93	0.54	0.81	0.93	0.93	0.37	0.64	0.87	0.91	—	0.39	0.76	0.86
I(mA)	1.0	2.0	2.0	1.1	4.4	8.7	8.7	7.2	16.5	30	30	35	—	57	57	70
$3I_2$(mA)	1.0	1.0	1.0	1.0	2.8	2.8	2.8	2.8	12	12	12	12	—	51	51	51
I_{CO_2}/I	—	—	—	—	0.92	0.73	0.58	0.10	0.90	0.78	0.60	0.09	—	0.81	0.61	0.13
$I_{CO_2}/3I_2$	—	—	—	—	1.4	2.3	1.8	0.26	1.2	1.9	1.5	0.26	—	0.91	0.68	0.18

The ratio I_{CO_2}/I in Tables 5 and 6 is less than 1 and decreases considerably with ${}_bc_M$ at a given potential (compare Fig. 4). The formation of largely [83] CH_2O cannot be disregarded at $X > 0.01$. This reaction occurs in spite of the result that the maximum value of $(I_1 + I_2)$ would be sufficient at $U \leq 0.5\,V$ to account for the net current on the cylinder. In contrast to the results for the cylinder, the values of $3\,I_2$ are smaller than I and I_{CO_2} on the foil. The latter difference may be due to the larger platinization of the cylinder. The decrease of $(1 - {}_SQ'_H/{}_SQ_H)$, which is a measure for the coverage of chemisorbed species with U at constant ${}_bc_M$, does not affect the results $I_{CO_2}/I < 1$ in Table 6. Carbon dioxide cannot solely be produced through the formation and oxidation of $C_sH_pO_q$ at $U \leq 0.65\,V$ for ${}_bc_M > 0.01\,M$ since $I_{CO_2}/3\,I_2 < 1$.

It was pointed out in section 4 of chapter IV that the definition of a rate-determining step for the total process of methanol oxidation is meaningful only if there is only one oxidation product. The results in Table 5 and 6 demonstrate clearly that there is more than one oxidation product on platinized platinum. Additional experimental evidence is required before the different rate-determining steps of the net reactions of the path involving CH_2O and HCOOH as intermediate products can be elucidated. It is not certain how much the efficiency for CO_2 production on smooth platinum differs from that on platinized platinum under equivalent conditions. However, the similarity of the shape of $i - U$ curves measured by cyclic voltammetry on smooth or platinized platinum in the same solution and at the same sweep rate, indicates that the situation is not too different. An experimental determination of the efficiency of CO_2 production on smooth platinum under steady-state conditions could be made by analyzing the CO_2 content of the stirring gas by mass spectrometry.

The inhibiting effect of the oxygen layer on the methanol oxidation at different platinum metals and its influence on the reactivity of these metals were discussed in section 6 of chapter VIII.

6. Oxidation of Higher Alcohols and Aldehydes

When introduced into the solution at potentials of the double layer region, alcohols and aldehydes containing more than one carbon atom cause [4, 39, 84, 85] the potential to become less positive. After passing through a minimum, the potential remains unchanged or becomes more positive. In contrast to methanol, the potential may move into the positive direction again. If the substances are added close to $U = 0$, gas bubble formation is observed. The evolution of gases proceeds actively at first and slows down with time. It is very small when a steady-state potential has been reached. The results [39] of the analysis of the initial gas samples are given in Table 7 for different fuels.

Table 7. *Composition of gases formed initially at open circuit during the interaction of different fuels with platinum in the hydrogen region*

Substance	${}_{b}c$(M)	Added at U_1(V)	Final Potential U_2(V)	Composition of gases in vol.% CH_4	C_2H_6	C_3H_8	C_4H_{10}
C_2H_5OH	2	0.057	0.159	13.4	86.6	—	—
C_2H_5OH	2	0.499	0.157	68.2	31.8	—	—
C_3H_7OH	1	0.054	0.140	—	19.3	76.3	4.4
C_4H_9OH	satd.	0.041	0.131	<1	<1	23	76
C_4H_9OH	satd.	0.511	0.138	—	2.8	69.4	27.8
CH_3CHO	0.7	0.038	0.184	54	46	—	—

When alcohols are brought into contact with the Pt electrode nearly covered with a hydrogen monolayer, the hydrocarbon that contains as many carbon atoms as the alcohol predominates in the gas evolved initially. In contrast, a predominant breaking along the C_1-C_2 bond is observed when the substances begin to react with an electrode initially free of H_{ad}. The appearance of hydrogenation products in the first case is evidence for the hydrogenation [86] of the alcohols by H_{ad}. In the latter case a self-hydrogenation takes place involving the oxidation of part of the alcohol and of the decomposition products as anodic process. Reactions of the type of the well-known Canizzaro reaction occur.

Charging curves taken after adequate washing procedures reveal that chemisorbed carbonaceous species are formed at open circuit. Curve 1a in Fig. 65 is the $U-t$ curve on platinized platinum in the absence of $C_sH_pO_q$. Curves 1 and 2 were obtained with 10^{-4} A/cm^2 after the introduction of 0.5 M C_2H_5OH, curves 3 and 4 after the addition of 0.5 M CH_3CHO to 0.05 M H_2SO_4 at about 0.5 V and 0.06 V respectively. The curves were taken after the potential had achieved about the same steady-state value for each of the two pairs of curves. When the fuel is added at 0:06 V, the arrest is longer. It was suggested [39] that the chemisorbed species formed after fuel addition at potentials of the double layer region are less rich in hydrogen than the species produced in the hydrogen region. It is the same distinction which was discussed as type I and II species or O and CH type species for hydrocarbons in section 10 of chapter IX. When hydrogen is bubbled [87] through the ethanol solution, about 50 % of the chemisorbed species is removed from the surface. This is evidence of at least two types of species, as in the similar hydrogenation experiments with hydrocarbons. The one type reacts with H_{ad} and desorbs. The other type does not interact.

The behavior of formaldehyde differs in some aspects from that of higher aldehydes. After adding formaldehyde, potentials below 0.1 V are established as for methanol. It was proposed [39] that the hydrogenation of formaldehyde proceeds in sulfuric acid solution only until methanol is formed. Chemisorbed species similar to those formed in the presence of methanol or formic acid are produced at open circuit [58] and by anodic oxidation [40] of formaldehyde.

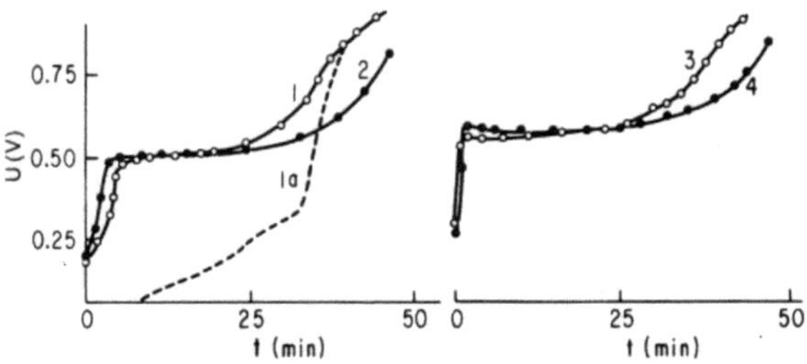

Fig. 65. Charging curves on platinized platinum in 0.05 M H_2SO_4 in the absence (curve 1 a) and presence (curves 1, 2, 3, 4) of chemisorbed carbonaceous species. The species were formed at open circuit after the addition of fuel at about 0.5 V (curves 1 and 3) or at 0.06 V (curves 2 and 4). Fuel was C_2H_5OH for curves 1 and 2 and CH_3CHO for curves 3 and 4

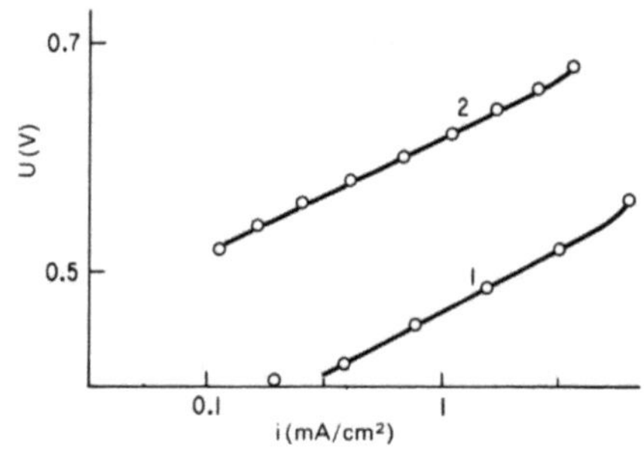

Fig. 66. The $i-U$ curve 1 was measured on platinized platinum in 0.5 M C_2H_5OH + 0.05 M H_2SO_4 by an anodic potential sweep with 5 mV/sec. Curve 2 corresponds to the oxidation of chemisorbed species formed in the presence of C_2H_5OH

Evolution of gases is not observed during the steady-state oxidation of alcohols and alhydes in acid and alkaline electrolytes. The $i-U$ curve 1 in Fig. 66 was measured [39] on platinized platinum during an anodic sweep with 5 mV/sec in 0.5 M C_2H_5OH + 0.05 M H_2SO_4. It represents nearly the steady-state $i-U$ curve. Curve 2 is the Tafel line resulting from the oxidation of $C_sH_pO_q$. Curve 2 is located above curve 1 as in the case of formic acid and methanol. The rate of oxidation of $C_sH_pO_q$ at a given potential, for instance 0.52 V, is about a hundred times smaller than that of C_2H_5OH. It is concluded that the chemisorbed species are not the intermediates of ethanol oxidation in the Tafel region. Product analysis [88] at 0.75 V showed that the primary product is acetaldehyde:

(29) $$CH_3CH_2OH = CH_3CHO + 2H^+ + 2e\,.$$

A small amount of acetic acid was also built up.

(30) $$CH_3CH_2OH + H_2O = CH_3COOH + 4H^+ + 4e^-\,.$$

Reactions 29 and 30 occur on sites free of $C_sH_pO_q$ and OH_{ad}. It is likely that weakly bonded species participate. The difficulties of interpreting the kinetic data of methanol oxidation were pointed out in the preceding chapter. The anodic oxidation [4, 39, 84, 85, 88, 89] of higher alcohols and aldehydes is equally complex. Further experimental results are needed for a quantitative interpretation.

7. Oxidation of Hydrocarbons

Although the interaction of hydrocarbons with platinum at open circuit, the formation and oxidation of chemisorbed carbonaceous species under different conditions and the oxidation of hydrocarbons under steady-state conditions have been studied [20, 21, 90—122] extensively in recent years, the oxidation mechanisms are not established. The difficulty in arriving at a quantitative interpretation arises from the complex task of obtaining information about the individual role of the different types of chemisorbed species under the prevailing experimental conditions. The situation is similar to that described for higher alcohols and aldehydes in the preceding section.

Niedrach [92] carried out volumetric and galvanostatic studies of the species formed at open circuit on Teflon-bonded platinum black in acid, neutral and alkaline electrolytes at 25 °C and 60° from saturated (CH_4, CH_3CH_3, $CH_3CH_2CH_3$, $(CH_3)_3CH$) and unsaturated (CH_2CH_2, CH_3CHCH_2, $CH_2CH_2CH_2$) hydrocarbons. Sulfuric acid was incorporated by equilibration into a zerolit C-20 cation exchange membrane. Zerolit A-20 membranes were used with potassium bicarbonate and potassium hydroxide electrolytes. The test electrode and the counter electrode were pressed against the two sides of the membrane. The electrode was brought to $U = 0.4$ V after a pretreatment procedure. The hydrocarbon was fed at open circuit from a burette through a small chamber to the other side of the porous electrode which was not in contact with the membrane. The system was allowed to equilibrate for a certain time. The potential and the change of gas volume in the burette were recorded during this period. At the end of the equilibration the gas was forced by electrolyte from the small chamber into a graduated collector tube. Samples of this gas were analyzed by gas chromatography. The species chemisorbed on the electrode were removed by galvanostatic oxidation or reduction. The anodic charging curves in Fig. 51 were obtained in this way. Fully hydrogenated hydrocarbons were desorbed during the reductions. These gases, together with hydrogen, replaced the electrolyte in the chamber. Samples were analyzed.

The potential decreased during the equilibration period and tended towards steady-state values between 0.15 V and 0.26 V for the saturated hydrocarbons in 2.5 M H_2SO_4 at 25 °C. It decreased initially, passed through a minimum and increased slowly to values between 0.23 V and 0.27 V in the case of unsaturated hydrocarbons. The charging curves in Fig. 51 show to what extent chemisorbed species are formed from the various hydrocarbons (compare section 10 of chapter IX). The corresponding charging curves reveal that formation of chemisorbed species from hydrocarbons is small with the exception of C_2H_4 in 6 M KOH.

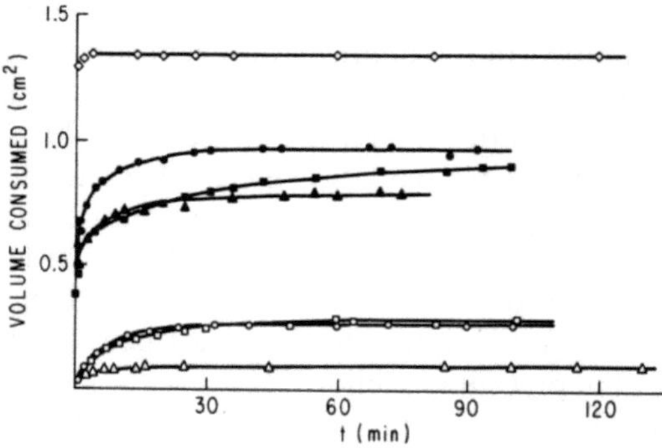

Fig. 67. Gases consumed during the equilibration of a Teflon-bonded platinum electrode at 25 °C.
◇: hydrogen, △: methane, ○: ethane, □: propane, ●: ethylene, ■: cyclopropane, ▲: propylene

The gases that were consumed during the equilibration are plotted as a function of time in Fig. 67. The volumes were corrected for the methane and ethane due to cracking. The hydrogen consumption corresponds to about a monolayer of H_{ad}. Methane is used up to a small extent. A small amount of chemisorbed species is formed in agreement [92, 112, 115, 116, 119] with the short arrest in anodic charging curves. The gas consumption of saturated hydrocarbons of higher molecular weight is larger than that of methane. All the saturated hydrocarbons form the type I species at open circuit [92, 99] and also at constant potential [107, 112, 115—119]. As the molecular weight of the saturated hydrocarbons above methane increases, the type II species are produced in greater quantities [119]. The length of the arrest due to the oxidation of $C_sH_pO_q$ above 0.8 V increases with the molecular weight of these fuels in Fig. 51. The behavior of the saturated hydrocarbons is consistent with results [123, 124] in the gas phase where less than monolayers are formed. Dissociative adsorption with the formation of H atoms and probably of both ethyl radicals and 1, 2 diadsorbed ethane occurs [124]. The gas consumption of unsaturated hydrocarbons is larger than that of the saturated ones. This was attributed [92] to self-hydrogenation as in the gas phase [125, 126]. The net result is a surface covered to a large extent by type I and type II species.

The analysis of the gases after an equilibration period of 100 minutes shows [92] that methane is formed from ethane and ethylene. Ethane is found besides methane from propane, propylene, and cyclopropane as original fuels. Saturated hydrocarbons are released [92] to a varying extent besides the original fuel on cathodic hydrogenation of the chemisorbed species that were previously formed during the equilibration with ethane, ethylene, propane, propylene, and cyclopropane.

It was demonstrated [101] that the adsorption of ethane on micro Teflon-bonded platinum black electrodes in 4.3 M $HClO_4$ is mechanistically and kinetically similar to that on smooth platinum after normalization to the active area defined by ${}_{S}Q_H$. The hydrocarbon is admitted to the one side of the porous electrode. The other side is in contact with the electrolyte. Steady-state $i-U$ curves and $Q_{Ox}-U$ curves, measured [119] point by point in 75 % H_3PO_4 on microelectrodes at 100 °C in the presence of four different saturated hydrocarbons, are shown in Fig. 68. The current density i, the anodic charge Q_t for the oxidation of all the chemisorbed species and the charge Q_1 for the oxidation of type I species are normalized. The sequence of potential steps employed was A to G in Fig. 52 with step D at 0.06 V, step E at the desired potential and $t_F = 0$. The sweep rate for step G was 0.1 V/sec. This rate is large enough to make the contribution from the oxidation of hydrocarbons, diffusing into the electrode from the gas phase, negligible. The charge Q_1 was determined approximately for ethane and propane from the area under the first wave of the $i-U$ curve of step G. The steady-state current was measured at the end of step E with $t_E = 10$ min. This time was sufficient for the achievement of steady-state values of i, Q_t, and Q_1.

The $i-U$ curves in Fig. 68 have a similar shape. The current begins to increase with potential at about 0.2 V. It reaches a maximum located between 0.5 V and 0.6 V. The subsequent decrease of i with U extends to about 1.0 V. The Q_t-U curves and Q_1-U curves are bell-shaped with the maximum at about 0.2 V. Only type I species are present for

CH_4. The amount of both type I and type II species is smaller for C_2H_6 than for C_3H_8. The Q_t value at 0.2 V increases with molecular weight. BRUMMER [107] distinguished between an α and β type of the type II species in the case of C_3H_8. The α type is cathodically desorbable. A mixture of alkyl radicals was proposed [118] for its composition which changes with potential. The β type species are unreactive to both reduction and oxidation and are probably a polymer.

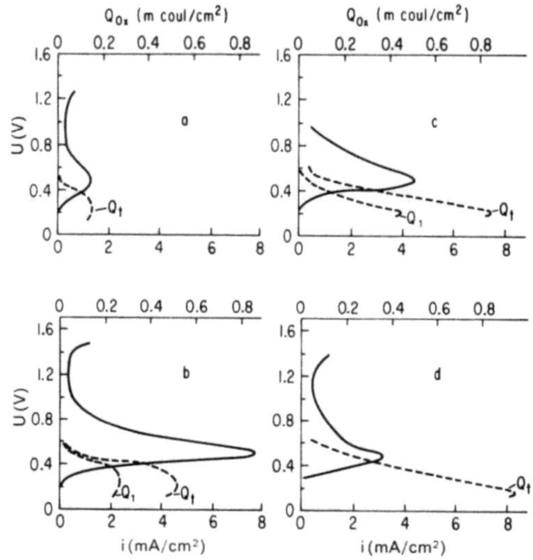

Fig. 68. Steady-state current densities and charges due to the oxidation of chemisorbed species as a function of potential in 75% H_3PO_4 at 100 °C in the presence of saturated hydrocarbons.
a: methane, b: ethane, c: propane, d: butane

The production of j types of chemisorbed species at constant potential may be expressed in a general way:

$$\frac{dQ_t}{dt} = \sum_j (i_{F,j} - i_{Ox,j} - i_{Hy,j}) . \tag{31}$$

Here $i_{F,j}$ designates the rate of formation of the species, j, $i_{Ox,j}$ the rate of oxidation, and $i_{Hy,j}$ the rate of hydrogenation. Studies in the ethane system [94, 95, 101] revealed that dQ_t/dt is greatest at potentials in the vicinity of the potential U_M of the maximum of the $Q_t - U$ curve. It was concluded [94] that $i_{F,j} \gg i_{Hy,j}$ there. The net rate dQ_t/dt is smaller at $U > U_M$ and $U < U_M$ because the contribution of $i_{Ox,j}$ and $i_{Hy,j}$ respectively is no longer negligible. Since $i_{F,j}$, $i_{Ox,j}$ and $i_{Hy,j}$ cannot be measured separately, the extent to which the decrease of Q_t at both sides of U_M results from the changes of the partial currents with potential is not known.

The total coverage decreases with U in the rising portion of the $i-U$ curves. The question arises how the type I and type II species are involved in the steady-state oxidation of saturated hydrocarbons to CO_2. The 100% efficiency [127] for the CO_2 production under steady-state conditions implies that all the intermediate products between the reactant and the final product do not leave the electrode. A likely reaction scheme [108, 109, 112, 128] is:

$$\text{sat. hydrocarbon} \longrightarrow \text{type II species,} \tag{32}$$

$$\text{type II species} \longrightarrow \text{type I species,} \tag{33}$$

$$\text{type I species} \longrightarrow CO_2. \tag{34}$$

The decrease of the reactivity of saturated hydrocarbons with molecular weight may result from an inhibition of the rate-determining step 34 by the type II species.

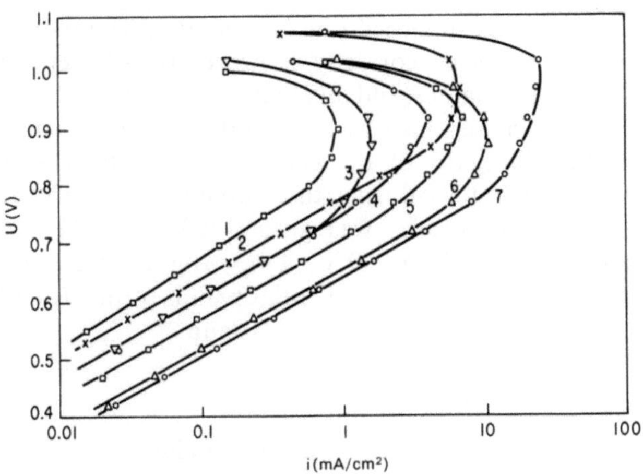

Fig. 69. Steady-state current-potential curves for the oxidation of unsaturated hydrocarbons on platinized platinum in 0.5 M H_2SO_4 at 80°C.
Curve 1: benzene, curve 2: butadiene, curve 3: 2-butadiene, curve 4: l-butene, curve 5: propylene, curve 6: ethylene, curve 7: allene

Fig. 69 shows a comparison of potentiostatic $i-U$ curves determined [100] on platinized platinum in 0.5 M H_2SO_4 at 80°C for several unsaturated hydrocarbons. Tafel lines with slopes of $2\,RT/F$ were obtained over three decades of current. The maximum of the $i-U$ curves at about 0.9 V was attributed [100] to the inhibition of the reactions by the oxygen layer. The similarity in the oxidation of the compounds can be expressed by the empirical relation:

$$I = k c_{\text{fuel}}^{\beta} c_{H^+}^{-\frac{1}{2}} \cdot \exp\left(\frac{UF}{RT}\right). \tag{35}$$

Where $0.1 < \beta < 0.25$ and k decreases about one order of magnitude from the oxidation of allene and ethylene to that of benzene and butadiene.

This similar behavior was taken as an indication that the same step is rate-controlling and that the position of this step in the reaction sequence is the same. The following mechanism which is written for ethylene as an example, was suggested [100]:

$$C_2H_4 = C_2H_{4\,ad}, \tag{36}$$

$$H_2O = OH_{ad} + H^+ + e^-, \tag{37}$$

$$C_2H_{4\,ad} + OH_{ad} = C_2H_{3\,ad}^{\cdot} + H_2O\,(\text{or } C_2H_4OH_{ad}), \tag{38}$$

$$C_2H_{3\,ad}\,(\text{or } C_2H_4OH_{ad}) + 4\,H_2O\,(\text{or } 3\,H_2O) = 2\,CO_2 + 11\,H^+ + 11\,e^-. \tag{39}$$

The water discharge is supposed rate-determining. The model does not predict the pH dependence correctly. However, the pH dependence was considered [100] anomalous, and different interpretations were advanced for the anomaly. The serious objection to the above mechanism comes [108] from the disregard of the chemisorbed carbonaceous species formed [98, 99, 108, 109] during the oxidation of ethylene. The net composition of these species was given [109] as C_2H_2 on smooth platinum in 1 M $HClO_4$ at 30 °C. The coverage with chemisorbed species decreases with U in the potential region of the Tafel lines and reaches zero at about 1.2 V on smooth platinum.

8. Oxidation of Hydrazine

The anodic oxidation of hydrazine which is of interest as a non-carbonaceous fuel has been studied on different transition metals in acid and alkaline media in recent years [8, 129—138].

The net reaction in acid electrolytes is:

$$N_2H_5^+ = N_2 + 5\,H^+ + 4\,e^-. \tag{40}$$

The net reaction in alkaline solutions

$$N_2H_4 + 4\,OH^- = N_2 + 4\,H_2O + 4\,e^- \tag{41}$$

may involve the following steps:

(42) $$N_2H_{4\,ad} = NH_{2\,ad} + NH_{2\,ad},$$

(43) $$NH_{2\,ad} = NH_{ad} + H_{ad},$$

(44) $$NH_{ad} = N_{ad} + H_{ad},$$

(45) $$N_{ad} + N_{ad} = N_2,$$

(46) $$H_{ad} + OH^- = H_2O + e^-.$$

The N—N bond is broken. Alternatively, the following scheme may be considered:

(47) $$N_2H_{4\,ad} = H_2H_{3\,ad} + H_{ad},$$

(48) $$N_2H_{3\,ad} = N_2H_{2\,ad} + H_{ad},$$

(49) $$N_2H_{2\,ad} = N_2H_{ad} + H_{ad},$$

(50) $$N_2H_{ad} = N_2 + H_{ad}.$$

Successive dehydrogenation without breaking of the N—N bond is assumed. Instead of the intermediate formation of H_{ad} followed by reaction 46, water may be produced directly:

(51) $$N_mH_{n\,ad} + OH^- = N_mH_{n-1\,ad} + H_2O + e^-.$$

Steady-state $i-U$ curves measured [133] potentiostatically on a rotating platinized Pt cylinder in 0.1 M KOH + 0.3 M K_2SO_4, and in 1 M KOH at 24° ± 1 °C are shown in Fig. 70. The $i-U$ curve for the H_2 oxidation at $p_{H_2} = 1$ atm is included for comparison. The shape of the $i-U$ curves was found to depend upon the ratio ${}_bc_{N_2H_4}/{}_bc_{OH^-}$. Well defined limiting currents are observed at the lower concentrations of hydrazine (curves b, c, e, f). The limiting currents were about 30% smaller than those computed for diffusion control on the rotating cylinder while good agreement was achieved in the case of H_2. Desorption of a reaction product was suggested [133] tentatively as the origin of the discrepancy between experimental and theoretical limiting currents. The resemblance between the $i-U$ curves for the oxidation of hydrogen and those for the oxidation of hydrazine for small values of ${}_bc_{N_2H_4}/{}_bc_{OH^-}$ extends from the rest potential at open circuit to oxygen evolution. The inhibition of the H_2 oxidation by the oxygen layer, which was discussed in section 6 of chapter VIII, is more pronounced than that of the hydrazine oxidation. The latter process does not involve O_{ad} or surface oxides. The rest potential was interpreted [133, 137] as a mixed potential. It was concluded [133] that the oxidation of hydrazine is not primarily by a hydrogen oxidation mechanism, as suggested by PAVELA [129], but by a sequential oxidation of intermediates. A detailed analysis of possible consecutive steps was carried out under certain assumptions on the adsorption of intermediate products in reference 135. It was suggested tentatively that a chemical reaction of the type

(52) $$N_2H_{n\,ad} + n\,OH_{ad} = n\,H_2O + N_2$$

with the discharge of OH^- ions initially as rate-determining step would give the correct Tafel slope (about 120 mV) and a negative reaction order with respect to ${}_bc_{N_2H_4}$ at small concentrations in alkaline media. At

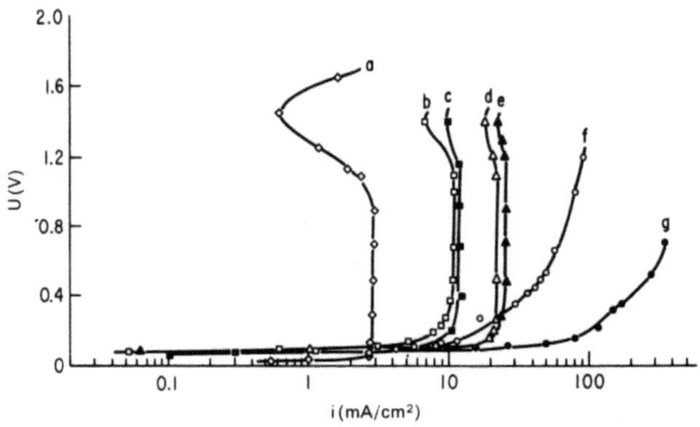

Fig. 70. Steady-state current-potential curves for the oxidation of hydrogen (curve 1) and hydrazine (curves 2 to 7) on platinized platinum at 24 °C.
Curves a, b, c: ${}_bc_{N_2H_4} = 0.005, 0.01, 0.1$ M in 0.1 M KOH + 0.3 M K_2SO_4,
curves d, e, f: ${}_bc_{N_2H_4} = 0.005, 0.01, 0.1$ M in 1 M KOH

higher concentrations of N_2H_4, a direct oxidation with an adsorbed species at full coverage seems the only possibility [135, 138].

(53) $$N_2H_{3\,ad} + OH^- = N_2H_{2\,ad} + H_2O + e^-.$$

In acid solutions, the reaction order is fractional and increases with the hydrazine bulk concentration. The reaction

(54) $$N_2H_{4\,ad} = N_2H_{3\,ad} + H^+ + e^-$$

was supposed rate-determining in reference [137].

Volumetric measurements [136] of the self-decomposition of hydrazine on platinized carbon electrodes and analysis of the gas mixture lead to the net reaction

(55) $$2\,N_2H_4 = 2\,NH_3 + H_2 + N_2$$

between 20 °C to 40 °C in alkaline solutions. Between 60° and 80 °C the net reaction

(56) $$N_2H_4 = N_2 + 2H_2$$

was involved. Reactions 55 and 56 have to be taken into account in the interpretation of the electrochemical oxidation of hydrazine on platinized carbon. In contrast, the contribution from reactions 55 and 56 is negligible [138] on platinized platinum.

9. Oscillatory Phenomena on Solid Electrodes

Oscillatory behavior observed as periodic potential transients at constant current or periodic current transients at constant potential is found frequently when more than two parallel electrode reactions are coupled. Usually, an upper and a lower current-potential curve limit the oscillation region. These two curves represent stable states [139] according to the theory of stability of electrode states [140]. Oscillatory phenomena occurring during the oxidation of certain fuels on solid electrodes are discussed in this section. The discussion is not extended to porous electrodes because the theory of the diffusion electrode has not been developed to the point to allow an adequate description of the complex coupling of parallel electrode reactions and mass transport processes in the liquid and gaseous phase.

Oscillatory behavior has been reported for the oxidation of methanol [49], formaldehyde [11, 49, 141], and formic acid [47]. The oscillations appear when a certain threshold current density is reached. Steady-state measurements are possible at currents below the threshold current and yield Tafel relations between I and U with regular slopes. The period of the oscillations becomes smaller with increasing current as shown by the $U-t$ curves in Fig. 71. These curves were obtained [141] on platinized platinum with 3 cm^2 of geometric area in 1 M $CH_2O + 3.75$ M H_2SO_4 at room temperature. The amplitude of the oscillations becomes larger with increasing current. The limits of the potential regions of oscillations at different currents form the upper and lower $I-U$ curve corresponding to stable states. A sudden transition from the potential region of oscillations to a steady-state potential of predominant oxygen evolution takes place at a certain current if the current is gradually increased under galvanostatic conditions.

The following interpretation of the oscillatory behavior observed during the galvanostatic oxidation of fuels like CH_3OH, CH_2O and $HCOOH$ is proposed by the author for large fuel concentrations when the influence of mass transport processes is negligible. The net reaction occurs mainly on sites free of chemisorbed carbonaceous species $C_sH_pO_q$ below the threshold current. The potential is sufficiently positive for the oxidation of $C_sH_pO_q$ when the threshold current is reached. The coverage with $C_sH_pO_q$ is depleted when the potential becomes more positive during the first part of one cycle because the rate of formation is smaller than that of oxidation of $C_sH_pO_q$. This situation results initially

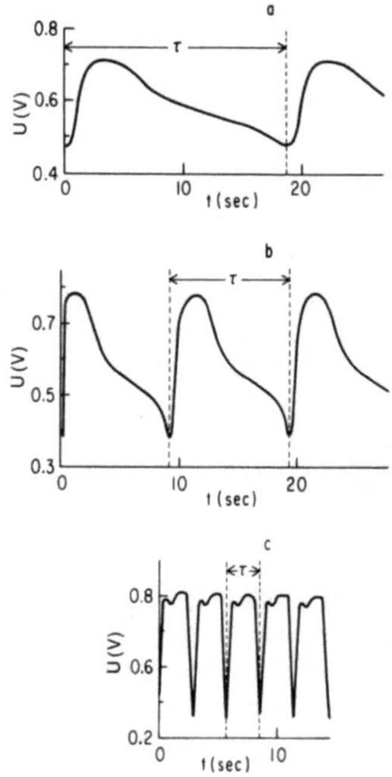

Fig. 71. Periodic potential transients on platinized platinum in 3.7 M H_2SO_4 + 1 M CH_2O. Curve a: $I = 20$ mA, curve b: $I = 30$ mA, curve c: $I = 400$ mA

from the small rate of formation at large coverage with $C_sH_pO_q$ and subsequently from a more rapid increase of the rate constant of the oxidation of $C_sH_pO_q$ than that of the formation with potential (compare the discussion of the respective rates for methanol in chapter IX). Deposition of the oxygen layer starts at potentials above 0.7 V. The oxidation of fuel to CO_2 at negligible coverage with $C_sH_pO_q$ is strongly inhibited by small amounts of O_{ad} while the tendency to the formation of $C_sH_pO_q$ is affected to a lesser extent. The change of the potential to less positive values in the second part of one cycle results at constant current because the deaccelerating effect of increasing coverage with

$C_sH_pO_q$ is compensated by the larger rate constant of the process of formation at less positive potentials. The oxygen layer is reduced when the appropriate potential for the given solution is passed during the second part of the cycle. The inhibition of the fuel oxidation to CO_2 is removed. The initial state of the surface for the beginning of another cycle is approached.

An analysis of oscillatory kinetics in the electrochemical oxidation of formate and ethylene was given recently by WOJTOWICZ, MARINCIC, and CONWAY [142]. In the formate case the oscillations are associated with electrochemical reactions. An autocatalytic step for the removal of the oxygen layer was considered consistent with the conditions for oscillations under certain circumstances. For ethylene oxidation it was shown that only imaginary solutions are possible under certain conditions of adsorption and mass transport control. Necessarily the derivations depend strongly upon the form of the kinetic expressions and upon the reaction mechanisms. The reader is referred for details to reference 142.

References

1. MÜLLER, E.: Z. Elektrochem. **29,** 264 (1923).
2. —, and S. TAKEGAMI: Z. Elektrochem. **34,** 704 (1928).
3. TANAKA, S.: Z. Elektrochem. **35,** 38 (1929).
4. FRUMKIN, A. N., and B. I. PODLOVCHENKO: Doklady Akad. Nauk S.S.S.R. **150,** 349 (1963).
5. AIKASYAN, E. A., and YU. V. PLESKOV: Zhur. Fiz. Khim. **31,** 205 (1957).
6. BAGOTZKY, V. S., and YU. V. VASILIEV: Electrochim. Acta **9,** 869 (1964).
7. VIELSTICH, W.: Chem. Ing. Tech. **35,** 362 (1963).
8. — Z. Instrumentenk. **71,** 29 (1963).
9. KUTSCHKER, A., and W. VIELSTICH: Electrochim. Acta **8,** 985 (1963).
10. LIANG, C., and T. C. FRANKLIN: Electrochim. Acta **9,** 517 (1964).
11. BUCK, R. P., and L. R. GRIFFITH: J. Electrochem. Soc. **109,** 1005 (1962).
12. MUNSON, R. A.: J. Electrochem. Soc. **111,** 372 (1964).
13. SCHWABE, K.: Z. Elektrochem. **61,** 744 (1957).
14. GERISCHER, H., and J. HELD: Z. Elektrochem. **64,** 1111 (1960).
15. SHLYGIN, A. I., and G. A. BOGDANOVSKY: 4th Conf. Electrochem., 1956, Acad. Sci. Moscow (published 1959), p. 282.
16. MARTINYUK, G. A., and A. I. SHLYGIN: Zhur. Fiz. Khim. **32,** 164 (1958).
17. BREITER, M. W., and S. GILMAN: J. Electrochem. Soc. **109,** 622 (1962).
18. GILMAN, S., and M. W. BREITER: J. Electrochem. Soc. **109,** 1099 (1962).
19. — J. Phys. Chem. **68,** 70 (1964).
20. WROBLOWA, H., B. J. PIERSMA, and J. O'M. BOCKRIS: J. Electroanalyt. Chem. **6,** 401 (1963).
21. JOHNSON, J. W., H. WROBLOWA, and J. O'M. BOCKRIS: J. Electrochem. Soc. **111,** 863 (1964).
22. GILMAN, S.: J. Phys. Chem. **66,** 2657 (1962).
23. — J. Phys. Chem. **67,** 78 (1963).

24. — J. Phys. Chem. **67**, 1898 (1963).
25. WARNER, T. B., and S. SCHULDINER: J. Electrochem. Soc. **111**, 992 (1964).
26. BRUMMER, S. B., and J. I. FORD: J. Phys. Chem. **69**, 1355 (1965).
27. STONEHART, P.: Proc. 5th International Symposium on Power Sources 1966, p. 509. Oxford: Pergamon Press 1967.
28. EISCHENS, R. P., and W. PLISKIN: Advan. Catalysis **1**, 18 (1958).
29. FASMAN, A. B., G. L. PODYUKOVA, and D. V. SOKOLSKY: Doklady Akad. Nauk S.S.S.R. **150**, 856 (1963).
30. PADYUKOVA, G. L., A. B. FASMAN, and D. V. SOKOLSKY: Electrokhim. **2**, 885 (1966).
31. BREITER, M. W.: J. Phys. Chem. **72**, 1305 (1968).
32. BINDER, H., A. KÖHLING, and G. SANDSTEDE: Advd. Energy Conversion **6**, 135 (1966).
33. BREITER, M. W.: Electrochim. Acta **12**, 1213 (1967).
34. TEMKIN, M.: Zhur. Fiz. Khim. **15**, 296 (1941).
35. MCKEE, D. W.: General Electric R and D Center, private communication, see also D. W. McKee, J. Catalysis **8**, 240 (1967).
36. DORLING, T. A., and R. L. MOSS: J. Catalysis **7**, 378 (1967).
37. BRENNAN, D., and F. H. HAYES: Phil. Trans. Roy. Soc. London **258**, 347 (1965).
38. PETRY, O. A., B. I. PODLOVCHENKO, A. N. FRUMKIN, and HIRA LAL: J. Electroanalyt. Chem. **10**, 253 (1965).
39. PODLOVCHENKO, B. I., O. A. PETRY, A. N. FRUMKIN, and HIRA LAL: J. Electroanalyt. Chem. **11**, 12 (1966).
40. BREITER, M. W.: J. Electroanalyt. Chem. **14**, 407 (1967).
41. STONEHART, P.: Electrochim. Acta **12**, 1185 (1967).
42. GILMAN, S.: J. Phys. Chem. **71**, 4330 (1967).
43. BRUMMER, K. D. N.: J. Catalysis **9**, 207 (1967).
44. NIEDRACH, L. W., D. W. MCKEE, J. PAYNTER, and I. F. DANZIG: Electrochem. Tech. **5**, 318 (1967).
45. MCKEE, D. W., L. W. NIEDRACH, J. PAYNTER, and I. F. DANZIG: Electrochem. Tech. **5**, 419 (1967).
46. MÜLLER, E.: Z. Elektrochem. **28**, 101 (1922).
47. —, and S. TANAKA: Z. Elektrochem. **34**, 256 (1928).
48. BRUMMER, S. B., and A. C. MAKRIDES: J. Phys. Chem. **68**, 1448 (1964).
49. PAVELA, T. O.: Ann. Acad. Sci. Fennicae, Series, A, II, Chemica **59** (1954).
50. BREITER, M. W.: Electrochim. Acta **8**, 447 (1963).
51. — Electrochim. Acta **8**, 457 (1963).
52. GINER, J.: Electrochim. Acta **9**, 63 (1964).
53. BRUMMER, S. B.: J. Phys. Chem. **69**, 562 (1965).
54. BREITER, M. W.: Electrochim. Acta **10**, 503 (1965).
55. BRUMMER, S. B.: J. Phys. Chem. **69**, 1363 (1965).
56. MINAKSHISUNDARAM, N., YU. B. VASILIEV, and V. S. BAGOTZKY: Elektrokhim. **3**, 193 (1967).
57. — — — Elektrokhim. **3**, 283 (1967).
58. BREITER, M. W.: J. Electroanalyt. Chem. **15**, 221 (1967).
59. FLEISCHMANN, C. W., G. K. JOHNSON, and A. T. KUHN: J. Electrochem. Soc. **111**, 602 (1964).
60. GOTTLIEB, M. H.: J. Electrochem. Soc. **111**, 465 (1964).
61. BINDER, H., A. KÖHLING, and G. SANDSTEDE: Adv. Energy Conversion **7**, 121 (1967).
62. BREITER, M. W.: J. Electrochem. Soc. **111**, 1298 (1964).
63. — Abhandl. Sächsische Akademie der Wissenschaften, **49**, 222 (1968).
64. ECKERT, J.: Electrochim. Acta **12**, 307 (1967).
65. ANDREW, M. R.: Electrochim. Acta **11**, 1425 (1966).
66. BINDER, H., A. KÖHLING, and G. SANDSTEDE: Nature **214**, 268 (1967).

67. BINDER, H., A. KÖHLING, K. RICHTER, and G. SANDSTEDE: J. Electrochem. Soc. **112,** 355 (1965).
68. — —, and G. SANDSTEDE: Advances in Chemistry, in print.
69. POLIAK, A. G., YU. VASILIEV, V. S. BAGOTZKY, and R. M. SMIRNOVA: Elektrokhim. **3,** 1076 (1967).
70. BODE, H.: Chem. Ing. Techn. **35,** 367 (1963).
71. KRUPP, H., H. RABENHORST, G. SANDSTEDE, G. WALTER, and R. MCJONES: J. Electrochem. Soc. **109,** 553 (1962).
72. JULIARD, A. L., and H. SHALIT: J. Electrochem. Soc. **110,** 1002 (1963).
73. OXLEY, J. E., G. K. JOHNSON, and B. T. BUZALSKI: Electrochim. Acta **9,** 897 (1964).
74. BREITER, M. W.: Electrochim. Acta **9,** 827 (1964).
75. TAKAMURA, T., and K. MINAMIYAMA: J. Electrochem. Soc. **112,** 333 (1965).
76. BAGOTZKY, V. S., and YU. B. VASILIEV: Electrochim. Acta **11,** 1439 (1966).
77. BESKOROVAINAYA, S. S., YU. B. VASILIEV, and V. S. BAGOTZKY: Elektrokhim. **2,** 167 (1966).
78. BINDER, H., A. KÖHLING, and G. SANDSTEDE: Advd. Energy Conversion **6,** 135 (1966).
79. BIEGLER, T., and D. F. A. KOCH: J. Electrochem. Soc. **114,** 904 (1967).
80. BAGOTZKY, V. S., and YU. B. VASILIEV: Electrochim. Acta **12,** 1323 (1967).
81. STENIN, V. F., and B. I. PODLOVCHENKO: Elektrokhim. **3,** 481 (1967).
82. BREITER, M. W.: Discussions Faraday Soc. **45**, 79 (1968).
83. ICENHOWER, D. E., and A. P. BOND: Ext. Abstr., Battery Div., Electrochem. Soc. Abstract **19** (1961).
84. PODLOVCHENKO, B. I., O. A. PETRY, and A. N. FRUMKIN: Doklady Akad. Nauk S.S.S.R. **153,** 379 (1963).
85. DRAZIC, D. M., and V. DRAZIC: Electrochim. Acta **11,** 1235 (1966).
86. SOKOLSKY, D. V.: Hydrogenation in Solutions, Akad. Nauk Kazakh. S.S.S.R., Alma-Ata 1962.
87. PODLOVCHENKO, B. I., and S. A. JOFA: Zhur. Fiz. Khim. **38,** 211 (1964).
88. RIGHTMIRE, R. A., R. L. ROWLAND, D. L. BOOS, and D. L. BEALS: J. Electrochem. Soc. **111,** 242 (1964).
89. BREITER, M. W.: J. Electrochem. Soc. **108,** 42 (1962).
90. GREEN, M., J. WEBER, and V. DRAZIC: J. Electrochem. Soc. **111,** 721 (1964).
91. DAHMS, H., and J. O'M. BOCKRIS: J. Electrochem. Soc. **111,** 728 (1964).
92. NIEDRACH, L. W.: J. Electrochem. Soc. **111,** 1309 (1964).
93. CAIRNS, E. J., and D. I. MCDONALD: Electrochem. Tech. **2,** 65 (1964).
94. GILMAN, S.: Trans. Faraday Soc. **61**, 2546 (1965).
95. — Trans. Faraday Soc. **61**, 2561 (1965).
96. BRUMMER, S. B., J. I. FORD, and M. J. TURNER: J. Phys. Chem. **69,** 3424 (1965).
97. GILEADI, E., B. T. RUBIN, and J. O'M. BOCKRIS: J. Phys. Chem. **69,** 3335 (1965).
98. BURSHTEIN, R. H., V. S. TYURIN, and A. G. PSHENICHNIKOV: Doklady Akad. Nauk S.S.S.R. **160,** 629 (1965).
99. —, A. G. PSHENICHNIKOV, V. S. TYURIN, and L. L. KNOTS: Elektrokhim. **1,** 1268 (1965).
100. BOCKRIS, J. O'M., H. WROBLOWA, E. GILEADI, and B. J. PIERSMA: Trans. Faraday Soc. **61,** 2531 (1965).
101. NIEDRACH, L. W., S. GILMAN, and I. WEINSTOCK: J. Electrochem. Soc. **112,** 1161 (1965).
102. BINDER, H., A. KÖHLING, H. KRUPP, K. RICHTER, and G. SANDSTEDE: Fuel Cell Systems, Advances in Chemistry Series **47,** 269 (1965).
103. — —, and G. SANDSTEDE: Fuel Cell Systems, Advances in Chemistry Series **47,** 283 (1965).

104. — —, H. KRUPP, K. RICHTER, and G. SANDSTEDE: J. Electrochem. Soc. **112,** 355 (1965).
105. SANDSTEDE, G.: Chem. Ing. Techn. **37,** 632 (1965).
106. HEILAND, W., E. GILEADI, and J. O'M. BOCKRIS: J. Phys. Chem. **70,** 1207 (1966).
107. BRUMMER, S. B.: J. Electrochem. Soc. **113,** 1041 (1966).
108. GILMAN, S.: J. Electrochem. Soc. **113,** 1036 (1966).
109. — Trans. Faraday Soc. **62,** 466 (1966).
110. — Trans. Faraday Soc. **62,** 481 (1966).
111. CAIRNS, E. J.: J. Electrochem. Soc. **113,** 1200 (1966).
112. NIEDRACH, L. W.: J. Electrochem. Soc. **113,** 645 (1966).
113. TYURIN, V. S., A. G. PSHENICHNIKOV, and R. H. BURSHTEIN: Elektrokhim. **2,** 948 (1966).
114. BELOSLYUDOVA, T. M., and D. V. SOKOLSKY: Elektrokhim. **2,** 704 (1966).
115. MARVET, R. V., and O. A. PETRY: Elektrokhim. **3,** 153 (1967).
116. PETRY, O. A., R. V. MARVET, and ZH. N. MALYSHEVA: Elektrokhim. **3,** 1141 (1967).
117. BRUMMER, S. B., and M. J. TURNER: J. Phys. Chem. **71,** 2825 (1967).
118. — — J. Phys. Chem. **71,** 3494 (1967).
119. NIEDRACH, L. W., and M. TOCHNER: J. Electrochem. Soc. **114,** 17 (1967).
120. GILMAN, S.: J. Phys. Chem. **71,** 2424 (1967).
121. BRUMMER, S. B., and M. J. TURNER: J. Phys. Chem. **71,** 3902 (1967).
122. BOGRACHEV, A. M., A. G. PSHENICHNIKOV, and R. H. BURSHTEIN: Elektrokhim. **4,** 358 (1968).
123. TRAPNELL, B. M. W.: Trans. Faraday Soc. **52,** 1618 (1956).
124. BOND, G. C.: Catalysis by Metals. New York: Academic Press (1962).
125. JENKINS, G. I., and E. RIDEAL: J. Chem. Soc. 2490 (1955).
126. STEPHENS, S. J.: J. Phys. Chem. **62,** 714 (1958).
127. GRUBB, W. T., and C. J. Michalske: J. Electrochem. Soc. **111,** 1015 (1964).
128. BRUMMER, S. B.: Fuel Cells, Advances in Chemistry Series, in print.
129. PAVELA, T. O.: Suomen Kemistilehti **30B,** 240 (1957).
130. KARP, S., and L. MEITES: J. Am. Chem. Soc. **84,** 906 (1962).
131. SUSBIELLES, G., and O. BLOCH: Compt. Rend. **255,** 685 (1962).
132. BARD, A. J.: Analyt. Chem. **35,** 1602 (1963).
133. SZPAK, S., P. STONEHART, and T. KATAN: Electrochim. Acta **10,** 563 (1965).
134. NESTEROV, B. P., and N. V. KOROVIN: Elektrokhim. **2,** 1296 (1966).
135. CONWAY, B. E., M. MARINIC, D. GILROY, and E. RUDD: J. Electrochem. Soc. **113,** 1144 (1966).
136. DANIEL-BEK, V. S., and G. V. VITVITSKAYA: Elektrokhim. **3,** 8 (1967).
137. GLASATOVA, T. N., and V. S. DANIEL-BEK: Elektrokhim. **3,** 396 (1967).
138. VITVITSKAYA, G. V., and V. S. DANIEL-BEK: Elektrokhim. **3,** 973 (1967).
139. BONHOEFFER, K. F.: Z. Elektrochem. **52,** 24 (1948).
140. FRANK, U. F.: Z. Phys. Chem. N.F. **3,** 183 (1955).
141. HUNGER, H. F.: J. Electrochem. Soc. **115,** 492 (1968).
142. WOJTOWICZ, J., N. MARINCIC, and B. E. CONWAY: J. Chem. Phys. **48,** 4333 (1968).

XI. The Oxygen Electrode at Low Temperatures

The reduction of molecular oxygen that is supplied either directly from containers or in a diluted form as air constitutes the reaction at the cathode in fuel cells. The use of air is preferable for economic reasons. Platinum metals and alloys of platinum metals are electrocatalysts for acid and alkaline electrolytes. Silver, silver alloys, nickel, carbon, and intermetallic compounds represent less expensive electrocatalysts for the oxygen electrode in alkaline solutions. In contrast to the hydrogen electrode, the overvoltage of the oxygen electrode is large at temperatures below 100 °C when a reasonable current is drawn.

The mechanism of the oxygen electrode is discussed for the above metals in this chapter. Depending upon the potential and the platinum metal, the oxygen reduction may occur on surfaces free of oxygen layers or on surfaces covered partially with layers. The two cases are treated separately since the reduction mechanism is affected by oxygen layers. As discussed in chapter VI, adsorbed impurities decrease the ability of platinum metals to adsorb hydrogen. Similarly, the O_2 reduction depends strongly upon the surface state. For two reasons, the emphasis is on clean surfaces of platinum metals, obtained for instance by previous anodic and subsequent cathodic pretreatment:

a) The influence of the oxygen layer on the O_2 reduction cannot be recognized clearly if a part of the surface is blocked by adsorbed impurities.

b) Disregarding the influence of mass transport processes in the pores, the kinetics of an oxygen diffusion electrode with large real surface are comparable to those of the smooth electrode of the same platinum metal in the active state.

1. Distinction of Reduction Mechanisms

Three types of mechanism have been postulated:

a) Oxygen layers are formed [1—6] by the interaction of O_2 with the electrode. The oxygen molecules are split into atoms during this initial process.

b) Oxygen molecules are not dissociated [7—12] in the initial steps involving O_2. The splitting of the O—O bond occurs after the formation of peroxide radicals. Adsorbed hydrogen atoms do not participate in the initial steps.

c) The formation of peroxide or peroxide radicals involves [13, 14] adsorbed hydrogen atoms on platinum and palladium.

The so-called [2, 15] oxide path

(1) $$2M + O_2 = 2MO,$$

(2) $$2M + 2MO + 2H_2O = 4MOH,$$

(3) $$4MOH + 4e^- = 4M + 4OH^-,$$

(4) $$O_2 + 2H_2O + 4e^- = 4OH^-$$

is an example of a mechanism of type a in alkaline electrolytes. M represents a metal atom on the surface. Complete reduction to OH^- is assumed. The hydrogen peroxide path

(5) $$M + O_2 + H^+ + e^- = MHO_2,$$

(6) $$MHO_2 + H^+ + e^- = MH_2O_2,$$

(7) $$M + MH_2O_2 = 2MOH,$$

(8) $$2MOH + 2H^+ + 2e^- = 2M + 2H_2O,$$

(9) $$O_2 + 4H^+ + 4e^- = 2H_2O$$

represents an example of a mechanism of type b in acid media. The reactions 10 and 11

(10) $$MH + O_2 = MHO_2,$$

(11) $$M + H_2O + e^- = MH + OH^-$$

belong to a mechanism of type c.

Mechanisms of type c are unlikely at $U > 0.4$ V on platinum metals since the hydrogen coverage is very small there (compare chapter VI). KRASIL'SHIKOV's argument [13] that hydrogen adsorption may exist for $U > 0.4$ V if molecular oxygen is present in the electrolyte is not supported by the measurements [16] of voltammetric $I-U$ curves and by impedance measurements [12] on platinum.

A distinction between mechanisms of type a and b is feasible on the basis of studies [12, 16—38] of the influence of oxygen layers at platinum metals. Let us assume that either reaction 2 or reaction 3 are rate-determining and that the equilibrium of step 1 is established [2—6]. If step 2 is rate-determining, the rate of the net process should increase with oxygen coverage. If step 3 is rate-determining, the pseudo-equilibrium of reaction 2 causes the coverage with MOH to increase with that of MO. In general, an increase of the oxygen coverage should lead to an increase in the rate of the process for mechanisms of type a. The experimental results of the O_2 reduction on platinum [12, 16, 21, 23, 24, 26—33, 35, 38] palladium [25], rhodium [34—37] and ruthenium [36, 37] demonstrate the opposite behavior. The net process is inhibited by oxygen layers.

The inhibition is illustrated by the voltammetric $i-U$ curves in Fig. 72. The curves were measured [39] with 100 mV/sec at 25 °C on smooth electrodes of Pt, Ir, Rh, and Pd in 0.5 M H_2SO_4 stirred vigorously with molecular oxygen. The sweep extended from 0.1 V to 1.4 V for Pt,

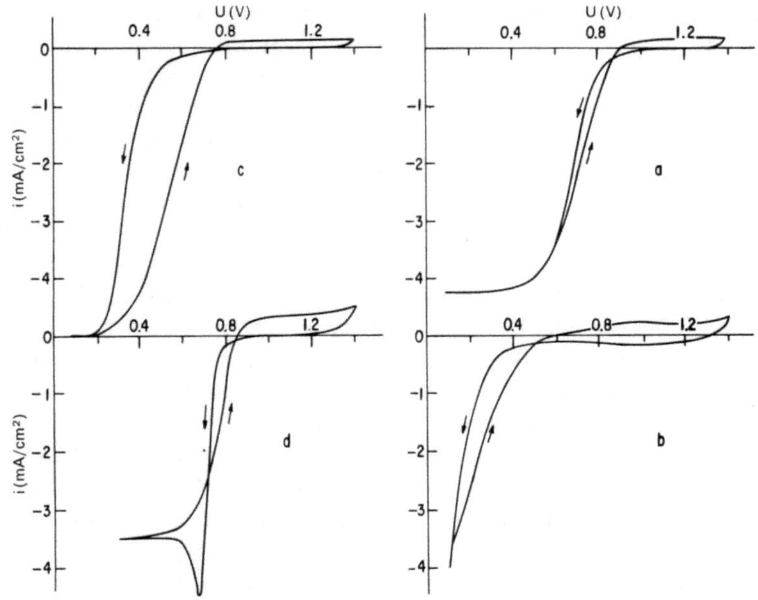

Fig. 72. Voltammetric $i-U$ curves taken with 100 mV/sec on platinum metals in 0.5 M H_2SO_4 stirred with O_2.
Curve a: platinum, curve b: iridium, curve c: rhodium, curve d: palladium

Ir, Rh and from 0.3 V to 1.4 V for Pd. The $i-U$ curves were recorded after ten minutes of continuous cycling. The sweep rate is sufficiently large for the formation of the oxygen layer to be clearly recognizable during the later portion of the anodic sweep. The comparison with the corresponding curves in Fig. 8 shows that the $i-U$ curves on Pt, Ir, and Rh have the same shape in the presence or absence of molecular oxygen at potentials of the oxygen region during the anodic sweep. The same statement holds for Pd. The formation of the oxygen layer is not influenced to a noticeable extent by the presence of O_2. This result was previously reported [12, 16, 32] for platinum.

The rate of reaction 9 is much smaller than that of the formation of the oxygen layer. After the sweep reversal at 1.4 V the $i-U$ curves on Pt, Ir, and Pd in Fig. 72 do not differ from the corresponding $i-U$ curves in the absence of O_2 until the reduction of the oxygen layer starts during the cathodic sweep. The current rises rapidly with decreasing potential and tends towards the limiting O_2 diffusion current. Average values of the current which displays fluctuations due to irregular stirring in the potential region of the limiting current are plotted in Fig. 72. The reduction of the relatively large oxygen layer on Pd is reflected by the overshoot of the cathodic current between 0.7 V and 0.6 V. On Ir the reduction of the oxygen layer is not finished before about 0.4 V. The rapid rise of current with decreasing potential starts so late on Ir that the limiting O_2 diffusion current is not yet reached at 0.1 V. Immediately after the sweep reversal at 0.1 V, the current decreases on Ir with increasing potential. The decrease starts at 0.2 V for Rh and at 0.4 V for Pt and Pd. There is a narrow potential range for Pt, Pd, and Rh in which a partial coverage with the oxygen layer exists during the cathodic sweep, but not during the anodic sweep. The comparison of the current at a given potential in this range demonstrates a larger current in the absence than the presence of the oxygen layer.

The overlapping of the potential regions of the O_2 reduction and of the reduction of the oxygen layers in Fig. 72 results because oxygen layers inhibit the O_2 reduction. The O_2 reduction occurs with a fast rate only on sites free of the oxygen layer. The pH dependence of the O_2 reduction cannot be used [24] as a diagnostic criterion for the elucidation of the rate-determining step since the pH dependence of the reduction of the oxygen layer is measured. Mechanisms of type a are not compatible with the experimental evidence of the O_2 reduction on platinum metals.

2. The Role of Hydrogen Peroxide in the Oxygen Reduction on Platinum Metals

The formation of hydrogen peroxide in the O_2 reduction on platinum, and also on the other electrocatalysts of interest, was not demonstrated with certainty [7, 40—42] until 1941. Two cathodic waves are observed during the O_2 reduction on mercury [9,43] or amalgams [44,45]. The wave at more positive potentials corresponds to net reaction 12 in acid electrolytes

(12) $$O_2 + 2H^+ + 2e^- = H_2O_2$$

or to net reaction 13 in alkaline solution.

(13) $$O_2 + H_2O + 2e^- = HO_2^- + 2OH^-.$$

The second wave results from reaction 14 or reaction 15:

(14) $$H_2O_2 + 2H^+ + 2e^- = H_2O,$$

(15) $$HO_2^- + H_2O + 2e^- = 3OH^-.$$

The reversible potential U_0 of reactions 12 and 13 is 0.713 ± 0.002 V at $\vartheta = 25\,°C$ and unit activity of the reactants. The corresponding U_0 value for reactions 14 and 15 is 1.744 V. The equilibrium potential of the reduction of H_2O_2 is more positive than that of the formation. From the thermodynamic point of view, hydrogen peroxide is an unstable intermediate of the O_2 reduction. Kinetic factors are responsible if hydrogen peroxide is found as a stable intermediate. Since the kinetic factors vary with the electrode material and the other experimental conditions, the extent of the formation of hydrogen peroxide is not the same.

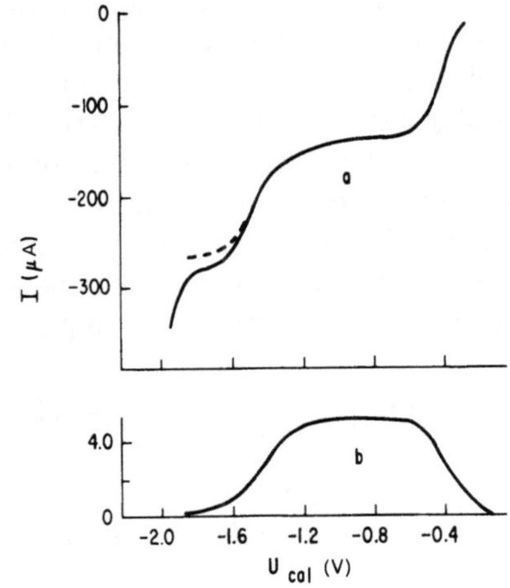

Fig. 73. Current of the disc (curve a) and of the ring (curve b) of a ring-disc assembly as a function of the potential of the disc during the reduction of molecular oxygen in 0.1 M KOH at 2150 rotations/min

The $I-U$ curve a in Fig. 73 is an example for the presence of two consecutive waves. It was obtained by FRUMKIN and NEKRASSOV [45] on the rotating disc of a ring-disc assembly [46] in 0.1 M KOH at 2150 rotations/min. Hydrogen peroxide formed during the oxygen reduction on an amalgamated gold disc at a given potential is carried away from the disc to a platinum ring electrode by forced convection. The ring and the disc are coplanar. The potential of the ring electrode is maintained at such a value that all the hydrogen peroxide transported to it is oxidized to O_2 while oxygen evolution from H_2O does not yet occur at a significant rate. Curve b represents the anodic current at the ring as a function of the potential of the disc. The anodic current becomes larger with decreasing potential in the rising portion of the first wave of curve a. It reaches the maximum value at the first plateau of curve a when the rate of O_2 reduction to H_2O_2 is controlled by the diffusion of O_2 to the disc. The ring current decreases with U when hydrogen peroxide is reduced according to reaction 15 on the disc. A second plateau is clearly recognizable in curve a if a correction for the residual current in the absence of O_2 is applied as indicated by the dotted line. The second plateau reflects the diffusion control of the reduction of O_2 to OH^-.

In contrast to mercury, the $I-U$ curves of the O_2 reduction on platinum metals consist only of one wave. The limiting diffusion current corresponds nearly to a four-electron process. Separate studies of the reduction of O_2 and H_2O_2 demonstrate in agreement with this result that the two species are reduced in the same potential range. Reaction 9 may occur as a parallel reaction to the sequence of reactions 12 and 14 in acid electrolytes. A similar statement holds for reaction 4 and the sequence of reactions 13 and 15 in alkaline media. Dissolved hydrogen peroxide is not formed as a stable intermediate in reactions 4 and 9. However, the intermediate production of H_2O_{2ad} or adsorbed peroxide radicals cannot be ruled out. For this reason the following scheme was proposed [12] previously for the O_2 reduction on active smooth platinum:

(16)
$$\begin{array}{l} O_2 \xrightarrow{I_1} H_2O_{2ad} \xrightarrow{I_4} H_2O_2 \xrightarrow{I_5} H_2O_{2,b} \\ \qquad\quad H_2O_{2ad} \xrightarrow{I_2} H_2O \text{ or } OH^- \xrightarrow{I_3} OH^-_b \end{array}$$

Eq. 17 to 22 hold under steady-state conditions with Eq. 22 applying to alkaline electrolytes only:

(17)
$$I = I_1 + I_2 ,$$

(18)
$$I_1 = \frac{2 D_{O_2} F A}{\delta_{O_2}} (c_{O_2} - {}_b c_{O_2}) ,$$

(19)
$$I_4 = I_5 = \frac{2 D_{H_2O_2} F A}{\delta_{H_2O_2}} (c_{H_2O_2} - {}_b c_{H_2O_2}) ,$$

(20)
$$I_1 = I_2 + I_4 ,$$

$$I_2 = -k_2 \cdot \theta_{H_2O_2}, \tag{21}$$

$$I_2 = I_3 = \frac{D_{OH^-} F A}{\delta_{OH^-}} (c_{OH^-} - {}_b c_{OH^-}). \tag{22}$$

The anodic partial current was neglected in Eq. 21 because the reduction of H_2O_2 to H_2O or OH^- is highly irreversible. If equilibrium between $\theta_{H_2O_2}$ and H_2O_2 and a small coverage $\theta_{H_2O_2}$ is assumed, it follows:

$$\theta_{H_2O_2} = K_{H_2O_2} \cdot c_{H_2O_2}. \tag{23}$$

With $K_2 = K_{H_2O_2} k_2$, Eq. 20 may be rewritten:

$$I_1 + K_2 c_{H_2O_2} = \frac{2 D_{H_2O_2} F A}{\delta_{H_2O_2}} c_{H_2O_2}. \tag{24}$$

As in previous work [45—47], it is assumed in Eq. 24 that ${}_b c_{H_2O_2} \ll c_{H_2O_2}$. Eq. 24 is equivalent to the boundary condition [45—47] at the surface of the disc in the case where a stable intermediate which is formed at rate I_1 undergoes a first order reaction besides diffusing away.

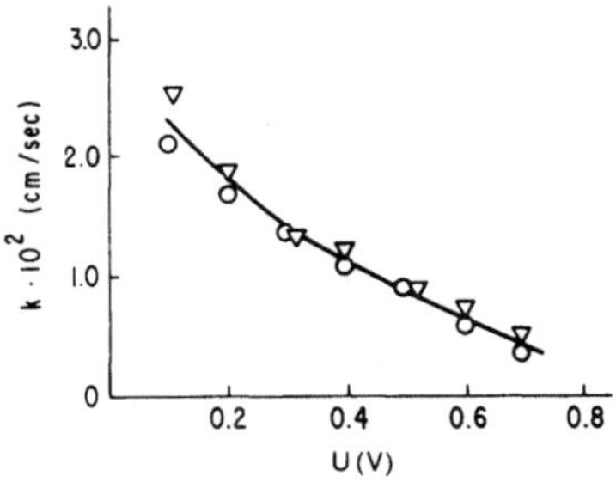

Fig. 74. Dependence of the rate constant for the reduction of H_2O_2 on rhodium in 0.1 M KOH upon potential.
○: computed according to Eq. 25, ▽: from measurements on the disc

The rate constant $k_2' = K_2/2FA$ for the reduction of hydrogen peroxide was computed [24, 25, 34, 36] according to Eq. 25

$$k_2' = \left(\frac{N |I_{13}|}{I_r} - 1 \right) \frac{D_{H_2O_2}}{\delta_{H_2O_2}} \tag{25}$$

from studies of the O_2 reduction with the ring-disc assembly. Here I_{13} designates the current due to reaction 13, I_r is the anodic current of H_2O_2 oxidation on the ring. N is the proportion of the material produced electrochemically on the disc, which reaches the ring. It is calculable

[47,48] from the dimensions of the assembly. Eq. 25 satisfies the boundary condition 24 at the surface of the disc. The rate constant of reaction 15 was also determined from direct measurements of the H_2O_2 reduction on the rotating disc in solutions containing known concentrations of H_2O_2. This rate constant is designated as $k_{H_2O_2}$. Satisfactory agreement was found between k_2' and $k_{H_2O_2}$ for platinum [24] and rhodium [34,36] in alkaline electrolytes. This is illustrated by Fig. 74 for rhodium. It was concluded [24,25,34,36,37] that the O_2 reduction in alkaline electrolytes occurs on Pt, Pd, Rh, and Ru in two successive stages with the intermediate formation of hydrogen peroxide. However, an agreement between k_2' and $k_{H_2O_2}$ is also to be expected for reaction scheme 16 if Eq. 24 holds. The latter assumption appears reasonable to the author. The experimental evidence, especially from the impedance measurements [12], points to a small total coverage with adsorbed intermediates of the O_2 reduction.

Bockris and coworkers [49—53] used Eq. 25 to develop a diagnostic criterion to determine whether the main path for the O_2 reduction is the sequence of reactions 12 and 14 or reaction 9 in acid electrolytes. In alkaline electrolytes, the sequence of reactions 13 and 15 or reaction 4 may be predominant. Their general reaction scheme

(26)

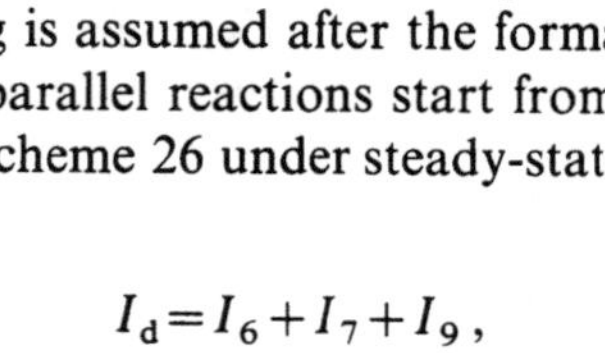

differs from scheme 16 with respect to the point where the two paths branch. The branching is assumed after the formation of H_2O_{2ad} in the scheme 16 while two parallel reactions start from O_2 in the scheme 26. Eq. 27 to 29 hold for scheme 26 under steady-state conditions at a given potential at the disc:

$$I_d = I_6 + I_7 + I_9 , \tag{27}$$

$$I_8 = \frac{I_r}{N} , \tag{28}$$

$$I_6 = I_8 + I_9 = I_9 \left(1 + k_2' \cdot \frac{\delta_{H_2O_2}}{D_{H_2O_2}} \right) . \tag{29}$$

Some transformations [49] lead to Eq. 30

$$\left| \frac{I_d}{I_r} \right| = \frac{x+1}{N} + \frac{1.61 (x+2) v^{1/6} k_2'}{N D_{H_2O_2}^{2/8} \omega^{1/2}} . \tag{30}$$

Here $x = I_7/I_6$, ω is the angular rate of rotation and ν the kinematic viscosity of the electrolyte.

The ratio $|I_d/I_r|$ obtained from measurements [53] on platinum in 0.1 M KOH is shown as a function of $\omega^{-1/2}$ at different potentials in Fig. 75. The electrode was free of oxygen layers in these experiments. The

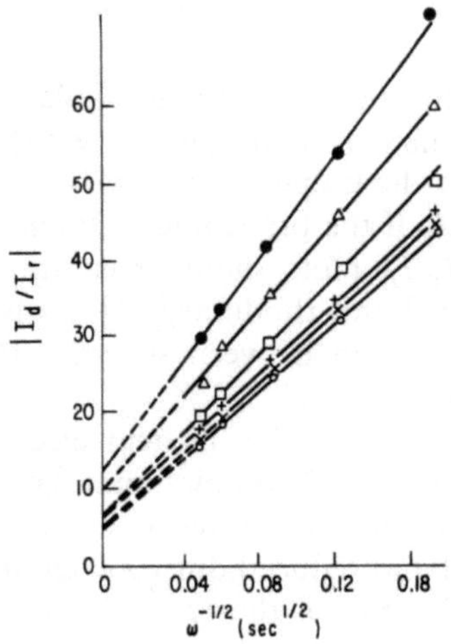

Fig. 75. Absolute value of the ratio between the current of the disc and the current of the ring as a function of $\omega^{-1/2}$ at different potentials on platinum in 0.1 M KOH. ●: 0.60 V, △: 0.55 V, □: 0.50 V, +: 0.45 V, ×: 0.40 V, ○: 0.35 V

intercept of the $|I_d/I_r|$ versus $\omega^{-1/2}$ plots with the $|I_d/I_r|$ axis allows the computation of x. The intercept was found to be greater than 1/N (≈ 3.3) on platinum [53] and rhodium [52] in 0.1 M KOH. Since $x > 0$, it was concluded that two paths of the O_2 reduction are present and that dissolved hydrogen peroxide is produced as an intermediate in one of them.

From the experimental point of view, the above conclusions depend in a critical way on maintaining the same surface state at different angular velocities, on the accuracy of the determination of $|I_d/I_r|$ and that of the extrapolation of the $|I_d/I_r|$ versus $\omega^{-1/2}$ plots. Contrary to the opinion of Bockris and coworkers [52, 53], the accuracy of the determination of the intercept is not considered sufficient by the author to separate the contribution of reaction 4 from that of the sequence of reactions 13 and 15 if the intercept is smaller than 2/N. Such values were reported for platinum [53] and rhodium [52] in a large potential range.

From the theoretical point of view, the possibility of a distinction between schemes 16 and 26 is of interest. If the validity of Eq. 24 is assumed for simplicity, the analysis leads to Eq. 31 with $y = I_2/I_1$ for the scheme 16.

$$\left| \frac{I_d}{I_r} \right| = \frac{y+1}{N} + \frac{1.61(y+1)k_2'}{N D_{H_2O_2}^{2/3} \omega^{1/2}}. \tag{31}$$

Schemes 16 and 26 are not distinguishable as long as the experimental intercept is smaller than 2/N (x, y<1). A maximum value of 1 is allowed for y in the frame of scheme 16. Such a restriction does not exist for scheme 26. If the branching is assumed from the peroxide radical O_2H_{ad}, the $|I_d/I_r|$ plots should pass through the origin. The results on platinum [53] and rhodium [52] in alkaline electrolytes do not allow a clear distinction between schemes 16 and 26 since x<1 in most cases.

The situation is more complex in acid electrolytes. Reproducible $I-U$ curves are obtainable [12] on platinum metals (see Fig. 72) under voltammetric conditions at sweep rates as low as 30 mV/sec when the oxygen layer is formed and reduced during each of the repetitive cycles. The decrease [5, 10, 52—54] of the O_2 reduction current at constant potential with time after an anodic pretreatment of the electrode does not yet affect shape of the $I-U$ curves. If the measurements are made more slowly, the shape of the $I-U$ curves depends [54] upon the way in which they were measured. The influence of time on the efficiency of the H_2O_2 production in 0.5 M H_2SO_4 after a standardized pretreatment [5] of smooth platinum is shown in Fig. 76. The efficiency was computed [5] by Bianchi and Mussini on the basis of an analytical determination of the bulk concentration of H_2O_2 after maintaining the electrode for a given time at constant potential. The highest efficiencies were found in the short time experiments in agreement with similar results of Winkelman [10]. The curves in Fig. 76 show a rapid increase of the H_2O_2 production with decreasing potential below 0.6 V. A maximum is reached at about 0.45 V. Then the efficiency becomes smaller again.

Winkelman [10] reported a smaller efficiency of H_2O_2 production at a given potential for platinized platinum than for smooth platinum. Similarly, less hydrogen peroxide was formed [30] on active than on inactive smooth platinum during the O_2 reduction at constant current. Hydrogen peroxide could not be detected [32] on smooth platinum in a cell with small electrolyte volume if a periodic voltage of triangular shape was applied between 0 and 1.4 V in 4 M H_2SO_4 at about 100 mV/sec. Apparently the electrode surface remained in the active state under these conditions.

BOCKRIS and coworkers determined the I_d/I_r values for platinum [51] and rhodium [53] from measurements with the ring-disc assembly in 0.05 M H_2SO_4. The $|I_d/I_r|$ versus $\omega^{-1/2}$ plots yielded [51] lines parallel to the abscissa in "insufficiently purified" solutions for platinum free of oxygen layers. Solutions that were not purified by a special electrochemical procedure [51] are designated as insufficiently purified by these authors. The x values increased from about 0.8 to 15 between 0.3 V and 0.75 V on platinum. It was concluded that reactions 9 and 12 occur simultaneously and that $k_2' \approx 0$ in these solutions. In "pure" solutions, the decrease of I with time at constant potential was found to be much smaller and I_r approached zero. Practically all the oxygen is reduced to water when the bulk concentration of residual impurities,

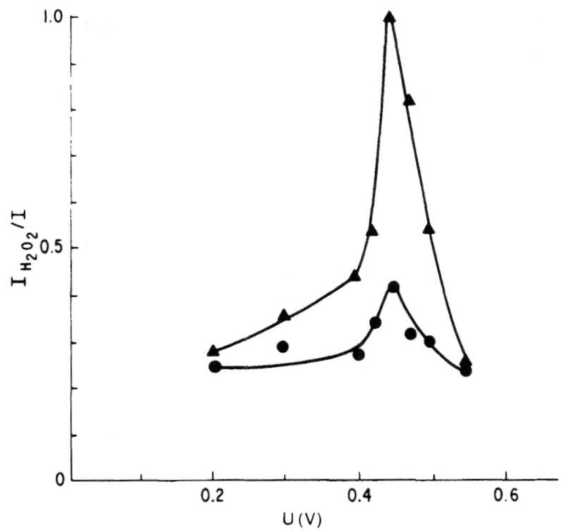

Fig. 76. Efficiency of platinum for the H_2O_2 production during the O_2 reduction in 0.5 M H_2SO_4 as a function of potential.
▲: after 30 min of electrolysis at a given potential, ●: after 360 min of electrolysis at a given potential

and, therefore, the adsorbed amount of impurities is small. The conclusion [51] of a very small H_2O_2 production on platinum in pure solutions is in agreement with the results on platinized platinum [10] and active smooth platinum [18, 27, 30, 32]. The appearance of H_2O_2 was attributed [51] to a change of the relative rates of the two parallel paths in the scheme 26 by adsorbed impurities. The result [51] that $k_2' = 0$ between 0.75 V and 0.3 V on platinum in the so-called insufficiently pure solutions is not in agreement with the evidence of other authors [5, 27, 31, 54] on the predominance of the reductions of O_2 to H_2O_2 in a limited potential range (compare Fig. 76). Further experimental work is required to establish with certainty that an increased production of H_2O_2 is solely due to the adsorption of impurities. The results on rhodium [52] were similar to those on platinum with the exception that $k_2' > 0$ in insufficiently pure acid solution. The possibility that oxygen layers are present at $U > 0.5$ V on electrodes that had been pretreated cathodically was not recognized.

NEKRASSOV and KRUSHCHEVA [36] compared k_2' and $k_{H_2O_2}$ obtained in the described way for rhodium or ruthenium in 0.5 M H_2SO_4. In general, k_2' was larger by at least an order of magnitude than $k_{H_2O_2}$. This result was attributed to the occurrence of the parallel reaction 9 which is not taken into account in the computation of k_2' according to Eq. 25. Therefore, NEKRASSOV and KRUSHCHEVA [36] used Eq. 25 with $k_{H_2O_2}$ instead of k_2' to compute I_{13} from measurements with the ring-disc assembly in 0.5 M H_2SO_4. Thus the plots in Fig. 77 were constructed. The curves in Fig. 77 demonstrate that the path 7 is

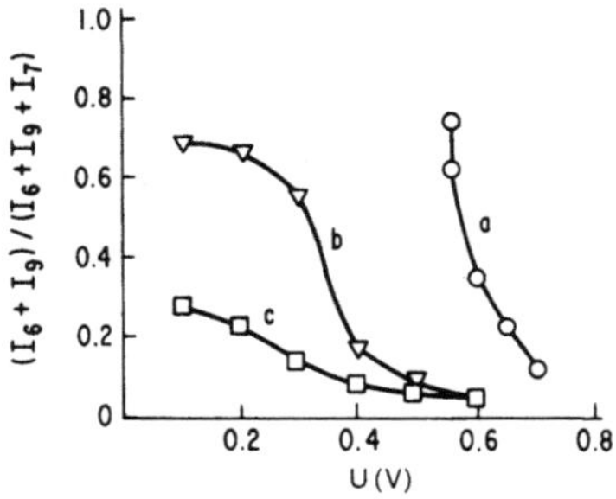

Fig. 77. Ratio of partial currents according to the scheme 26 as a function of potential for the O_2 reduction on rhodium and ruthenium in 0.5 M H_2SO_4.
Curve a: active rhodium, curve b: passive rhodium, curve c: active ruthenium

predominant for $U \geq 0.65$ V on active rhodium, for $U > 0.4$ V on inactive rhodium, and in the whole potential range for ruthenium (see the reaction scheme 26). The contribution of path 7 becomes rapidly smaller with decreasing potential below 0.70 V on active rhodium. Curve b for inactive rhodium shows the transition from the predominance of reaction 9 to the sequence of reactions 12 and 14 at much less positive potentials than curve a. The possible influence of the oxygen layer on the shape of curve a was not discussed. The said transition was considered responsible [36] for the appearance of dissolved H_2O_2 during the O_2 reduction on platinum metals below a certain potential.

The reduction of hydrogen peroxide (reaction 14 or 15) may start with either an electrochemical step [10, 55, 56] or a chemical step [23, 24, 57—59]. Molecular oxygen should be formed if the Haber-Grinberg mechanism [60] of the catalytic decomposition of H_2O_2 on platinum applies:

(32) $$Pt + H_2O_2 = PtO + H_2O\,,$$

(33) $$PtO + H_2O_2 = Pt + H_2O + O_2\,,$$

(34) $$2H_2O_2 = 2H_2O + O_2\,.$$

The formation of molecular oxygen during the H_2O_2 reduction on platinum in alkaline [23, 24] and acid media [59] was demonstrated by measurements with the ring-disc assembly. Both the limiting anodic and the limiting cathodic current were determined on the ring at a given potential of the disc. The absolute value of the limiting cathodic current on the ring was found larger than the anodic one. This is evidence

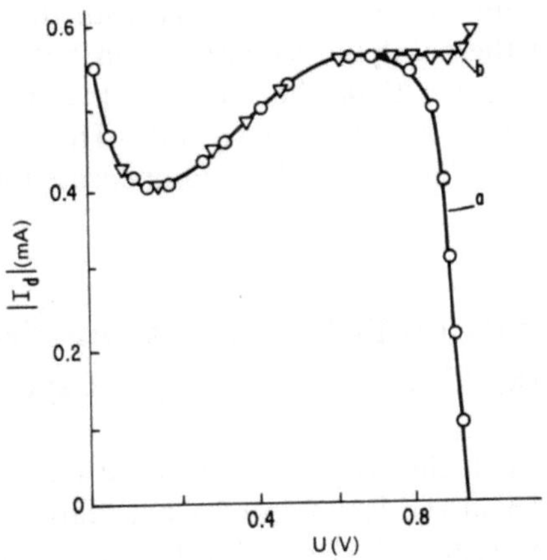

Fig. 78. Steady-state current-potential curve of the H_2O_2 reduction on the platinum disc (curve a) in 0.125 M KOH and curve corrected for the catalytic decomposition of H_2O_2 (curve b)

for the formation of O_2 since molecular oxygen that diffuses from the disc to the ring can only be reduced at potentials of the cathodic limiting current.

Curve a in Fig. 78 is the steady state $I-U$ curve for the reduction of H_2O_2 on the platinum disc in 0.125 M KOH. The current rises with decreasing potential below 0.9 V and reaches the limiting H_2O_2 diffusion current between 0.8 V and 0.6 V. There is a minimum in the $I-U$ curve between 0.6 V and 0 V. The amount of O_2 reaching the disc was computed from the difference between the absolute value of the cathodic limiting current and the anodic limiting current on the ring. At $I_d=0$ the amount of O_2 was found to be equivalent to the value expected for the complete decomposition of every arriving H_2O_2 molecule according to Eq. 34. The amount of O_2 becomes smaller with decreasing potential in the rising portion of curve a. It reaches zero at the potential U_M of the maximum of curve a and remains zero at $U<U_M$. If the disc current is corrected for the amount of O_2, curve b results. It was concluded [23, 24, 59] that the reduction of H_2O_2 involves the initial decomposition to O_2 and the subsequent reduction of O_2 to OH^- or H_2O at $U>U_M$ on platinum. A similar conclusion was reached in ref. 30.

It was demonstrated by BIANCHI and coworkers [61] that the catalytic decomposition of H_2O_2 is increased by oxygen layers on platinum. This observation gives an explanation [54] for the decrease of the disc current with decreasing potential after the maximum in curve a. The surface is partially covered with an oxygen layer in the potential region of the broad maximum. When the coverage with the oxygen layer decreases, the rate of the catalytic decomposition becomes smaller. The latter influence is greater than the effect of the increasing cathodic overvoltage. A negative characteristic results. The current rises again when the overvoltage is large enough to compensate the other effect.

3. Mechanism of the O_2 Reduction on Active Platinum Metals in the Absence of the Oxygen Layer

Electrodes on which the oxygen layers were removed by adequate cathodic polarization in the last step of a pretreatment procedure are frequently called "prereduced" electrodes. If a steady state $I-U$ curve is measured in the direction of increasing potential, the O_2 reduction occurs on a surface free of oxygen layers up to the potential U_F at which the respective oxygen layer begins to form on the given platinum metal. These potentials are known from voltammetric and similar studies (see chapter VIII). Once the studies of the O_2 reduction are made at $U>U_F$ the inhibiting effect of the oxygen layer has to be taken into account in the elucidation of the rate-determining step from current-potential curves. Insufficient attention has been paid to the latter point in recent work, especially on rhodium and iridium [6]. The Tafel slope, which is already a questionable diagnostic criterion for the O_2 reduction in the absence of oxygen layers (see section 3 in chapter IV), is often used as a criterion for prereduced electrodes in potential ranges where oxygen layers are present.

WINKELMAN [10] found that steady-state $I-U$ curves of the O_2 reduction could be approximated on prereduced platinized platinum in the potential region where the oxygen layer is absent by the relation:

$$(35)\qquad i = -k_2\left(1 - \frac{i}{i_{d,O_2}}\right)\,{}_{b}c_{O_2}\,{}_{b}c_{H^+}^{\frac{2}{3}}\exp\left(-\frac{(1-\alpha)UF}{RT}\right).$$

Since the fit between experimental and theoretical curves was made in the transition region to the limiting O_2 diffusion current, it is likely that the $i-U$ dependence is not influenced by the inhibiting effect of the oxygen layer. However, the pH dependence may not be correct since the equilibrium of reactions 36 or 37

$$(36)\qquad H_2O = OH_{ad} + H^+ + e^-,$$

$$(37)\qquad OH^- = OH_{ad} + e^-$$

appears to be established at rather small coverages with OH_{ad} (see chapter VIII).

On smooth platinum, Eq. 35 does not contain [10] the term ${}_{b}c_{H^+}^{2/3}$. The results on platinized and smooth platinum with $\alpha < 1$ suggest that one of the first discharge steps in the O_2 reduction is rate-determining. On smooth platinum, it may be the reaction

$$(38)\qquad O_{2ad} + e^- = O_{2ad}^- .$$

The experimental results do not allow a more quantitative analysis at present.

4. The O_2 Reduction on Platinum Metals in the Presence of Oxygen Layers

Electrodes on which an oxygen layer is present because of a preceding anodic pretreatment are frequently designated [3] as "preoxidized" electrodes. Since the reduction of the oxygen layers on platinum metals is a process with kinetic hindrance, the layers are stable during the time of the measurements in a certain potential range which is different for each of the metals. The O_2 reduction can be studied on surfaces having a constant coverage with the layer in the said range. The electrodes with oxygen layers are free of adsorbed impurities in most cases (compare chapter VIII).

The $|I_d/I_r|$ versus $\omega^{-1/2}$ plots yielded [52] the same lines which are parallel to the abscissa for preoxidized rhodium both in "pure" and "insufficiently pure" solutions of sulfuric acid at $U<0.60$ V in agreement with the above comment on the adsorption of impurities. The oxygen layer was formed at 1.0 V. The intercept of the lines with the $|I_r/I_d|$ axis was larger than 8. The reaction paths 6 and 7 of scheme 26 are effective simultaneously while path 9 does not contribute because of $k_2' \approx 0$ in acid electrolytes. The Tafel slope was about 100 mV for the preoxidized electrode with a large oxygen layer and 50 mV for the prereduced rhodium electrode with a smaller oxygen layer.

The oxygen layer on ruthenium is reduced at less positive potentials than the one on rhodium (see chapter VIII). Thus the curves a and c in Fig. 77 indicate that the preference of path 7 over the sequence of paths 6 and 9 in the scheme 26 is due to the oxygen layer. Such a preference would be understandable [36] if the path 6 involves the intermediate formation of oxygen layers from O_2. However, the inhibition of the O_2 reduction by the oxygen layer and the intermediate formation of the layer from O_2 are not reconcilable at present. The elucidation of the rate-determining step of the O_2 reduction in the presence of oxygen layers requires the quantitative description of the inhibition as a function of coverage in the kinetic equations.

5. The O_2 Reduction on Silver, Nickel, and Silver Alloys

Steady-state $i-U$ curves of the O_2 reduction in 6 M KOH at 25 °C are compared for Ag, C, and Ni in Fig. 79. The curves were taken [62] by VIELSTICH on diffusion electrodes prepared according to the techniques of JUSTI and coworkers [63]. The potential U_{cal} was measured versus a saturated calomel electrode. The silver electrode shows the best performance. Carbon is a better electrocatalyst than nickel in 6 M KOH. The limiting current due to transport processes was not reached in the measurements.

The O_2 reduction on Ag [28, 64, 65], Ni [66], and Ni-Ag alloys [66] was studied with the ring-disc assembly in recent years. The $i-U$ curves in Fig. 80 were measured [66] at 1000 rotations/min on the disc in 0.1 M KOH. The disc was made of silver for curves a and a′ and of nickel for curves b and b′. The disc surface was polished mechanically with quartz powder and subsequently polarized cathodically for curves a and b (prereduced electrodes). Anodic polarization with 10 mA/cm^2 for 2 min served as the pretreatment for curves a′ and b′ (preoxidized electrodes). The curves a, a′, and b are similar in shape to the $i-U$ curves on platinum metals in Fig. 72. The net process is limited by the O_2 diffusion at $U<0.4$ V. The decrease of $|i|$ with increasing potential occurs on prereduced nickel in the potential range where the surface

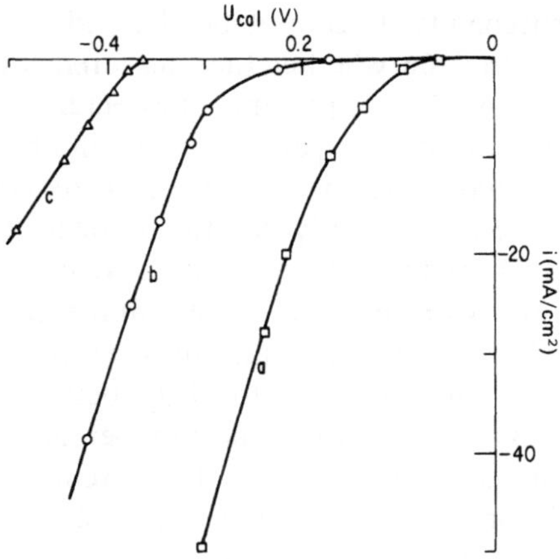

Fig. 79. Steady-state current-potential curves of the O_2 reduction on silver (curve a), carbon (curve b) and nickel (curve c) in 6 M KOH at 25 °C

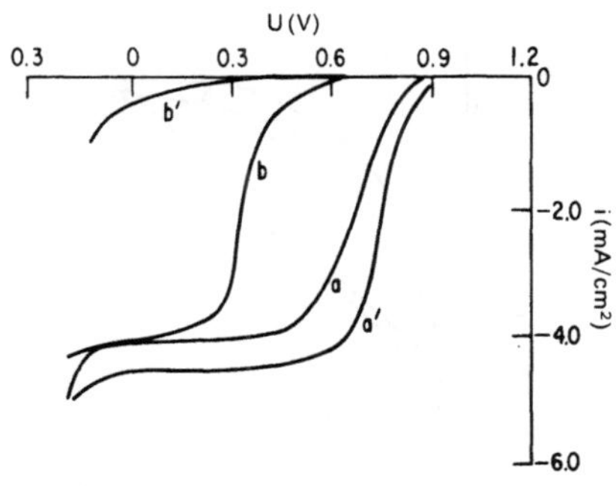

Fig. 80. Steady-state current-potential curves of the O_2 reduction on prereduced (curves a and b) and preoxidized (curves a', b') disc electrodes of silver (curves a, a') and nickel (curves b, b') in 0.1 M KOH

coverage with a thin film of Ni $(OH)_2$ increases (see chapter VIII). This film inhibits the O_2 reduction. The higher nickel oxide, probably NiOOH, that had been formed during the anodic pretreatment is not converted completely to $Ni(OH)_2$ even at $U < 0.3$ V during the measurement of curve b'. The small reactivity of the preoxidized electrode is attributed to this circumstance. In contrast to nickel, the silver surface displays a slightly larger reactivity in the preoxidized than in the prereduced state. However, the reduction of Ag_2O that is mainly formed during the anodic pretreatment should be completed [62] at $U < 0.8$ V if curve a' represents truly steady-state measurements. It cannot be decided on the basis of the information in reference 66 whether the measurement of curve a′ was made so rapidly that Ag_2O was present. The decrease of i with U above 0.6 V on silver was interpreted [64,65] as an inhibition effect of the chemisorbed oxygen (compare Fig. 33) on the O_2 reduction.

Delahay [42] reported already that the number of electrons for the net process of the O_2 reduction on Ag and Ni increases from values between 2 and 3 to 4 with increasing overvoltage in neutral solutions. Studies [64,65] with the ring-disc assembly revealed for prereduced silver that the reduction leads only to HO_2^- between 0.8 V and 0.9 V (two electron process). Below 0.8 V, the number of electrons becomes larger with overvoltage. The portion of the current that is used for the hydrogen peroxide production is smaller than 20% below 0.5 V. If a correction is applied for the formation of HO_2^-, the I_d versus $\omega^{1/2}$ plot for the disc coincides [64, 65] with the theoretical line of a four-electron process. The simultaneous occurrence of a four- and a two-electron process was also suggested [44] on the basis of measurements of the O_2 reduction on prereduced silver electrodes in slightly acid electrolytes.

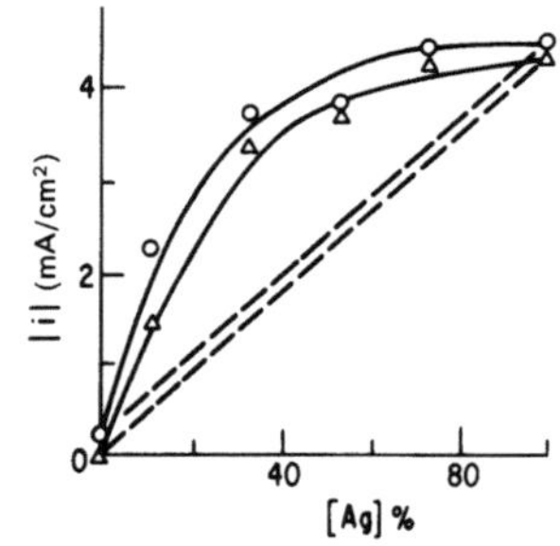

Fig. 81. Rate of the O_2 reduction in 0.1 M KOH on disc electrodes of silver-nickel as a function of the silver content of the bulk at 0.5 V (△) and 0.1 V (○)

Measurements [28] of $i-U$ curves in 1 M KOH with additions of H_2O_2 in the vicinity of the open circuit potential showed a smooth transition of the cathodic branch into the anodic branch on silver. A reasonable exchange current density can be computed from the slope $(\Delta U/\Delta i)_{i=0}$ of the $i-U$ curve. Such behavior is characteristic for the vicinity of the reversible potential of an electrode reaction. At sufficient distance from the reversible potential, Tafel lines with slopes of 25 mV to 30 mV were obtained [28]. The value of $(\partial U/\partial pH)_{i,p_{O_2}}$ was about 30 mV. These results suggest that the open circuit potential is identical with the reversible potential of reaction 13 on silver in alkaline electrolytes at ${}_bc_{H_2O_2} > 10^{-6}$ M. Silver behaves like mercury [9] in this respect. When the hydrogen peroxide concentration becomes smaller than 10^{-6} M, the open circuit potential has to be considered [67] a mixed potential.

The O_2 reduction was investigated [66] on silver-nickel alloys prepared by pressing appropriate amounts of silver and nickel powder under heat. The porosity of the pressed electrodes was between 3 to 5%. The current measured [66] in 0.1 M KOH at 0.5 V or 0.1 V with 1000 rotations/min is plotted in Fig. 81 as a function of the silver content for preoxidized discs made of different alloys. The increase in the reactivity of the alloys with silver content is greater below 40% than would be expected on the assumption of an additive superposition of the currents of the pure components according to the bulk composition. The currents for additive superposition are shown by dotted lines in Fig. 81. Above 50% the reactivity of the alloys is comparable to that of silver. The cause of the shape of the curves in Fig. 81 is not known with certainty. It is conceivable that the surface layers are richer in silver than the bulk.

Current-potential curves of the O_2 reduction were measured [68] in concentrated KOH on alloy electrodes prepared from silver and small amounts of elements that form oxides of low electron affinity. The reactivity of a silver −1.7 w/o (weight percent) magnesium alloy was better than that of pure silver. In contrast, no improvement was found for foils of silver −1 w/o thorium, silver −1 w/o radium, and silver −1 w/o barium.

Silver-palladium alloys showed [69] a performance comparable to that of Pd at relatively small additions (10%) of palladium in 5.3 M KOH at room temperature.

Although the mechanism of the O_2 reduction has been studied extensively for the said electrocatalysts in recent years, the experimental evidence is not sufficient to establish the rate-determining steps with certainty.

6. The O_2 Reduction on Carbon

Carbon can be used [7, 11, 62, 70—73] as an electrocatalyst for the O_2 reduction in alkaline electrolytes (compare also section 5 in chapter VIII). The performance which is not so good as that of silver (see Fig. 79) appears adequate for certain purposes, for instance, in small zinc-air cells. Activation procedures [72, 73] which are not of an electrochemical nature improve the performance of carbon oxygen electrodes. The performance rapidly becomes poor with decreasing pH below pH < 14. In acid solution, the impregnation of carbon with platinum metals or other electrocatalysts is required. The data [73] in Table 8 show that the number of electrons involved in the net process depends [7] strongly upon the type of carbon from which the electrode is prepared.

Table 8. *Efficiency of Different Carbon Electrodes for the* H_2O_2 *Production at Current Densities between* 10 to 20 mA/cm² *in* 1 M KOH at 5 °C.

	Initial efficiency in %	Efficiency in % after 20 to 50 hours
Ringsdorff-Kohle	70	40
Aktiv-Kohle	35	30
Elorit DA	10	7
Elorit TD	3	2

Carbon electrodes of the type Elorit TD are desirable for practical oxygen electrodes.

In his classical studies, Berl [7] demonstrated the establishment of the equilibrium of reaction 13 on activated carbon for which the efficiency of H_2O_2 production amounted to 100%. He suggested that the O—O bond is not disrupted in the reduction of O_2 to HO_2^-. The latter suggestion was tested by Yeager and coworkers [11]. Isotopic techniques involving O^{18} were employed in studies of the anodic and cathodic behavior of the oxygen-peroxide couple on active carbon electrodes exhibiting a 100% efficiency for the H_2O_2 production from O_2 in alkaline solution. The isotopic analyses were made with an analytic mass spectrometer. It was established by tracer and equilibrium experiments that all of the oxygen in the peroxide originates from gaseous oxygen and that the O—O bond is not actually broken in the formation of H_2O_2. The tracer experiments indicated that the molecular oxygen which is produced during the anodic oxidation of HO_2^- is oxygen from the peroxide.

The results in Table 6 were obtained on carbon electrodes for which the efficiency of H_2O_2 production is smaller than 100%. Experiments which allow to decide whether the main path of the O_2 reduction is the sequence of reactions 13 and 15 or reaction 4 have not yet been carried out on carbon electrodes. As demonstrated by Frumkin and coworkers [74—77], the adsorption of oxygen on carbon occurs in two different forms. The formation of the first type requires the dissociation of the O_2 molecules into O atoms. The second type is formed without dissociation. It was suggested [36] that the ratio I_{13}/I_4 of the rates of reactions 13 and 4 depends upon the ratio between the rate constant of the electrochemical production of H_2O_2 and the rate constant for the formation of the first type of adsorbed oxygen. Thus the mechanism of the O_2 reduction on carbon will vary with the type of carbon and the pretreatment of the surface.

7. The O_2 Reduction on Intermetallic Compounds

The search for effective electrocatalysts led to investigations of the corrosive resistance and of the reactivity for the O_2 reduction of compounds of metals with other elements like boron, carbon, nitrogen, oxygen, etc. The cathode [78] of the Bacon cell represents the classical example. The porous cathodes were made by pressing and sintering mixtures of carbonyl nickel and ammonium bicarbonate. Subsequently, they were coated with a layer of lithiated nickel oxide.

Many of the intermediate compounds are semiconductors. The conductivity may be improved by adding appropriate ions during their preparation. Solid solutions result in certain cases. DAVTYAN and coworkers [79] interpreted the mechanism of the oxygen electrode on semiconductors of the type $L_xT_{1-x}O$ by postulating the intermediate formation and dissociation of complexes of the type $L-O^*-T^{3+}-O$. There L^+ designates an alkali cation and T a transition metal like Ni, Cu, Co, Mn, and $x\cdot 100$ gives the atomic percent of L. The chemisorption of oxygen leads to the formation of the complex:

$$L^+ + T^{2+} - O + \tfrac{1}{2}O_2 = L - O^* - T^{3+} - O. \tag{39}$$

Reaction of the complex with water leads to OH^-

$$L - O^* - T^{3+} - O + H_2O = L^+ + [T^{3+} - O]^+ + 2\,OH^-. \tag{40}$$

For the O_2 reduction on NiO, the reactions 38 and 41

$$O_{ad} + e^- = O_{ad}^-. \tag{41}$$

were postulated [80] as rate-determining steps. The electrochemical behavior of tungsten, tantalum, and titanium carbides and of titanium nitride was studied [81] in acid and alkaline electrolytes. Tungsten carbide showed the best corrosion resistance among the compounds in acid electrolytes at potentials below oxygen evolution. Tantalum carbide proved acceptable for alkaline solutions. Tungsten carbide was the best electrocatalyst among the above materials for O_2 reduction in acid electrolytes. However, the large overvoltage (>0.7 V for tungsten carbide at $-10\,\mu A/cm^2$) rules these compounds out for practical purposes.

The O_2 reduction was reported [82] to occur in a potential region where the passive films are stable on Ni, NiAs, NiSi, and NiSb. The order of reactivty for the O_2 reduction is Ni > NiAs > NiSi in acid solution, and NiS > Ni > NiSi ≈ NiSb > NiAs in alkaline solution. However, the performance of these compounds is not satisfactory for their use as electrocatalysts in a practical oxygen electrode.

Sodium tungsten bronzes of the general formula Na_xWO_3 with $0.2 \leq x \leq 0.93$ were stable [83] in sulfuric acid solutions even at potentials for the oxygen evolution. The compound $Na_{0.6}WO_3$ was found to be better electrocatalyst for the O_2 reduction than prereduced platinum at potentials above 0.9 V ($|i| < 10\,\mu A/cm^2$). Below 0.9 V, however, the reactivity of platinum rapidly becomes superior to that of $Na_{0.6}WO_3$ with increasing overvoltage. If compounds with a reactivity comparable to that of platinum at high current densities exist and if their electrical conductivity is acceptable, tungsten bronzes may become useful electrocatalysts for the oxygen electrode.

8. The Reversible Oxygen Electrode

The equilibrium potential of net reaction 4 in alkaline solution or reaction 9 in acid solution is generally not established because the exchange currents of the rate-determining steps are so small that other reactions interfere. However, the equilibrium potential can be determined [62, 84, 85] from kinetic data as demonstrated first by HOAR [84]. The intersection of the extrapolated Tafel line for the O_2 evolution with the extrapolated Tafel line for the O_2 reduction is determined. As illustrated by the data [6] for Ir and Pt in Fig. 82, the intersection is close to the theoretical value of the reversible potential $U_{rev} = 1.23$ V in acid electrolytes. However, the extrapolation over several orders of magnitude of the current raises doubts about the accuracy of the procedure.

BOCKRIS and HUQ [85] reported the attainment of the reversible oxygen potential on preoxidized platinum electrodes in purified sulfuric acid solution, saturated with O_2. WATANABE and DEVANATHAN [22] pointed out that the surface has to be prepared in such a way that the exchange currents of the reactions involving the formation and reduction of oxygen layers are much smaller than the exchange current of the rate-determining step of reaction 9. The latter exchange current density can be estimated from the intersection of the Tafel lines (see Fig. 82). It is about 10^{-10} A/cm². Owing to this small value, the suppression of other reactions appears extremely difficult. Nevertheless, the intermediate attainment of an open-circuit potential close to 1.23 V was interpreted [22, 85] as establishment of the equilibrium potential of the oxygen electrode. When recorded [22, 85] as a function of time, the open-circuit potential approaches a constant value close to 1.23 V within 30 minutes. After staying there for 1 to 2 hrs., it becomes less positive with time. The pressure dependence of the open-circuit potential in the constant portion of the potential-time curve corresponds [22, 85] roughly to the theoretical one (Eq. II, 10).

HOARE found [86—88] that the potential of a bright platinum electrode which had been pretreated in concentrated nitric acid became steady at 1.23 V in purified 1 M H_2SO_4, saturated with O_2, for more than 24 hrs. Traces of hydrogen peroxide were absent. He suggested that the pretreatment in nitric acid produced a film similar to the oxygen

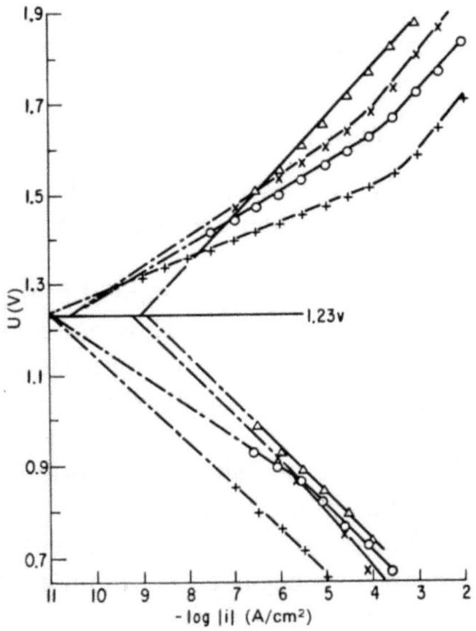

Fig. 82. Anodic Tafel plots for the O_2 evolution and cathodic plots for the O_2 reduction on rhodium (○), iridium (+), platinum-40at.% rhodium (×), and platinum (△) in 1 M $HClO_4$

layer formed by anodic pretreatment [22, 85] in sulfuric acid. The film covers the surface completely and suppresses the establishment of a mixed potential. The mixed potential was attributed [19] to the simultaneous occurence of reaction 4 and reaction 42

$$H_2O = O_{ad} + 2H^+ + 2e^- . \tag{42}$$

It was concluded [19] that the film is unstable in sulfuric acid solution and eventually becomes incomplete, whereupon a mixed potential process becomes effective. The open-circuit potential falls to less positive values. In phosphoric acid solutions, the film is more stable [89]. A potential close to 1.24 V was established for longer periods in phosphoric acid solutions (about 72 hrs.) than in sulfuric acid solution (about 24 hrs.). It is difficult to judge whether an open-circuit potential that stays close to the theoretical value of the equilibrium potential for a certain time is a transient phenomenon or results from the equilibrium of an electrochemical reaction.

SCHULDINER and coworkers [90] studied the dependence of the open-circuit potential of smooth platinum upon the partial pressure of oxygen in a gas-tight system with negligible oxygen leak. The rest potential was found to increase with 0.06 V/decade of pressure between $p_{O_2}=10^{-6}$ atm and $p_{O_2}=10^{-2}$ atm in agreement with earlier results [91,92]. The open-circuit potential was considered as an equilibrium potential of the reaction

(43) $$O_2+H^+ +e^- =HO_{2ad}.$$

Constant activity of HO_{2ad} was assumed. Since the coverage with HO_{2ad} is small in the potential range under consideration, the latter assumption is questionable.

As is to be expected, the reversible potential of the oxygen electrode is established at higher temperatures. Thus good agreement between measured and computed U_{rev} values was reported [93] for the Bacon cathode in concentrated potassium hydroxide solution between about 100 °C and 240 °C.

Although it is difficult to establish the equilibrium of reactions 4 and 9, reaction 13 is reversible on mercury [9], silver [28], carbon [7], and platinum [20] in alkaline electrolytes. Measurements [94] of the H_2O_2 oxidation and reduction on a degassed platinum electrode indicate the reversibility of reaction 12 in acid solution. It is clear from the preceding comments that small traces of H_2O_2 exert a considerable influence on the open-circuit potential in solutions saturated with O_2.

References

1. ROITER, W., and R. JAMPOLSKAJA: Physicochim. Acta. U.R.S.S. **7**, 247 (1937).
2. RIDDIFORD, A. C.: Electrochim. Acta **4**, 170 (1961).
3. SAWYER, D. T., and L. V. INTERRANTE: J. Electroanalyt. Chem. **2**, 310 (1961).
4. —, and R. J. DAY: Electrochim. Acta **8**, 589 (1963).
5. BIANCHI, G., and T. MUSSINI: Electrochim. Acta **10**, 445 (1965).
6. DAMJANOVIC, A., A. DEY, and J. O'M. BOCKRIS: J. Electrochem. Soc. **113**, 739 (1966).
7. BERL, W. G.: Trans. Electrochem. Soc. **83**, 253 (1943).
8. KRASIL'SHIKOV, A. I.: Zhur. Fiz. Khim. **23**, 332 (1949) and **26**, 216 (1952).
9. BAGOTZKY, V. S., and I. E. YABLOKOVA: Zhur. Fiz. Khim. **27**, 1663 (1953).
10. WINKELMANN, D.: Z. Elektrochem. **60**, 731 (1956).
11. DAVIES, M. O., M. CLARK, E. YEAGER, and F. HOVORKA: J. Electrochem. Soc. **106**, 56 (1959).
12. BREITER, M. W.: Electrochim. Acta **9**, 441 (1964).
13. KRASIL'SHIKOV, A. I., and V. A. ANDREJEVA: Zhur. Fiz. Khim. **27**, 389 (1953).

14. — Proceedings of the 4th Conference on Electrochemistry, Moscow 1956, p. 272. Moscow: Academy of Sciences 1959.
15. Bockris, J. O'M.: J. Chem. Phys. **24,** 817 (1956).
16. Vielstich, W.: Z. Instrumentenk. **71,** 29 (1963).
17. Giner, J.: Z. Elektrochem. **63,** 386 (1956).
18. Lingane, J. J.: J. Electroanalyt. Chem. **2,** 296 (1961).
19. Hoare, J. P.: J. Electrochem. Soc. **109,** 858 (1962).
20. Nekrassov, L. N., and L. Müller: Doklady Akad. Nauk S.S.S.R. **149,** 1107 (1963).
21. Kozawa, A.: J. Electroanalyt. Chem. **8,** 20 (1964).
22. Watanabe, N., and M. A. Devanathan: J. Electrochem. Soc. **111,** 615 (1964).
23. Müller, L., and L. N. Nekrassov: Doklady Akad. Nauk S.S.S.R. **157,** 416 (1964).
24. — — J. Electroanalyt. Chem. **9,** 282 (1965).
25. Sobol, V. V., E. I. Krushcheva, and V. A. Dagaeva: Elektrokhim. **1,** 1332 (1965).
26. Frumkin, A. N., E. I. Krushcheva, M. R. Tarasevich, and N. A. Shumilova: Elektrokhim. **1,** 17 (1965).
27. Krushcheva, E. I., N. A. Shumilova, and M. R. Tarasevich: Elektrokhim. **1,** 730 (1965).
28. Bagotzky, V. S., L. N. Nekrassov, and N. A. Shumilova: Uspekhi Khim. **34,** 1697 (1965).
29. Oshe, A. I., V. I. Tikhomirova, and V. S. Bagotzky: Elektrokhim. **1,** 688 (1965).
30. Peters, D. C., and R. A. Mitchell: J. Electroanalyt. Chem. **10,** 306 (1965).
31. Sandler, Y. L., and E. A. Pantier: J. Electrochem. Soc. **112,** 928 (1965).
32. Liang, C. C., and A. L. Juliard: J. Electroanalyt. Chem. **9,** 390 (1965).
33. Damjanovic, A., and J. O'M. Bockris: Electrochim. Acta **11,** 376 (1966).
34. Nekrassov, L. N., E. I. Krushcheva, N. A. Shumilova, and M. R. Tarasevich: Elektrokhim. **2,** 363 (1966).
35. Hoare, J. P.: Electrochim. Acta **11,** 203 (1966).
36. Nekrassov, L. N., and E. I. Krushcheva: Elektrokhim. **3,** 677 (1967).
37. — — Elektrokhim. **3,** 166 (1967).
38. Gnanamuthu, D. S., and J. V. Petrocelli: J. Elektrochem. Soc. **114,** 1036 (1967).
39. Breiter, M. W.: to be published.
40. Laitinen, H. A., and I. M. Kolthoff: J. Phys. Chem. **45,** 1061 (1941).
41. Delahay, P.: J. Electrochem. Soc. **97,** 198 (1950).
42. — J. Electrochem. Soc. **97,** 205 (1950).
43. Heyrovsky, J.: Cas. cesk. Lekarn. **7,** 242 (1927).
44. Siver, Yu. G., and B. N. Kabanov: Zhur. Fiz. Khim. **22,** 53 (1948).
45. Frumkin, A. N., and L. N. Nekrassov: Doklady Akad. Nauk S.S.S.R. **126,** 115 (1959).
46. — —, V. Levich, and Yu. Ivanov: J. Electroanalyt. Chem. **1,** 84 (1960).
47. Ivanov, Yu. B., and V. G. Levich: Doklady Akad. Nauk S.S.S.R. **126,** 1029 (1959).
48. Albery, W. J., and S. Bruckenstein: Trans. Faraday Soc. **62,** 1920 (1966).
49. Damjanovic, A., M. A. Genshaw, and J. O'M. Bockris: J. Chem. Phys. **45,** 4057 (1966).
50. — — — J. Phys. Chem. **70,** 3761 (1966).
51. — — — J. Electrochem. Soc. **114,** 466 (1967).
52. — — — J. Phys. Chem. **71,** 3722 (1967).
53. — — — J. Electrochem. Soc. **114,** 1107 (1967).
54. Müller, L., and L. N. Nekrassov: Doklady Akad. Nauk S.S.S.R. **154,** 437 (1964).
55. Weiss, J.: Trans. Faraday Soc. **31,** 1547 (1935).
56. Gerischer, R., and H. Gerischer: Z. Physik. Chem. N. F. **6,** 178 (1965).
57. Bockris, J. O'M., and L. F. Oldfield: Trans. Faraday Soc. **51,** 249 (1955).
58. Lingane, J. J., and P. J. Lingane: J. Electroanalyt. Chem. **5,** 411 (1963).
59. Müller, L., and L. N. Nekrassov: Zhur. Fiz. Khim. **38,** 3028 (1964).

60. HABER, F., and S. GRINBERG: Z. Anorg. Allgem. Chem. **18,** 37 (1898).
61. BIANCHI, G., F. MAZZA, and T. MUSSINI: Electrochim. Acta **7,** 457 (1962).
62. VIELSTICH, W.: Z. Physik. Chem. N. F. **15,** 409 (1958).
63. JUSTI, E., M. PILKUHN, W. SCHEIBE, and A. WINSEL: Hochbelastbare Wasserstoff-Diffusions-Elektroden für Betrieb bei Umgebungstemperatur und Niederdruck, Abh. Mainzer Akad. Nr. **8.** Wiesbaden: Steiner 1959.
64. ZHUTAEVA, G. V., N. A. SHUMILOVA, and M. R. TARASEVICH: Doklady Akad. Nauk S.S.S.R. **161,** 151 (1965).
65. SHUMILOVA, N. A., G. V. ZHUTAEVA, and M. R. TARASEVICH: Electrochim. Acta **11,** 967 (1966).
66. LIN-IN, GU, N. A. SHUMILOVA, V. S. BAGOTZKY, and P. F. BELMER: Elektrokhim **3,** 1279 (1967).
67. HURLEN, T., Y. L. SANDLER, and E. A. PANTIER: Electrochim. Acta **11,** 1463 (1966).
68. BEER, S. Z., and Y. L. SANDLER: J. Electrochem. Soc. **112,** 1133 (1965).
69. MAZITOV, YU. A., K. I. ROSENTAL, V. I. VESELOVSKY: Zhur. Fiz. Khim. **38,** 697 (1964).
70. KORDESCH, K., and A. MARKO: Öster. Chem. Ztg. **53,** 125 (1951).
71. JACQ, J., and O. BLOCH: Electrochim. Acta **9,** 551 (1964).
72. SCHWABE, K., R. KÖPSEL, K. WIESENER, and E. WINKLER: Electrochim. Acta **9, 413** (1964).
73. VIELSTICH, W.: Brennstoffelemente, Chap. III. 6 and III. 7. Weinheim: Verlag Chemie 1965.
74. BRUNS, B., and A. FRUMKIN: Z. Physik. Chem. **141,** 141 (1929).
75. FRUMKIN, A., R. BURSTEIN, and P. LEVIN: Z. Physik. Chem. **157,** 442 (1932).
76. KUCHINSKI, E., R. BURSTEIN, and A. FRUMKIN: Zhur. Fiz. Khim. **14,** 441 (1940).
77. BURSTEIN, R. H., and A. N. FRUMKIN: Doklady Akad. S.S.S.R. **32,** 327 (1941).
78. ADAMS, A. M., F. T. BACON, and R. G. H. WATSON: Fuel Cells. W. Mitchell, Jr., ed., New York: Academic Press 1963.
79. MISYUK, E. G., O. K. DAVTYAN, R. N. STUPICHENKO, and E. A. KALYUZHNAJA: Elektrokhim. **2,** 788 (1966).
80. GOLOVKIN, YU. I., N. I. VASILISTOV, and N. A. FEDOTOV: Elektrokhim. **3,** 805 (1967).
81. MAZZA, F., and S. TRASSATTI: J. Electrochem. Soc. **110,** 847 (1963).
82. SHAMSUL HUQ, A. K. M., A. J. ROSENBERG, and A. C. MAKRIDES: J. Electrochem. Soc. **111,** 278 (1964).
83. SEPA, D. B., A. DAMJANOVIC, and J. O'M. BOCKRIS: Electrochim. Acta **12,** 746 (1967).
84. HOAR, T. P.: Proc. Roy. Soc. London **A 142,** 628 (1933).
85. BOCKRIS, J. O'M., and A. K. M. SHAMSUL HUQ: Proc. Roy. Soc. London **A 237,** 277 (1956).
86. HOARE, J. P.: J. Electrochem. Soc. **110,** 1019 (1963).
87. — J. Electrochem. Soc. **112,** 849 (1965).
88. — Nature **204,** 71 (1964).
89. THACKER, R., and J. P. HOARE: J. Electrochem. Soc. **113,** 862 (1966).
90. SCHULDINER, S., B. J. PIERSMA, and T. B. WARNER: J. Electrochem. Soc. **113,** 573 (1966).
91. —, and R. M. ROE: J. Electrochem. Soc. **110,** 332 and 1142 (1963).
92. WARNER, T. B., and S. SCHULDINER: J. Electrochem. Soc. **112,** 853 (1965).
93. ROCKET, J. A., and R. BROWN: J. Electrochem. Soc. **113,** 207 (1966).
94. TIKHOMIROVA, V. I., V. I. LUK'YANYCHEVA, and V. S. BAGOTZKY: Elektrokhim. **1,** 645 (1965).

XII. Corrosion of Electrodes at Low Temperatures

The corrosion of electrodes in fuel cells leads to a decrease in performance with time. It may determine the life of the fuel cell. The rate of corrosion depends upon the electrode material, the electrolyte and the temperature. The presence of anions which form complexes with the respective electrocatalyst (for instance Cl^- for platinum metals) has to be avoided since complex formation tends to increase the rate of corrosion. In general, corrosion problems become more serious with increasing temperature. While the corrosion of electrocatalysts is discussed in this chapter, the degradation of electrode structures due to mechanical strain or to the disintegration of the bonding material is not dealt with.

1. Predictions from Potential-pH Diagrams

The use of potential-pH diagrams in the discussion of corrosion was introduced by POURBAIX [1]. The reversible potential and the pH are considered independent variables for reactions of the type:

(1) $$\mathrm{xX + yH_2O + ne^- = wW + mH^+}$$

Here X represents the oxidized form and W the reduced form of the substance in question. The equilibrium of reaction 1 may be expressed at 25 °C:

(2) $$U_{rev} = U_0 - \frac{0.0591\,\mathrm{m}}{\mathrm{n}}\,\mathrm{pH} + \frac{0.0591}{\mathrm{n}}\log\frac{a_\mathrm{X}^\mathrm{x}}{a_\mathrm{W}^\mathrm{w}}\,.$$

Here U_{rev} refers to the normal hydrogen electrode ($a_{\mathrm{H}^+}=1$, $p_{\mathrm{H}_2}=1$ atm). If $a_\mathrm{X}^\mathrm{x}/a_\mathrm{W}^\mathrm{w}=1$, Eq. 2 simplifies to Eq. 3.

(3) $$U_{rev} = U_0 - \frac{0.0591\,\mathrm{m}}{\mathrm{n}}\,\mathrm{pH}\,.$$

If x = w, the simplification requires only equal activities ($a_\mathrm{X}=a_\mathrm{W}$). The drawing of lines according to Eq. 3 in the potential-pH diagram allows [1,2] to establish domains of relative predominance of dissolved species or domains of relative stability of solid substances at equilibrium. By superimposing the diagrams for dissolved species and solid substances, the conditions of potential and pH are found under which a given solid substance and a given dissolved substance can be simultaneously stable. In addition the lines

$$U_{rev} = 0.000 - 0.0591\,\mathrm{pH}, \tag{4}$$

$$U_{rev} = 1.228 - 0.0591\,\mathrm{pH} \tag{5}$$

are plotted. These lines correspond to the equilibrium of reaction 6 or reaction 7 respectively at a pressure of 1 atm for the gas in question.

$$H_2 = 2H^+ + 2e^-, \tag{6}$$

$$2H_2O = O_2 + 4H^+ + 4e^-. \tag{7}$$

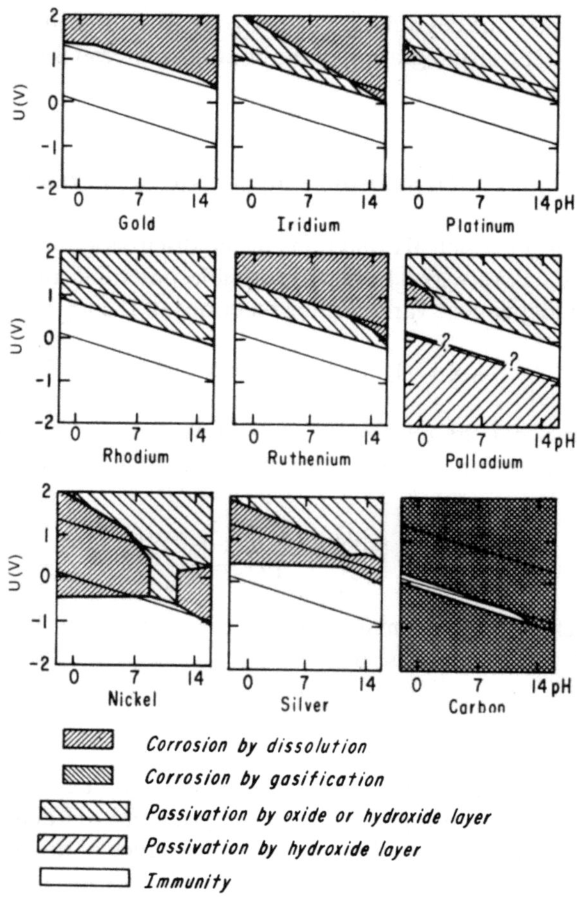

Fig. 83. Potential-pH diagrams for different electrocatalysts

Pourbaix [1,2] suggested as a first approximation that a metal is corrodible in an aqueous solution if the equilibrium concentration of its ions exceeds 10^{-6} M. The lines designating this solubility in the potential-pH diagrams allow to distinguish between domains where corrosion is possible and domains where corrosion is negligible. When the corrosion of the substance occurs by the formation of volatile hydrides or oxides, a partial pressure of 10^{-6} atm was chosen [1,2] as a rough guide to establish the domains of corrosion by gasification. The different domains are shown in the diagrams of Fig. 83, according to the data in reference 3 for materials of interest as electrocatalysts in the low-temperature fuel cell. The two parallel lines represent reactions 6 and 7.

The following points should be considered when the diagrams in Fig. 83 are used:

a) The diagrams are reliable to the same extent as the thermodynamic data from which the values of U_0 were computed.

b) The metal may dissolve in the form of more complex ions than assumed [2] in the construction of the diagrams.

c) If the dissolution is strongly irreversible, the domain of immunity will be different from the one predicted thermodynamically.

d) If a thick film is present the corrosion may be impeded by it. The possibility that chemisorbed layers inhibit the metal dissolution is disregarded.

2. Dissolution of Platinum Metals

While the dissolution of platinum [4—11], iridium [5, 12], rhodium [5, 14, 15], ruthenium [16—18], and palladium [5] has been investigated in aqueous solutions containing chloride ions, there are few corrosion studies in acid or alkaline electrolytes without complexing ions. Undoubtedly the scarcity of information results from the difficulty of measuring extremely small rates of corrosion. The development [7,11,18] of sensitive techniques in recent years is likely to improve the situation.

Vetter and Berndt [19] maintained a constant potential in the region between 0.8 V and 2.0 V on a smooth Pd wire in 0.5 M H_2SO_4 at 25 °C for 48 hr. Spectroscopical analysis of the solution at the end of the experiment showed that palladium corrodes. The rate of palladium dissolution computed from the solution analysis according to reaction 8

$$Pd = Pd^{++} + 2e^- \qquad (8)$$

increased with potential between 0.9 V and 1.0 V. After reaching a maximum of about 1 μA/cm² at 1.0 V, the rate decreased with U. It was found to be very small between 1.2 V and 1.7 V. The dissolution rate became larger with U again in the potential region of oxygen evolution.

The sensitivity of the analysis [19] was not sufficient to determine traces of Pt in the potential range between 0.8 V and 2.14 V in 0.05 M H_2SO_4 and 0.5 M H_2SO_4 at 25 °C. If the hump at about 1.2 V in anodic charging curves taken with a small current density on smooth platinum in 0.5 M H_2SO_4 is attributable to platinum corrosion, the corrosion rate can be estimated to be smaller than 0.2 μA/cm². However, this estimate [19] depends in a critical way upon the interpretation of the hump which may be due to other processes (compare section 1 in chapter VII).

Another type of evidence for the dissolution of platinum in alkaline electrolytes was reported by MALACHESKY, JASINSKI, and BURROWS [20] recently. If a Teflon-bonded platinum electrode was used as the anode in the preelectrolysis of the solution, traces of platinum were deposited afterwards from the solution onto cadmium or gold electrodes. The presence of platinum on cadmium was established by the considerable increase in the reactivity of the electrode with respect to the O_2 reduction. The decrease of the overvoltage for the hydrogen evolution on gold is evidence for the deposition of platinum on gold. The latter deposition was also observed in 1 M $HClO_4$ by BRUMMER [21]. If the potential of the Teflon-bonded platinum electrode was kept at $U < 0.5$ V, any dissolution of platinum was not detected. The test for the increased reactivity of cadmium with respect to the O_2 reduction and for the decreased hydrogen overvoltage on gold was negative.

The preceding results demonstrate the corrosion of palladium and platinum at potentials of the oxygen layer and in the region of oxygen evolution. A similar situation may be expected for the other platinum metals. Since the potential of the anode of a fuel cell remains below those of the oxygen region, the corrosion of the electrocatalyst will be absent. However, corrosion of the platinum metals may be expected for the cathode. The precise assessment of this corrosion is difficult since it may be accompanied by other degradation effects.

3. Dissolution of Nickel, Silver, and Carbon in Alkaline Electrolytes

The corrosion of nickel in alkaline solutions is small [2,22,23] in the potential region between the two parallel lines representing the equilibrium of reaction 6 and 7 respectively in Fig. 83. This is the potential range of interest for the operation of fuel cells. Studies [23] of the dissolution of $Ni(OH)_2$ in NaOH up to 15 M suggest an upper limit of the amount of dissolved bivalent nickel of less than 10^{-5} M. Recent work [24] in 0.2 M KOH indicates that less than 0.1 ppm of nickel are dissolved during anodic polarization. Since some of the dissolved material may be dispersed solid, it is likely that the solubility of nickel in its different valence states is not significant in alkaline solutions.

Silver is stable in alkaline solutions in the potential range of interest since the formation of Ag_2O, which is soluble, occurs at more anodic potentials.

The diagram for carbon in Fig. 83 displays a very small domain of stability. It is thermodynamically possible for carbons to be easily oxidized to carbon dioxide, carbonic acid, and carbonates. Reduction of carbon may lead to the formation of methane, methyl alcohol and other organic substances. However, the energetically possible reactions are strongly irreversible [2] and do not occur under normal conditions of pressure and temperature. SCHMIDT [24] reported a corrosive destruction of carbon electrodes when a critical potential was exceeded during the reduction of O_2. The carbon electrodes were not impregnated with metallic electrocatalysts. The critical potential depended upon the extent to which an oxygen layer was present (compare section 5 in chapter VIII).

References

1. POURBAIX, M.: Thesis, Delft 1945.
2. — Atlas of Electrochemical Equilibria in Aqueous Solutions. Oxford: Pergamon Press 1966.
3. —, and G. GOVAERTS: First International Congress on Metallic Corrosion, London, 1961, p. 96. London: Butterworths 1962.
4. TRONEV, V. G.: Doklady Akad. Nauk S.S.S.R. **15**, 555 (1937).
5. RAUB, E., and G. BUSS: Z. Elektrochem. **46**, 195 (1940).
6. ERSHLER, B.: Acta Physicochim. U.S.S.R. **19**, 139 (1944).
7. MITUYA, A., and T. OBAYASHI: Zhur. Fiz. Khim. **34**, 679 (1960).
8. LLOPIS, J., and A. SANCHO: J. Electrochem. Soc. **108**, 720 (1961).
9. BREITER, M. W., and J. L. WEININGER: J. Electrochem. Soc. **109**, 1135 (1962).
10. LLOPIS, J., and M. VAZQUEZ: Electrochim. Acta **8**, 163 (1963).
11. LOSSEW, W. W., M. A. DEMBROWSKI, A. I. MOLODOW, and W. W. GORODEZKI: Electrochim. Acta **8**, 387 (1963).
12. LLOPIS, J., and L. JORGE: J. Electrochem. Soc. **110**, 947 (1963).
13. — — Electrochim. Acta **9**, 103 (1964).
14. —, and M. VAZQUEZ: Electrochim. Acta **9**, 1655 (1964).
15. —, I. M. TORDESILLAS, and M. MUNIZ: Electrochim. Acta **10**, 1045 (1965).
16. — —, and J. M. ALFAYATE: Electrochim. Acta **11**, 623 (1966).
17. —, and M. VAZQUEZ: Electrochim. Acta **11**, 633 (1966).
18. —, J. M. GAMBOA, and J. M. ALAFAYATE: Electrochim. Acta **12**, 57 (1967).
19. VETTER, K. J., and D. BERNDT: Z. Elektrochem. **62**, 378 (1958).
20. MALACHESKY, P., R. JASINSKI, and B. BURROWS: J. Electrochem. Soc. **114**, 1104 (1967).

21. Brummer, S. B.: J. Electrochem. Soc. **112,** 633 (1965).
22. MacGillavry, D., J. J. Singer, Jr., and J. H. Rosenbaum: J. Chem. Phys. **1195,** 195 (1951).
23. Gayer, K. H., and A. B. Garret: J. Am. Chem. Soc. **71,** 2973 (1949).
24. Schmidt, S.: Mitteilungsblatt Chem. Ges. DDR **14,** 23 (1967).

XIII. Processes in Fuel Cells with Molten Carbonate Electrolytes

Fuel cells with molten carbonate electrolytes have received attention [1—10] in recent years for two reasons:

a) They offer the prospect of small material costs.

b) They are well suited to natural gas as fuel.

The sections of this chapter deal with basic problems encountered in carbonate fuel cells.

1. General Considerations

Various eutectics of molten alkali carbonates have been chosen as electrolytes for cells in which carbonaceous fuel or hydrogen are oxidized. The invariance of the electrolyte (see section 4 of chapter II) and an acceptable electrolyte resistance between anode and cathode are generally not attainable with other molten compounds such as hydroxides, chlorides, nitrates, and sulfates. Binary and ternary mixtures of Li_2CO_3 (727 °C), Na_2CO_3 (860 °C), K_2CO_3 (900 °C) are favored because they have lower melting points than the pure carbonates the melting points of which are given in parenthesis behind the respective compound.

The deterioration of the performance due to the loss of electrolyte from the porous matrix in which the carbonates were kept in early models and the requirement of leak tightness led to the replacement of matrix electrolytes by paste electrolytes [4]. The latter electrolytes are prepared by blending a finely divided inert solid, for instance MgO, with the appropriate amount of the carbonates. The mixture turns into a stiff paste when the carbonates melt. The paste and the contacts between electrodes and paste are gas-tight. If silver is used as material for the oxygen electrode, the paste electrolyte has the important property of preventing the migration of silver to the anode. Silver migration is highly undesirable because of the formation of silver dendrites on the anode.

Porous anodes of nickel, stainless steel, stainless steel with iron-iron oxide mixtures, lithiated manganous oxide, and silverized catalysts (Rh, Co, Zn, ZnO, MnO_2, Co_2O_3, lithiated NiO) have been tested. The difference in the reactivity of these electrodes for the H_2 oxidation is not large at temperatures between 500 °C and 700 °C. The presence of oxides improves the performance of the anode for hydrogen gas containing some carbon monoxide. Stainless steel, nickel, silver, lithiated nickel oxide, and mixtures of copper and copper oxide have been tried as electrocatalysts for the O_2 reduction. Silver displays a good performance. An attempt to classify some of the important metals according to their reactivity for the H_2 oxidation and the O_2 reduction was made by DEGOBERT and BLOCH [2].

The initial structure, pore size, and pore size distribution is of minor importance for the long-term performance of diffusion electrodes in carbonate fuel cells because of changes due to sintering, oxidation and reduction, dissolution and precipitation of metals into and from the electrolyte during the operation at temperatures around 700 °C. The cells age with time and approach a natural operating level. The situation differs greatly from that of low-temperature cells for which properly chosen manufacturing techniques are important. Hydrogen-carbon dioxide mixtures resulting from steam-reforming of hydrocarbons or shifting of carbon monoxide appear to be the most suitable fuels for molten carbonate cells. Reforming may be accomplished by bringing [7] a small amount of reforming catalysts into the anode space or by using [6] a double-duty anode in which the fuel gas is introduced at one edge of the electrode and forced through a coarse layer parallel to the electrolyte-gas interface. Hydrocarbons and carbon monoxide are not used directly because coke is frequently formed. The net reactions at the anode and cathode are for these mixtures:

(1) $$\tfrac{1}{2}H_2+\tfrac{1}{2}CO_3^{--}=\tfrac{1}{2}H_2O+\tfrac{1}{2}CO_2+e^-,$$

(2) $$\tfrac{1}{4}O_2+\tfrac{1}{2}CO_2+e^-=\tfrac{1}{2}CO_3^{--}.$$

The occurence of net reaction 1 was established experimentally by BROERS and SCHENKE [4]. A mixture of oxygen and carbon dioxide or of air and carbon dioxide has to be fed to the anode to guarantee the invariance of the electrolyte with respect to carbonate ions. The formation of water vapor as one of the products of reaction 1 allows steam-reforming in the anode chamber or in a double-duty anode without external introduction of steam.

The electromotive force of the hydrogen-oxygen cell with carbonate electrolytes is:

(3) $$E=-\frac{\Delta G_0}{2F}-\frac{RT}{2F}\ln\frac{p_{H_2O}\,p_{a,CO_2}}{p_{H_2}\,p_{O_2}^{1/2}\,p_{c,CO_2}}.$$

Eq. 3 takes into account the possibility of different pressures of CO_2 at the anode (a) and cathode (c). It was found [7] that mixtures of $1\,M\,H_2+1\,M\,CO_2$ or $1\,M\,CO+M\,H_2O$ established the same potential at the anode and that the electromotive force of the cell was equal within experimental error to the value obtained from Eq. 3 at 700 °C. This implies that reactions 1 and 2 are reversible and that the equilibrium of reaction 4 is established.

(4) $$H_2 + CO_2 = H_2O + CO.$$

2. Properties of Molten Carbonate Electrolytes

Ionic melts consist [11] of a highly disordered but quasicrystalline arrangement. Defects in this arrangement include holes, interstitial ions, paired holes and paired ions, and cooperative defects similar to dislocations extending over more than two ions. Strong electrostatic interactions lead to nearest neighbor anionic-cationic distances that are smaller in the molten alkali halides than in the crystalline state. Carbonate melts have approximately the same kinetic energy content as the corresponding molten chlorides but differ in the aspect that the anions are disc-shaped polyatomic species rather than the rare-gas type spherical ions of the molten halides. The physical and electrochemical properties of molten carbonates were thoroughly studied by JANZ and coworkers [12—14] in recent years.

Measurements of the surface tension, density, and electrical conductance of Li_2CO_3, Na_2CO_3, K_2CO_3 and $Na_2CO_3 + K_2CO_3$ mixtures in the temperature range of 750 °C to 1000 °C were reported in reference 12. The molten carbonates showed the first appreciable dissociation and weight loss just above their respective melting points if the partial pressure of carbon dioxide in the surrounding atmosphere was smaller than the equilibrium dissociation pressure of reaction 5:

(5) $$CO_3^{--} = CO_2 + O^{--}.$$

The measurements [12] were made with a pressure of CO_2 in large excess to the equilibrium dissociation pressures.

Relative to the $Na_2CO_3 - K_2CO_3$ mixtures, the surface tensions and partial molal volumes indicated little deviation from the predictions for thermodynamically ideal mixtures. The mechanism of conductance was discussed under the assumption of complete dissociation with the species M^+ and CO_3^{--} as the predominant ionic species [14] for two limiting cases:

a) The species M^+ and CO_3^{--} contribute mainly to the conductance.

b) The contact ion pair $(M^+CO_3^{--})$ is the important species in the mechanism of conductance.

The experimental results seem to favor a cooperative jump-rotation mechanism (case b). The rate-determining step in this mechanism is the jump of M^+ from one anionic species to the other, followed by a rotation of the anionic pair. However, an extra mode of diffusion for carbon has to be postulated [12] in the diffusion [16] of CO_2 molecules formed by reaction 4 in order to explain the large ratio of self-diffusion coefficients $D_{CO_3^{--}}/D_{Na^+} = 1.45$ in molten Na_2CO_3. It appears that further experimental work is required before the mechanism of conductance can be established with certainty.

Measurements of the electrical conductance, density and surface tension of binary and ternary carbonate mixtures, in particular of the eutectic compositions

$$Li_2CO_3(53.3\,\text{mol}\,\%)\text{—}Na_2CO_3(46.7\,\text{mol}\,\%),$$

$$Li_2CO_3(42.7\,\text{mol}\,\%)\text{—}K_2CO_3(57.3\,\text{mol}\,\%),$$

$$Li_2CO_3(43.5\,\text{mol}\,\%)\text{—}Na_2CO_3(31{,}5\,\text{mol}\,\%)\text{—}K_2CO_3(25.0\,\text{mol}\,\%),$$

$$Na_2CO_3(58\,\text{mol}\,\%)\text{—}K_2CO_3(42\,\text{mol}\,\%),$$

are reported in reference 14. Density and surface tension properties are approximately additive for these systems. Calculations of the conductance by the simple principle of additivity or by the more general MARKOV [17] equation from the conductances of the single-salt alkali carbonates deviated by 4 to 19% from the measured conductance of the mixtures. If the cooperative conductance mechanism of the carbonates as single-salt melts is also assumed for the binary systems, the deviations from the Markov equation may be attributed to an increase in the time period for which the cation resides in the neighborhood of the anion.

3. Thermal Stability of Molten Carbonates and Corrosion of Metals

The rate of corrosion of metal electrodes in molten carbonates has to be small if the fuel cells are to perform satisfactorily for a reasonable length of time. LITTLEWOOD [18] extended the thermodynamic treatment of corrosion in aqueous electrolytes developed by POURBAIX [19] (compare section 1 in chapter XII) to molten chlorides by constructing plots of the reversible potential U_{rev} of possible reaction versus pO^{--}. The acidity of the melt is expressed by Eq. 6.

$$pO^{--} = -\log c_{O^{--}} \tag{6}$$

in analogy to the pH of aqueous solutions, assuming the activity coefficient as 1.

The function

$$pCO_2 = \log p_{CO_2} \tag{7}$$

was introduced in the thermodynamic discussion of stability and corrosion by INGRAM and JANZ [20] since the pressure of CO_2 is usually fixed by the experimental conditions. Diagrams showing redox and acid-base equilibria were constructed [20] for molten carbonates. The standard potentials U_0 were computed according to Eq. 8.

$$U_0 = -\frac{\Delta G_0}{zF} \tag{8}$$

as in aqueous electrolytes. Since the equilibrium of the oxygen electrode (reaction 2) is established [7,21,22] in molten carbonates, the potential at $a_{CO_3^{--}} = 1$ and 1 atm total pressure ($p_{CO_2} = \frac{2}{3}$ atm, $p_{O_2} = \frac{1}{3}$ atm) was set equal to zero. The equilibrium potential of the oxygen electrode is at 600 °C:

$$U_{rev} = 0.0359 - 0.086\,pCO_2 + 0.0433 \log p_{O_2}\,. \tag{9}$$

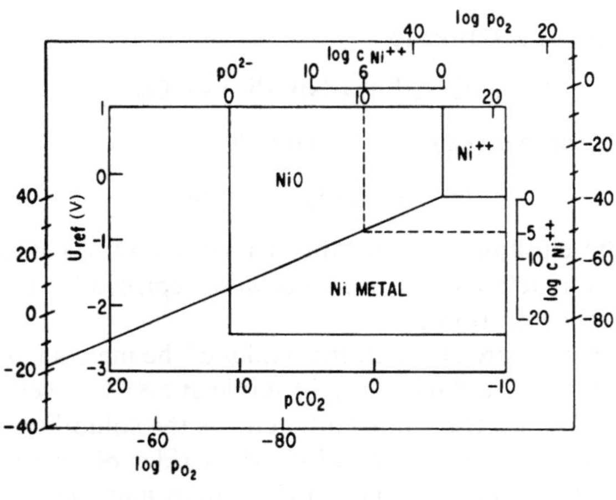

Fig. 84. Redox and acid-base equilibria for the system nickel-ternary eutectic at 600 °C

The diagram for the system nickel-ternary eutectic at 600 °C is shown in Fig. 84 as an example. It is divided into three regions in which only one nickel phase may be present at unit activity. The boundary between the Ni^{++} and NiO areas is the value of pCO_2 at which solid NiO precipitates from a pure nickel carbonate melt. When the oxygen pressure is reduced beyond the dissociation pressure of NiO ($10^{-19.2}$ atm), nickel metal may appear. The line for $\log p_{O_2} = -19.2$ is the boundary between the NiO and Ni domains. The standard potential of the electrochemical reaction

(10) $$Ni = Ni^{++} + 2e^-$$

is obtained as -0.34 V from the free energy change of the reaction:

(11) $$NiCO_3 = Ni + CO_2 + \tfrac{1}{2}O_2 .$$

The three boundaries intersect at -0.34 V. The equilibrium of reaction 11 may be expressed:

(12) $$U_{rev} = -0.34 + 2.3\frac{RT}{2F}\log c_{Ni^{++}} .$$

Eq. 12 establishes the $\log c_{Ni^{++}}$ scale in the domain of Ni metal; this scale is fixed in the NiO domain by the equilibrium of reaction 13.

(13) $$NiCO_3(liq) = NiO(sol) + CO_2 .$$

It is:

(14) $$\log c_{Ni^{++}} = -5.24 - pCO_2 .$$

The equilibrium of reaction 15

(15) $$NiCO_3(liq) = NiO(liq) + CO_2$$

gives a correlation between pO^{--} and pCO_2:

(16) $$pO^{--} + pCO_2 = -\log K_d .$$

Here K_d is the equilibrium constant of reaction 15. As it is usual, the activities of the reactants and products were replaced by the respective pressures or concentrations.

The oxygen pressure at which the oxides of the metals Co, Fe(II), Zn, Mn, Mg, Na, K, Ca, Li, Ba(II) are still stable at 600 °C is very small [19] ($<10^{-20}$ atm). Hence, the important factor is the solubility of the oxide in the melt. The pCO_2 values at which the oxides of alkali metals and barium precipitate from the melt are positive and large. Carbonates with these cations display good thermal stability while considerable pressures of carbon dioxide are required to guarantee stability for the other carbonates at 600 °C. Acid-base considerations make it also understandable [19] that the basic oxides have excellent properties as container material for carbonates.

For aqueous systems Pourbaix [19] made the arbitrary assumption that a metal would corrode if the equilibrium concentration of its ions was greater than 10^{-6} M. The same assumption was used as a guideline by Ingram and Janz [20]. If nickel metal is immersed under 1 atm total pressure ($p_{O_2}=\frac{1}{3}$ atm, $p_{CO_2}=\frac{2}{3}$ atm) in a carbonate melt, the point in Fig. 84 corresponds to a potential close to zero and a nickel ion concentration slightly smaller than 10^{-5} M. The thermodynamic treatment indicates two possibilities:

a) The metal is converted completely to NiO.

b) The coherent NiO film leads to passivation.

Experiments are required to distinguish between these two possibilities. Corrosion of Ni to NiO was observed [20]. A comparison [20] between the thermodynamic predictions and experiments showed good agreement for Au, Pt, Ag, Fe, and Ni within the limitation of the thermodynamic method. Only gold and platinum were found to be immune to attack. Protection of other metals will depend mainly on passivation. Cathodic protection is undesirable because of the loss in performance due to electrode polarization.

4. Formation of Carbon Deposits

The formation [23—29] of carbon deposits on the anode has to be avoided since it decreases the performance of carbonate fuel cells. Carbon may be deposited electrochemically by reactions 17 to 20:

$$3CO_2+4e^-=C+2CO_3^{--}, \qquad U_0=-1.02\ \text{V}, \tag{17}$$

$$3CO+2e^-=2C+CO_3^{--}, \qquad U_0=-0.87\ \text{V}, \tag{18}$$

$$CO+CO_2+2e^-=C+CO_3^{--}, \qquad U_0=-0.97\ \text{V}, \tag{19}$$

$$CO_3^{--}+4e^-=C+3O^{--}, \qquad U_0=-1.02\ \text{V}. \tag{20}$$

The standard potentials [30] referred to a reversible oxygen electrode ($p_{CO_2}=\frac{2}{3}$ atm, $p_{O_2}=\frac{1}{3}$ atm) are given for 600 °C. The standard state corresponds to a total gas pressure of 1 atm. Reactions 17 and 20 are thermodynamically equivalent [24] because of the equilibrium of reaction 5. Carbon may also result from the disproportionation of carbon monoxide

$$2CO=C+CO_2. \tag{21}$$

Mixtures of hydrogen and carbon dioxide are more suitable fuels than hydrogen because they suppress the formation of carbon by reaction 21, the equilibrium constant of which is equal [29] to 2.5 at 650 °C.

The values of the standard potentials indicate that some of the reactions 17 to 21 are feasible on thermodynamic grounds as side reactions on the anode under certain conditions determined by the partial pressures of the participating gaseous species. The extent of the participation is not known. CAIRNS, TEVEBAUGH, and HOLM [26] gave a detailed discussion of the conditions under which carbon deposition by reaction 21 is not favorable thermodynamically. In general, carbon deposition can be prevented by raising the temperature and by increasing the content of CO_2 or H_2O in the fuel.

5. Processes at the Anode

Steady-state $i-U$ curves taken [31] by ARKHIPOV and STEPANOV on a smooth platinum foil with $A=0.5\ cm^2$ in a eutectic of Li_2CO_3, K_2CO_3, and Na_2CO_3 are shown at three temperatures in Fig. 85. The electrolyte around the anode was stirred with molecular hydrogen. The $I R_{el}$ drop was eliminated by interrupting the current and measuring the

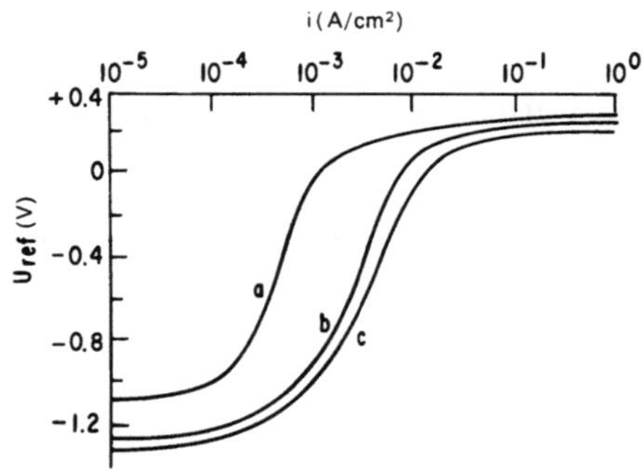

Fig. 85. Current-potential curves of the H_2 oxidation in a ternary eutectic of carbonates on smooth platinum.
Curve a: 500 °C, curve b: 600 °C, curve c: 700 °C

potential immediately after the interruption. The potential is referred to the reversible oxygen electrode ($p_{CO_2}=\frac{2}{3}$ atm, $p_{O_2}=\frac{1}{3}$ atm) in the same melt.

Reaction 1 occurs with negligible polarization at $i \leq 0.1\ mA/cm^2$. A limiting current is reached when the potential becomes less negative. It is reasonable to assume that the limiting current is due to diffusion of H_2. The limiting current of H_2 increases with temperature. The increase is larger from 500 °C to 600 °C than from 600 °C to 700 °C. It is uncertain whether this effect results from an overcompensation of the increase of the diffusion coefficient with temperature by a decrease of the solubility of H_2 in the eutectic. In aqueous electrolytes, the limiting current of the H_2 diffusion has a maximum [32] at about 60 °C for this reason. At positive potentials carbon dioxide and oxygen are evolved [27]. A similar shape of the $i-U$ curve was found [31] on porous platinum rods when the hydrogen was pressed through the pores from the interior of the rod.

The potentials at $i=10^{-5}$ A/cm^2 are more negative than the equilibrium values for which thermodynamic calculations give 1.07 V, 1.05 V, and 1.03 V at 500 °C, 600 °C, and 700 °C respectively. The influence of secondary reactions [33] was not considered [31]. The deviation between measured and theoretical values was attributed [31] to the influence of the partial pressures p_{CO_2} and p_{H_2O} upon the potential.

Although the oxidation of carbonaceous fuels supplied to the anode chamber without external or internal reforming was reported [34], it is likely that the cells are running on thermally produced hydrogen rather than by the direct electrochemical oxidation of carbonaceous fuels as occurs in the low temperature cells (see chapter X). Steady-state $i-U$ curves of the oxidation of different fuels measured by KRONENBERG [34] on a catalyzed anode in the ternary eutectic are shown in Fig. 86. A

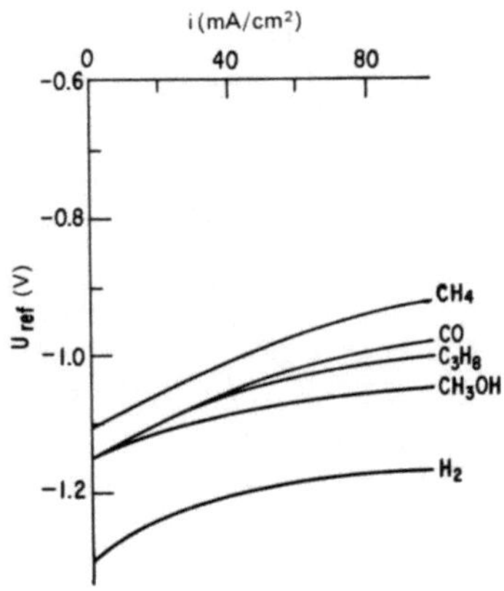

Fig. 86. Current-potential curves for the oxidation of various fuels at 610 °C after correction for the IR_{el} drop

powdered silver electrode ($p_{CO_2}=\frac{2}{3}$ atm, $p_{O_2}=\frac{1}{3}$ atm) served as reference electrode in the eutectic. The considerable IR_{el} drop was eliminated by the interrupter technique. In general, there is little polarization. The results obtained [7,8,34,35] by transient techniques suggest that the polarization results largely from concentration gradients. The nature of these gradients, which exist on porous electrodes in carbonate melts as well as in aqueous electrolytes, is discussed in chapter XVI.

At elevated temperature, the smallness of the overvoltages due to the hindrance of discharge steps makes it very difficult to determine the predominant step of net reaction 1. Possible reaction schemes are not given here because they would be based on guesses. The performance of fuels in Fig. 86 decreases in the order H_2, CH_3OH, C_3H_8, CO, CH_4. The cause for this decrease is not known. It may result from a decreasing ability of the anode for the thermal cracking and steam-reforming of the fuels.

6. The Oxygen Electrode in Molten Carbonates

The reversibility of reaction 2 was established by thorough studies [21,22] of the dependence of the open-circuit potential upon the concentration [21] of CO_3^{--}, upon the partial pressures [22] p_{CO_2} and p_{O_2} and upon the temperature [22]. In all cases, the electrode potential displayed the dependence predicted thermodynamically for the given carbonate melt.

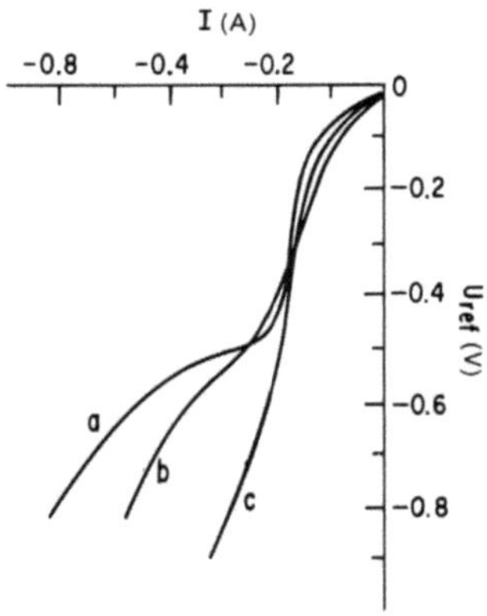

Fig. 87. Current-potential curves of the O_2 reduction on porous silver electrodes in a ternary eutectic of carbonates.
Curve a: $p_{O_2}=0.79$ atm, curve b: $p_{O_2}=0.45$ atm, curve c: $p_{O_2}=0.13$ atm

Steady-state $i-U$ curves of the reduction of O_2 on silver diffusion electrodes are represented in Fig. 87. The curves were taken [36] by STEPANOV and TRUNOV in a ternary eutectic of Li_2CO_3, K_2CO_3, and Na_2CO_3. The potential is given with respect to the reversible oxygen electrode ($p_{CO_2}=\frac{2}{3}$ atm, $p_{O_2}=\frac{1}{3}$ atm) in the same electrolyte. The interrupter technique was employed to eliminate the IR_{el} drop. Gas mixtures with $p_{CO_2}=0.2$ atm, $p_{N_2}+p_{O_2}=0.8$ atm were passed through the diffusion electrode.

The three curves (p_{O_2}=0.79 atm, 0.45 atm, 0.13 atm) in Fig. 87 consist of two portions. The portion between 0 and about −0.05 V was attributed [36] to reaction 2. When the limiting current of reaction 2 is reached, part of the oxygen reacts [36] according to Eq. 22.

(22) $$O_2 + 4e^{--} = 2O^{--}$$

As was to be expected on the basis of the stoichiometry of reaction 2, the limiting current determined [36] as a function of p_{CO_2}/p_{O_2} was found to have a maximum at $p_{CO_2}/p_{O_2} \approx 2$. The limiting current of curve b in Fig. 87 is the largest one for this reason. The currents above the limiting current of reaction 2 increase with the oxygen pressure.

Similar $i-U$ curves with two segments were obtained [37,38] in mixtures of 50% Li_2CO_3 and 50% Na_2CO_3 at different temperatures on platinum or palladium foil. The shape of the curves depended strongly on the length of the heat pretreatment of the electrodes at various temperatures in air. The initial segment virtually disappeared after palladium had been heated for 30 to 60 minutes at 700 °C in air. It was suggested that the two portions are due to the involvement of two different types of oxides in reaction 2 on platinum and palladium and not to the successive occurrence of reaction 2 and reaction 22.

The following mechanism was suggested [35] for reaction 2:

(23) $$2M + O_2 = 2MO,$$

(24) $$2MO + 4e^- = 2M + 2O^{--},$$

(25) $$2O^{--} + 2CO_2 = 2CO_3^{--}.$$

Here MO designates an oxide or surface oxide. Since most metals are covered with oxide at elevated temperatures, another mechanism was proposed [39]:

(26) $$2M + O_2 = 2MO,$$

(27) $$2MO + 2CO_2 = 2MCO_3,$$

(28) $$2MCO_3 = 2M^{++} + 2CO_3^{--},$$

(29) $$2M^{++} + 4e^- = 2M.$$

The decay of the overvoltage of reaction 2 with time after the current interruption is small [7,35]. This implies that the hindrance of discharge steps of net reaction 2 is small. Therefore, it is very difficult to establish the mechanism of the oxygen electrode in carbonate melts. A distinction between the two preceding mechanisms does not appear feasible at present.

References

1. BROERS, G. H. J., and J. A. A. KETELAAR: Fuel Cells, Vol. **1**, p. 78. G. J. Young, ed., New York: Reinhold Publ. 1960.
2. DEGOBERT, P., and O. BLOCH: Bull. Soc. Chim. France, 1887 (1962).
3. GORIN, E., and H. L. RECHT: Fuel Cells, Chap. 5. W. Mitchell, Jr., ed., New York: Academic Press 1963.
4. BROERS, G. H. J., and M. SCHENKE: Fuel Cells, Vol. **2**, p. 6. G. J. Young, ed., New York: Reinhold 1963.
5. TANTRAM, A. D. S., A. C. C. TSEUNG, and B. S. HARRIS: Hydrocarbon Fuell Cell Technology, p. 187. B. S. Baker, ed., New York: Academic Press 1965.
6. HARDY, R. W., W. E. CHASE, and J. MCCALLUM: Hydrocarbon Fuell Cell Technology, p. 213. B. S. Baker, ed., New York: Academic Press 1965.
7. BROERS, G. H. J., and M. SCHENKE: Hydrocarbon Fuel Cell Technology, p. 225. B. S. Baker, ed., New York: Academic Press 1965.
8. TRACHTENBERG, I.: Hydrocarbon Fuel Cell Technology, p. 251. B. S. Baker, ed., New York: Academic Press 1965.
9. MILLET, J., and R. BUVET: Hydrocarbon Fuel Cell Technology, p. 285. B. S. Baker, ed., New York: Academic Press 1965.
10. BAKER, B. S., L. G. MARIANOWSKI, J. ZIMMER, and G. PRICE: Hydrocarbon Fuel Cell Technology, p. 293. B. S. Baker, ed., New York: Academic Press 1965.
11. UBBELOHDE, A. R.: Proc. Chem. Soc., London, 332 (1960).
12. JANZ, G. J., and M. R. LORENZ: J. Electrochem. Soc. **108,** 1052 (1961).
13. —, and F. SAEGUSA: J. Electrochem. Soc. **110,** 452 (1963).
14. WARD, A. T., and G. J. JANZ: Electrochim. Acta **10,** 849 (1965).
15. JANZ, G. J., and D. W. JAMES: J. Chem. Phys. **35,** 739 (1961).
16. DJORDJEVIC, S., and G. J. HILLS: Trans. Faraday Soc. **56,** 269 (1960).
17. DELIMARSKII, YU. K., and B. F. MARKÓV: Electrochemistry of Fused Salts, Eng. Transl. Washington, D. C.: Sigma Press 1961.
18. LITTLEWOOD, R.: J. Electrochem. Soc. **109,** 525 (1962).
19. POURBAIX, M.: Atlas of Electrochemical Equilibria in Aqueous Solutions. Oxford: Pergamon Press 1966.
20. INGRAM, M. D., and G. J. JANZ: Electrochim. Acta **10,** 783 (1965).
21. FLOOD, H., T. FÖRLAND, and K. MOTZFELT: Acta Chem. Scand. **6,** 257 (1952).
22. STEPANOV, G. K., and M. A. TRUNOV: Doklady Akad. Nauk S. S. S. R. **142,** 886 (1962).
23. BROERS, G. H. J.: Dissertation, Univ. of Amsterdam, 1958.
24. DANIEL-BEK, V. S., M. Z. MINTZ, V. V. SYSOEVA, and M. V. TIKHONOVA: Zhur. Priklad. Khim. **32,** 649 (1959).
25. DOUGLAS, D. L.: Ind. Eng. Chem. **52,** 308 (1960).
26. CAIRNS, E. J., A. D. TEVEBAUGH, and G. J. HOLM: J. Electrochem. Soc. **110,** 1025 (1963).
27. DELIMARSKII, YU. K., A. V. GORODYRSKII, and V. F. GRICHSHENKO: Doklady Akad. Nauk S. S. S. R. **156,** 650 (1964).
28. INGRAM, M. D., B. BARON, and G. J. JANZ: Electrochim. Acta **11,** 1629 (1966).
29. WEBB, A. N., W. B. MATHER, JR., and R. M. SUGGITT: J. Electrochem. Soc. **112,** 1059 (1965).
30. Private Communication, Prof. G. J. JANZ: Renselaer Polytechnic Institute, Troy, N. Y.
31. ARKHIPOV, G. G., and G. K. STEPANOV: Electrochemistry of Molten and Solid Electrolytes, Vol. **3**, p. 73. A. N. Baraboshkin, ed., N. Y.: Consultants Bureau 1964.
32. BREITER, M., and K. HOFFMANN: Z. Elektrochem. **64,** 462 (1960).
33. SANDLER, Y. L.: J. Electrochem. Soc. **109,** 1115 (1962).
34. KRONENBERG, M. L.: J. Electrochem. Soc. **109,** 753 (1962).
35. TRACHTENBERG, I.: J. Electrochem. Soc. **111,** 110 (1964).

36. Stepanov, G. K., and A. M. Trunov: Reports of the Institute of Electrochemistry, Ural'sk. Fil. Akad. Nauk S.S.S.R. **3,** 67 (1962).
37. Silakov, A. V., G. S. Tyurikov, and N. V. Vasilistov: Elektrokhim. **1,** 613 (1965).
38. — — — Elektrokhim. **2,** 205 (1966).
39. Trunov, A. M.: Elektrokhim. **1,** 1482 (1965).

XIV. Processes in Fuel Cells with Solid Electrolytes

The desirability of constructing a fuel cell with a solid electrolyte that is stable during the time of operation was already recognized by HABER [1], NERNST [2], and BAUR [3]. Since the problem of electrocatalysis is not serious [4,5] at temperatures around 1000 °C, the main emphasis has been on the improvement of solid electrolytes and on the development of suitable cell designs in recent years.

1. General Considerations

The hydrogen-oxygen fuel cell with solid electrolyte may be represented schematically by Fig. 1 again. Porous electrodes cover the two sides of a thin flat disc of solid electrolyte. A thin-walled tube is formed [3] by solid electrolyte in another arrangement. The electrodes are attached to parts of the external and internal wall respectively. The electrolyte has to be chemically and physically stable at the high temperatures and impermeable to the fuel gases. In order to reduce ohmic losses, the piece of solid electrolyte has to be thin. The thinness is limited by the requirement of mechanical strength. Contact resistance between electrode and electrolyte should be small. This may be achieved by flamespraying of metals and oxides as electrocatalysts or by metal evaporation onto the electrolyte in vacuo. The basic requirement in the choice of the electrocatalyst is resistance to gaseous corrosion in the presence of the fuel at elevated temperatures. Corrosion limits the utility of Ni and Co as cathodes, but not as anodes [6—8]. The high vapor pressure of Ag and Pd is the principal limitation to their use. Platinum is also subject to a small evaporation loss. The ceramic compound $PrCoO_3$ is a promising electrocatalyst material for the cathode [7]. Its electronic conductivity is sufficient for it to be employed without embedding metallic current collectors.

Oxide ions are formed at the interface cathode/electrolyte from oxygen molecules according to the net reaction:

$$O_2 + 4e^- = 2O^{--}. \tag{1}$$

The oxide ions move to the anode and react with the fuel, for instance molecular hydrogen:

$$H_2 + O^{--} = H_2O + 2e^-. \tag{2}$$

Ideally the charge transport through the electrolyte should involve oxide ions only. Let us designate the current due to O^{--} by I_{ion}. If a sufficient number of electrons is present, a parasitic electronic current I_{El} results. The electronic current flows in the opposite direction to the ionic current. The subscripts "ion" and "El" are also used to distinguish between the values of the resistance (R_j), the conductivity (σ_j) and the transference number (t_j) for ionic and electronic conductance. It is:

$$-U = E_{rev} - I_{ion} R_{ion}, \tag{3}$$

$$I = I_{ion} - I_{El}, \tag{4}$$

$$\sigma = \sigma_{ion} + \sigma_{El}. \tag{5}$$

It follows from Eq. 3 to 5:

$$-U = \frac{\sigma_{ion}}{\sigma_{ion} + \sigma_{El}} E_{rev} - \frac{1}{\sigma_{ion} + \sigma_{El}} I = E_{rev} t_{ion} - \frac{I}{\sigma}. \tag{6}$$

Here E_{rev} is the thermodynamic value of the electromotive force. The voltage between the terminals 1 and 2 of the fuel cell is a linear function of the total current I, even in the case when the electrolyte has some electronic conductance. The slope of the current-voltage characteristic gives the total resistance ($R_{ion} + R_{El}$) of the electrolyte. The absence of any polarization due to electrochemical reactions at the interfaces is assumed with Eq. 6. Measurements of the cell voltage at open circuit allow to obtain the transference numbers t_{ion} and t_{El} for solid electrolytes. WEISSBART and RUKA [4] determined the transference number t_{El} for $Zr_{0.85}Ca_{0.15}O_{1.85}$ from such measurements and reported the small values of 0.002 and 0.006 for $p_{O_2} = 0.2$ and 10^{-17} atm respectively at 1000 °C. Electronic conductance is negligible for $Zr_{0.85}Ca_{0.15}O_{1.85}$.

The efficiency of fuel cells with solid electrolytes was discussed in detail by TAKAHASHI, ITO, and IWAHARA [9] under the assumption of negligible losses due to polarization. Defining the efficiency as the ratio between the power output W of the fuel cell and the sum W_s of all the partial energies generated, the following expression was derived [9]:

$$\frac{W}{W_s} = \frac{\sigma E_{rev} t_{ion} I - I^2}{\sigma E_{rev}(t_{ion}(1 - t_{ion}) \sigma E_{rev} + t_{ion} I)}. \tag{7}$$

The above definition of the efficiency is different from the one in section 2 of chapter II. If $t_{ion} < 1$, the efficiency W/W_s is zero at $I = 0$ and $I = \sigma E_{rev} t_{ion}$. It has a maximum value

$$\left(\frac{W}{W_s}\right)_{max} = \frac{1 - \sqrt{1 - t_{ion}}}{1 + \sqrt{1 - t_{ion}}}, \tag{8}$$

at an intermediate current I_m:

$$I_m = \sigma E_{rev}\left(\sqrt{1-t_{ion}} - (1-t_{ion})\right). \tag{9}$$

Eq. 8 demonstrates that the maximum efficiency depends only upon t_{ion} and tends towards 1 with increasing t_{ion}. The relation [6] between power output and efficiency

$$W = \frac{1}{2}\sigma E_{rev}^2 \frac{W}{W_s} t_{ion}\left(\sqrt{\left(1-\frac{W}{W_s}\right)^2 t_{ion}^2 - 4\frac{W}{W_s}(1-t_{ion})} + \left(1-\frac{W}{W_s}t_{ion}\right) + (1-t_{ion})\right) \tag{10}$$

approaches asymptotically the quadratic formula:

$$W_{t_{ion}\to 1} = \sigma E_{rev}^2 \frac{W}{W_s}\left(1-\frac{W}{W_s}\right). \tag{11}$$

In general the power output and the efficiency increase at first with increasing current until the maximum efficiency is reached. With further increase in current, the efficiency decreases while the power output tends towards its maximum value. The power output at maximum efficiency does not coincide with the maximum value of the power output. Since the slope of the W versus W/W_s curves has large absolute values [9] in the vicinity of the maximum efficiency, a higher power output can be achieved at a slight expense of efficiency. Both the power output and the efficiency decrease with further increase of current afterwards. This results from the large ohmic losses.

The electromotive force of the hydrogen-oxygen cell with solid electrolyte is given by Eq. II, 10. If the dissociation equilibrium 12

$$H_2 + \tfrac{1}{2}O_2 = H_2O \tag{12}$$

is assumed to be established at 1000 °C, Eq. II, 10 can be rewritten in terms of the oxygen pressure p_{c,O_2} in the cathode compartment and the oxygen pressure p_{a,O_2} in the anode compartment:

$$E_{rev} = \frac{RT}{4F}\ln\frac{p_{c,O_2}}{p_{a,O_2}}. \tag{13}$$

The solid electrolyte fuel cell displays the behavior of an oxygen concentration cell under this assumption. WEISSBART and RUKA [4] demonstrated agreement within ± 5 mV between cell voltages computed according to Eq. II, 10 and measured values over a wide range of ratios $p_{H_2}/(p_{H_2}+p_{O_2})$ for the hydrogen-oxygen cell with $Zr_{0.85}Ca_{0.15}O_{1.85}$ as electrolyte and platinum electrodes between 900° and 1000 °C. The oxygen pressure was 1 atm.

2. Properties of Solid Electrolytes

NERNST [10] proposed before 1900 the replacement of the metallic filaments that are poor light sources in lamps by ionic conductors. The important discovery in the development of the Nernst glower was that the low conductivity of pure solid oxides could be improved by adding other oxides. Mixtures containing zirconia as the main constituent and a rare earth, for instance 85% ZrO_2 + 15% Y_2O_3, were the most promising ones. WAGNER [11] established that vacant oxygen lattice sites are responsible for the mobility of the oxide ions in solid electrolytes with an imperfect fluorite lattice of the type $(ZrO_2)_{1-x}(CaO)_x$ or $(ZrO_2)_{1-x}(Y_2O_3)_x$. The replacement of zirconium ions by ions of lower vacancy results in a proportionate number of oxide ion lattice sites beeing vacant. After HUND confirmed [12] WAGNER's conclusions, the correlation between high ionic conductivity and mixed oxide crystals with vacancies became certain. The conductivity and structure of calcia-stabilized zirconia were thoroughly investigated by CARTER and ROTH [13] recently. TANNENBERGER, SCHACHNER, and KOVACS [14] studied the electrolytic properties of ZrO_2—Nd_2O_3, ZrO_2—Gd_2O_3, and ZrO_2—Yb_2O_3 systems, having 6 to 20% rare-earth content. MÖBIUS reviewed [15] solid electrolytes with oxide ion conductivity.

RUFF and EBERT [16] found that zirconia crystallizes at room temperature in a monoclinic form. This form is transformed at 1150 °C into a tetragonal one. The transformation is accompanied by a volume contraction of about 9%. The reverse transformation occurs during cooling at 900 °C. Objects made of zirconia are prone to cracking because of the density changes. Since both modifications differ only slightly from a cubic form, a cubic structure can be obtained by addition of other oxides which have a cubic structure and form solid solutions with zirconia. Effective additives include yttria, calcia, and magnesia. Thus mixed oxides that are stable in the temperature range of interest and possess satisfactory ionic conductivity are available.

Measurements of the conductivity of zirconia-calcia solutions [5, 12, 13, 17—20] or zirconia-yttria solutions [20, 21] when plotted [22] as a function of the concentration of CaO or Y_2O_3 at 1000 °C show a conductivity maximum between 10 and 15% CaO or between 5 and 10% Y_2O_3 respectively. The maximum conductivity occurs close to the concentration of CaO or Y_2O_3 which is the lower limit for the cubic phase stabilization of zirconia. It is likely that the conductivity maximum is correlated to the number and distribution of vacancies. Plots of the logarithm of the resistivity versus $1/T$ at constant composition are linear for CaO contents between 12 and 15% and for Y_2O_3 between 9 and 30% in the range between 800 °C and 1400 °C. Resistivities of about 25 Ωcm and 8.5 Ωcm are achievable [6] at 1000 °C for $ZrO_2 + 11-14$ mol % CaO and $ZrO_2 + 7-10$ mol % Y_2O_3. The temperature-dependence of the resistivity is typical for ionic conductance. The negligible contribution of electronic conductance in $Zr_{0.85}Ca_{0.15}O_{1.85}$ was demonstrated [18] by the independence of the conductivity upon the oxygen pressure between 1 and 10^{-10} atm at temperatures between 1400° and 1600 °C. As discussed by KIUKKOLA and WAGNER [23], the electronic conductivity is pressure-dependent because of the equilibria:

(14) $$O^{--} = \square + \tfrac{1}{2}O_2 + 2e^-,$$

(15) $$\tfrac{1}{2}O_2 + \square = O^{--} + 2p^+.$$

Here □ designates an oxygen ion vacancy, and p^+ is a hole. Since the concentration of vacancies and oxide ions is fixed by the composition of the oxide, the concentration of electrons and holes depends upon the oxygen pressure. CARTER and RHODES [24] found good agreement between the oxygen diffusion coefficients determined experimentally and those calculated from the electrical conductivity with an oxygen transference number of unity. The cation diffusion is not strongly influenced [25, 26] by the concentration of oxygen vacancies in $ZrO_2 + 12-16$ mol % CaO.

Both types of zirconia undergo a slow increase of the resistivity with time at elevated temperatures. The resistivity increase is associated [6] with a complex rearrangement of the anions in the lattice. The rate of this aging process is parabolic with time and depends on temperature and grain size. It can be reduced to an acceptable level with sufficiently fine grain size. Moreover, any previous aging can be annealed out by momentary heating to temperatures above 1250 °C.

3. Current-Voltage Characteristics

The voltage U between the terminals 1 and 2 of a hydrogen-oxygen fuel cell with $Zr_{0.85}Ca_{0.15}O_{1.85}$ as solid electrolyte is plotted as a function of the current in Fig. 88. The oxygen pressure was about 0.73 atm and $p_{H_2}/(p_{H_2}+p_{H_2O}) = 0.973$. The measurements were made [4] at different temperatures. Two porous platinum films served as electrodes. The current-voltage curves in Fig. 88 are linear in the range studied. Points measured with decreasing or increasing current fall on the same line. The slope of the curves decreases with temperature. As would be expected in the absence of significant polarization on the basis of Eq. 6, the resistance computed from the slope of each of the curves agreed within 5% with the resistance of the electrolyte determined with ac current of 1000 Hz at the given temperature. The lines intercept the U axis at $-E_{rev} t_{ion}$. The intercept depends [4] upon $p_{H_2}/(p_{H_2}+p_{H_2O})$ at constant oxygen pressure in the way predicted by thermodynamics.

Similar current-voltage curves were reported [27] for the hydrogen-oxygen fuel cell with different solid solutions of CeO_3—La_2O_3 as electrolyte. However, the solid electrolytes possessed some electronic conductance. Although the addition of ThO_2 reduced the electronic conductance, it led to an undesirable decrease in performance.

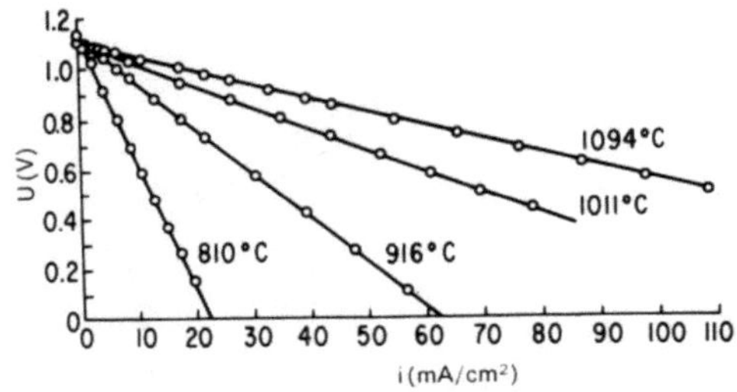

Fig. 88. Cell voltage of a fuel cell with $Zr_{0.85}\,Ca_{0.15}\,O_{1.85}$ as solid electrolyte versus current density at different temperatures with hydrogen and oxygen as fuels

It was pointed [5] out by BINDER, KÖHLING, KRUPP, RICHTER, and SANDSTEDE that a defined potential is not established at the anode for hydrocarbons since these fuels are partially or completely dehydrogenated at elevated temperatures. The ratio $p_{H_2}/(p_{H_2}+p_{H_2O})$ is determined by the hydrogen generated by cracking and by the water vapor which is accidentally present. The ratio $p_{H_2}/(p_{H_2}+p_{H_2O})$ becomes defined if the hydrocarbon is converted with water or carbon dioxide in a reformer before the effluent gas reaches the anode. The theoretical open-circuit voltage of a hydrocabon-oxygen fuel cell was computed [5] as a function of the number of carbon atoms in the hydrocarbon molecule and the composition of the gas mixture prior to conversion. The maximum difference between experimental values, measured as a function of temperature and gas composition, and the theoretical values was [5] one percent.

Current-voltage curves (a, b) for the oxidation of converted propane are shown in Fig. 89 together with a curve (c) for the oxidation of hydrogen. The curves were taken [5] at 1000 °C in a cell with $Zr_{0.85}\,Ca_{0.15}\,O_{1.85}$ as solid electrolyte and porous platinum electrodes. The oxygen pressure was 1 atm. The curves for the oxidation of converted propane are linear and parallel to the curve for the oxidation of hydrogen in a wide potential range. The ohmic potential drop in the solid electrolyte is responsible for the decrease of U with i. The curves in Fig. 89 display a small curvature at small currents in contrast to the curves in Fig. 88. This initial behavior of the curves in Fig. 89 was considered [5] as an indication for the presence of a concentration gradient of water vapor.

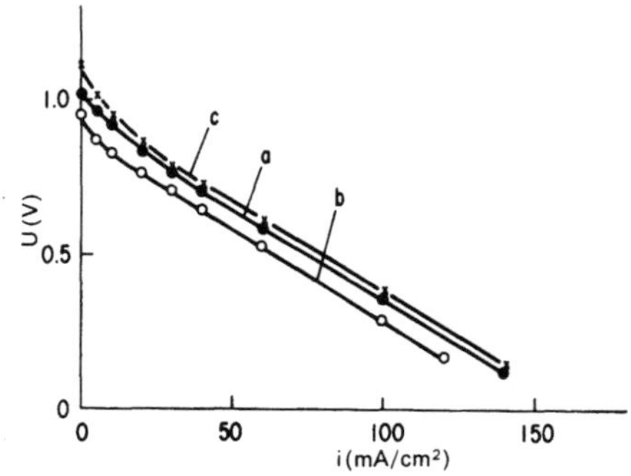

Fig. 89. Current-voltage characteristic of a fuel cell with $Zr_{0.85}\,Ca_{0.15}\,O_{1.85}$ as solid electrolyte at 1000 °C for different fuel mixtures at the anode and oxygen at the cathode. a: $C_3H_8:H_2O:CO_2=1:3:1$, b: $C_3H_8:H_2O:CO_2=1:5:4$, c: $H_2:H_2O=35:1$

The properties of the interface at which the formation of oxide ions occurs have been of special interest [6, 7, 28—35]. While solid electrocatalysts, Pt [28, 29, 31, 32] and C [30], were studied mainly, a molten silver cathode was employed in another type of zirconia-electrolyte fuel cell developed [34, 35] at the General Electric Research and Development Center in Schenectady. Since the hindrance of the electrochemical steps of the O_2 reduction at the cathode surface is small [28, 32] on platinum around 1000 °C, it is hard to elucidate the reaction mechanism beyond the net reaction 1. Analysis [33] of the potential distribution curves inside $Zr_{0.9}Y_{0.2}O_{2.1}$ in contact with two platinum electrodes showed at 1380 °C that the electronic hole contribution to the conductivity in the bulk of the specimen depended upon $p_{O_2}^{1/4}$, as would be expected from the equilibrium of reaction 15. The partial oxygen pressure had values between 10^{-4} and 10^{-2} atm. However, if the production of oxide ions is assumed to occur at the cathode solely by reaction 15, the rate of production is much lower than the rate of loss at the anode. A cathodic reaction of the type

$$\tfrac{1}{2}O_2+2e^-=O^{--} \qquad (16)$$

predicts [33] a cathode space charge variation with p_{O_2} of the form observed experimentally.

References

1. HABER, F., and A. MOSER: Z. Elektrochem. **11,** 593 (1905).
2. NERNST, W.: German Patents 259. 241 (1912).
3. BAUR, E., and H. EHRENBERG: Z. Elektrochem. **18,** 1002 (1912).
4. WEISSBART, J., and R. RUKA: J. Electrochem. Soc. **109,** 723 (1962).
5. BINDER, H., A. KÖHLING, H. KRUPP, K. RICHTER, and G. SANDSTEDE: Electrochim. Acta **8,** 781 (1963).
6. WHITE, D. W.: Revue Energie Primaire EPE, Vol. **III,** p. 10. Brussels 1965.
7. — 134th Meeting of the Electrochemical Society 1968, Extended Abstract No. **354.**
8. SCHACHNER, H., and H. TANNENBERGER: Revue Energie Primaire EPE, Vol. **III,** p. 49. Brussels 1965.
9. TAKAHASHI, T., K. ITO, and H. IWAHARA: Electrochim. Acta **12,** 21 (1967).
10. NERNST, W.: Z. Elektrochem. **6,** 41 (1899).
11. WAGNER, C.: Naturwiss. **31,** 265 (1943).
12. HUND, F.: Z. Elektrochem. **55,** 363 (1951).
13. CARTER, R. E., and W. L. ROTH: Proceedings of a Symposium held by the Nuffield Research Group Imperial College London, 1967, p. 125. Published by the Institution of Mining and Metallurgy, London.
14. TANNENBERGER, H., H. SCHACHNER, and P. KOVACS: Revue Energie Primaire EPE, Vol. **III,** p. 19. Brussels 1965.
15. MÖBIUS, H.: Z. Chem. **2,** 100 (1962) and **4,** 81 (1964), Z. Silikattechnik **17,** 358 (1966).
16. RUFF, O., and F. EBERT: Z. anorg. Chem. **180,** 19 (1929).
17. TROMBE, F., and M. FOËX: Compt. rend. **236,** 1783 (1953).
18. KINGERY, W. D., J. PAPPIS, M. E. DOTTY, and D. C. HILL: J. Am. Ceram. Soc. **42,** 393 (1959).
19. VOLCHENKOVA, Z. S., and S. F. PALGUEV: Electrochemistry of Molten and Solid Electrolytes, Vol. **1,** p. 97. A. N. Baraboshkin, ed., New York: Consultants Bureau 1961.
20. DIXON, J. M., L. D. LAGRANGE, U. MERTEN, C. F. MILLER, and J. T. PORTER: J. Electrochem. Soc. **110,** 276 (1963).
21. NODDACK, W., H. WALCH, and W. DOBNER: Z. Physik. Chem. **211,** 180 (1959).
22. SMITH, J. G.: An Introduction to Fuel Cells, Chap. 8. K. R. Williams, ed., Amsterdam: Elsevier Publishing Company 1966.
23. KIUKKOLA, K., and C. WAGNER: J. Electrochem. Soc. **104,** 379 (1957).
24. CARTER, R. E., and W. H. RHODES: J. Am. Ceram. Soc. **49,** 244 (1966).
25. MÖBIUS, H. H., H. WITZMANN, and D. GERLACH: Z. Chem. **4,** 154 (1964).
26. SIMPSON, L. A., and R. E. CARTER: J. Am. Ceram. Soc. **49,** 139 (1966).
27. TAKAHASHI, T.: J. Electrochem. Soc. Japan **34,** 60 (1966).
28. NEUIMIN, A. D., S. V. KARPACHEV, and S. F. PALGUEV: Doklady Akad. Nauk S. S. S. R. **141,** 402 (1961).
29. FILYAEV, A. T., S. V. KARPACHEV, and S. F. PALGUEV: Doklady Akad. Nauk S. S. S. R. **149,** 909 (1963).
30. KLEITZ, M., J. BESSON, and C. DEPORTES: Revue Energie Primaire EPE, Vol. **III,** p.35. Brussels 1965.
31. PERFIL'EV, M. V., S. F. PALGUĖV, and S. V. KARPACHEV: Elektrokhim. **1,** 90 (1965).
32. HARTUNG, R., and H. H. MÖBIUS: Z. Chem. **7,** 325 (1967).
33. CASSELTON, R. E. W., and J. C. SCOTT: Phys. Letters **25A,** 264 (1967).
34. "GE Demonstrates New Fuel Cell", Chem. and Eng. News **47** (1963).
35. TRAGERT, W. E.: U.S. Patent **3,** 138487 and **3,** 138488 (1964).

XV. Properties of Porous Electrodes

Practical units of fuel cells could not operate without porous electrode structures. Porous electrodes with their large electrochemically active surface allow reasonable currents to be supplied at acceptable losses due to polarization (see section 2 of chapter II). Although a few properties, like maximum available surface of electrocatalyst and hydrogenation and dehydrogenation of carbonaceous species for Teflon-bonded platinum black electrodes, and formation of oxygen layers for Raney nickel electrodes, have been discussed in preceding chapters, a discussion of the parameters that determine the operation of porous electrodes had to be offered in a separate chapter. While the empirical aspects concerning the operation of porous electrodes are covered in this chapter, theoretical aspects are dealt with in chapter XVI.

According to the mode of operation the porous electrodes may be classified in two groups:

a) Electrodes that are completely immersed in electrolyte. All the pores are filled with electrolyte.

b) Electrodes with one side in contact with electrolyte and the other side exposed to a gas phase (fuel or air). A meniscus represents the interface between electrolyte and gas in certain pores.

Electrodes of type a are called "flooded" or "drowned" in the terminology of fuel cell technology. Porous electrodes that are strongly wet-proof, such as Teflon-bonded platinum black electrodes, have to be vacuum-impregnated with electrolyte to assure complete flooding. Flooded electrodes are required as anodes for liquid fuels like methanol or hydrazine. Electrodes of type b are designated as gas-diffusion electrodes. These electrodes are of interest as anodes for gaseous fuels and as cathodes for the oxygen electrode in low-temperature fuel cells. Certain conditions that will become clear in the subsequent sections have to be met structurally to prevent the gas passing through the porous structure like a frit or all the pores becoming filled with electrolyte.

1. Porosity

The pore size distribution of porous electrodes is one of the fundamental parameters determining the extent to which the available surface of electrocatalyst participates in electrochemical reactions. The following techniques have been employed for the characterization of porous structures:

a) Direct observation with the optical microscope [1—4] or the electron microscope [3—5].
b) Mercury intrusion [6—8].
c) Capillary condensation [9—11].
d) Permeability measurements [12].
e) X-ray absorption [13—15].

A photomicrograph of a sectioned electrode of the Niedrach-Alford type is shown in Fig. 90. The micrograph was taken at 500 X. The electrodes were prepared by bonding mixtures of Teflon and platinum black (34 mg Pt/cm^2 and 3.1 mg Teflon/cm^2) to the central areas of a nickel screen. A part of a nickel wire of the screen is visible on the left side of the photomicrograph. A porous Teflon film (2.4 mg/cm^2) was deposited on the gas side of the diffusion electrode to make this side strongly hydrophobic. The Teflon film is recognizable at the top. It is evident from Fig. 90 that a porous structure consisting of Teflon agglomerates, Teflon fibers and platinum aggregates is present. The binding is ascribed to the Teflon fibers.

The distribution of the platinum catalyst in the body of the electrode is displayed in Fig. 91. Only platinum clusters are visible in the transmission micrograph obtained at 100000 X from microtomed cross sections. The platinum aggregates vary in size. Some of the clusters are barely connected with the others in the plane shown. If present, isolated clusters are electrochemically inactive because of the large contact resistance between them and other aggregates. The fine spaces between the platinum clusters represent largely micropores since they are smaller than 0.2 μ on the average. The wide spaces between the platinum clusters are partly filled with Teflon agglomerates [3]. It is unlikely that tight contact exists between different Teflon agglomerates and between Teflon agglomerates and the platinum aggregates which form the walls of the wide spaces. The surface of these walls is very irregular with an estimated roughness of 100 Å. The wide spaces may be considered as macropores of Teflon-bonded platinum black electrodes since the transport of gas occurs through these pores.

Micrographs of the type in Fig. 90 and Fig. 91 give a general picture of the porous structure and allow the size of the different agglomerates to be estimated. A resolution of 1 μ or about 15 Å is achievable by the optical microscope and the electrom microscope respectively. The porosity cannot be determined with certainty if plastic binders like Teflon are used. Although the determination of the total porosity is in principle possible for porous metals, the procedure of integrating the areas of cross sections of pores in micrographs is very cumbersome for complex structures.

The determination of the pore size distribution by mercury intrusion introduced by RITTER and DRAKE [16] is based on Eq. 1 which represents the condition for the existence of a stable meniscus:

$$p = \frac{2\gamma \cos \beta}{r}. \tag{1}$$

Here r denotes the radius of a cylindrical pore, γ the surface tension of mercury and β the contact angle between mercury and the material forming the pore wall. Mercury intrudes at a given pressure p into the pores with radii $r > (2\gamma \cos \beta)/p$ and fills them. The volume of mercury that penetrates into the pores with increasing pressure is determined experimentally. The pore size distribution function F(r) follows from Eq. 2:

$$\mathrm{F(r)} = \frac{dV}{d\mathrm{p}} \frac{p}{r}. \tag{2}$$

If the $F(r) - r$ curve displays N maxima, each maximum at r_j(j = 1,2,..., N) is assigned to a large number of pores with the radius r_j. It is said that the porous structure consists largely of N types of pores.

The mercury intrusion technique requires large pressures if the pore radius becomes smaller than 100 Å. Since Eq. 1 is based on a cylindrical pore, the complex structure is approximated by an idealized structure in which each of the pores has a constant diameter. In addition a uniform contact angle is assumed. The intrusion technique should not be applied to composite electrodes without special precautions for the latter reason.

While the mercury intrusion technique is suitable for the determination of an idealized distribution of pores with radii ≥ 100 Å, the capillary condensation technique is restricted to $r < 600$ Å if nitrogen gas is used. As it is frequently employed, the method is based on the assumption of cylindrical, noninteracting pores and Wheeler's derivation [17] of Eq. 3:

$$V_{\mathrm{p}} - V_{\mathrm{ad}} = \pi \int_{r_c}^{\infty} (r-d) L\, dr. \tag{3}$$

Here V_{p} is the total pore volume and V_{ad} the volume of the adsorbed layers on the pore walls. The thickness d of the adsorbed layer is a function of the relative gas pressure of the adsorbate. The total length of all

Fig. 90. Photomicrograph of a sectioned electrode of the Niedrach-Alford type

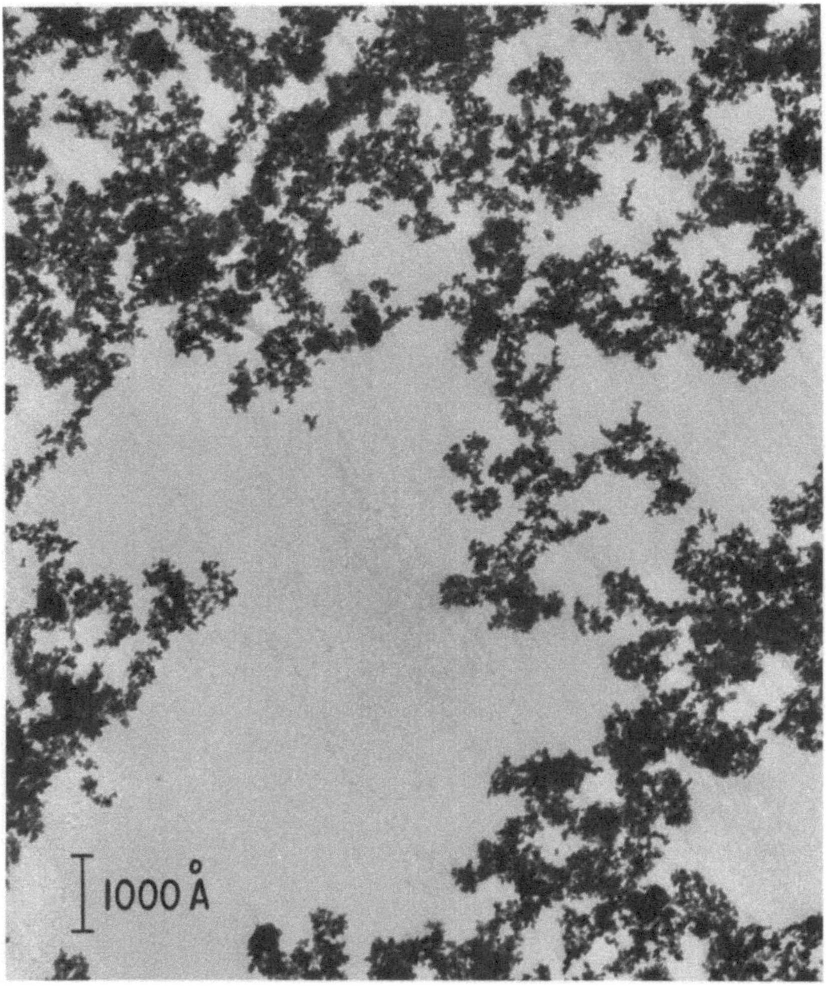

Fig. 91. Electron micrograph of a microtomed section of the interior of a Teflon-bonded platinum black electrode

pores with radius r per unit weight of adsorbent is designated by L. The critical pore radius r_c below which all the pores are completely filled by adsorbed or condensed gas is a function of the relative pressure. Extensive numerical calculations are required to obtain the pore size distribution from adsorption-desorption isotherms of nitrogen. WEISSMAN [11] always observed a maximum in the pore size distribution curve of Teflon-bonded platinum black electrodes at 20 to 50 Å. Since this peak was also present in platinum black, it was considered representative of the intra-particle geometry of the electrocatalyst. The peaks at larger radii had different locations for Teflon-bonded platinum black than for platinum black. They reflect the idealized pore size distribution of the porous electrodes.

The basic assumption of constant radius of every single pore is the same for the capillary condensation technique as for the mercury intrusion technique. Thus the satisfactory agreement of the pore size distribution of carbon obtained [18] by mercury intrusion and capillary condensation in an overlapping region of pore radii between 50 and 300 Å implies only that the same idealized structure is derived. A method of capillary condensation which is not based on the assumption of the shape of the pores was suggested by BRUNAUER, MIKHAIL and BODOR [19] recently. The application of this method to porous electrodes has not been reported so far.

Techniques d and e do not allow a determination of the pore size distribution. Mean hydraulic pore diameters can be estimated from permeability measurements using the flow of an inert gas through the porous structure. The total porosity P_t given by Eq. 4

$$\mathrm{P}_t = 1 - \frac{V_t}{V_{ap}} \tag{4}$$

is obtainable from the absorption of the K_α radiation according to Eq. 5:

$$\mathrm{P}_t = 1 - \frac{\ln \dfrac{J_0}{J}}{\lambda x_{ap}}. \tag{5}$$

Here V_{ap} is the apparent volume of the porous structure and V_t the true volume (volume of the same amount of material without pores). The intensity of the X-ray beam without the sample is designated by J_0, the intensity with porous sample by J. The linear absorption coefficient is denoted by λ, and the measured thickness of the porous sample by x_{ap}. Good agreement was reported by APPELT and NOWACKI [15] for P_t values determined for the same sample according to Eq. 4 and 5.

The porosity that results from open-ended pores, accessible to a liquid phase, is available according to Eq. 6

$$(6) \qquad P_0 = \frac{W_2 - W_1}{\rho V_{ap}}$$

from the difference $(W_2 - W_1)$ in weight of the porous structure before and after impregnation with a liquid of density ρ. Ranges [11] of the ratio of open to total porosity are given in Table 9 for different electrodes of the Niedrach-Alford type before and after use.

Table 9. *Comparisation of the ratio* P_0/P_t *for different electrodes*

Electrode No.	1	2	3	4	5
new	0.39–0.55	0.57–0.58	0.44–0.47	0.49–0.58	0.49–0.57
used	0.78–0.89	0.86–0.87	0.78–0.87	0.77–0.87	0.79–0.86

The data in Table 9 indicate that about half the porosity of the above electrodes is due to open-ended pores before use. The ratio P_0/P_t becomes significantly larger for electrodes that have been put on load even for only a few hours. Since the cumulative pore volume obtained from pore size distribution functions at $r < 600$ Å decreased simultaneously, the added porosity is mainly due to open macropores. It is likely that the pore size distribution of porous nickel electrodes [20] or carbon electrodes [8, 10] changes also with the time of operation. However, statements on the time of the pore size distribution are lacking in most publications in which a correlation between porous structure and performance is attempted.

Another parameter of importance for the characterization of porous structures is the tortuosity factor K_t. If the expression for the resistance to a flow process such as diffusion or conduction has a factor $L/\pi r^2$ for straight-through pores of length L and radius r, it will have a factor $K_t L/\pi r^2$ for a complex porous structure. The tortuosity factor K_t is greater than 1 for the following reasons:

a) All the pores do not go straight through.
b) There are blind alleys and pockets.
c) There may be molecular restrictions for the moving particles.
d) Alteration of solvent properties occurs in very narrow pores.

The tortuosity factor of a cubical lattice of pores is 3 if they represent only a small fraction of the volume. As the porosity increases, K_t approaches 1. The tortuosity factor can be determined experimentally by fluid flow of solvents with sufficiently small molecules, by diffusion with radioactive species under conditions of zero flow, and by conductance measurements.

2. Determination of Different Surface Areas

The electrochemical technique for determining the ${}_{S}Q_{H}$ values as a measure for the maximum available surface area of platinum in porous electrode structures was already discussed in section 5 of chapter VI. It is evident from the introductory remarks to this chapter, that the electrodes have to be in the flooded state for this type of measurement. Determinations of the maximum available surface of the electrocatalyst from the adsorbed amount of hydrogen or carbon monoxide in the gas phase yield values that are too large if many isolated clusters of electrocatalyst are present.

The quantity A_B found as specific surface area by the BET technique [21] in m^2/g is a measure of the maximum available surface of porous metal structures if metal and electrocatalyst are the same. The situation becomes more complex when the porous structure consists of a binder and an electrocatalyst. It is considered likely that the parameter A_B is proportional to the maximum available surface under the following conditions:

a) The binder is conducting like carbon.

b) The ratio of electrocatalyst surface to binder surface is small. This avoids much of the electrocatalyst not being in contact with the electrolyte because of crowding of catalyst crystallites.

Conditions a and b are not met by Teflon-bonded platinum black electrodes, so that a correlation between A_B and the performance appears questionable. SCHWABE [10] suggested that the number of active sites for the O_2 reduction on porous carbon electrodes is independent of A_B, as in the case of numerous catalytic reactions.

Mean specific areas A_S in m^2/g may be computed as

$$A_S(r) = \frac{2\,V_p(r)}{r} \tag{7}$$

from the cumulative pore volume $V_p(r)$ between r_1 and r_2. Here r designates a mean value for the given interval. Each of the pores is assumed to have a constant radius and the same length as the other pores with radii between r_1 and r_2. It is evident that A_S corresponds to an idealized structure of the porous electrode.

The fraction of the maximum available surface of electrocatalyst that participates in electrochemical reactions under conditions of gas diffusion will depend upon numerous parameters like conductivity of electrolyte, temperature, gas pressure, pore size distribution, hydrophobicity, rate-determining step of the electrochemical reaction, electrode potential, etc. An order of magnitude may be obtained [12] for this fraction by the ratio of the values of the double layer capacity under diffusion conditions and flooded conditions.

3. Experimental Current-Potential Curves for Porous Electrodes

A comprehensive study of the H_2 oxidation on four different gas-diffusion electrodes with platinum as electrocatalyst in 0.5 M H_2SO_4 at temperatures between 0 and 60 °C was published by AUSTIN and ALMAULA [12] recently. Gas mixtures corresponding to partial hydrogen pressures of 1.0, 0.5, and 0.1 atm respectively passed the gas side of the electrode at flow rates at least 30 times the maximum rate of the H_2 oxidation. The H_2 oxidation is an example of an electrode reaction with large i_0 values of the heterogeneous steps on platinum (compare chapter VII). If the quantity U is plotted versus the ratio $i/i_{l,H_2}$ after correction for the IR_{el} drop, the curves for the three partial pressures coincide for each of the electrodes in Fig. 92. Here i_{l,H_2} is the limiting H_2 current density. It was established [12] that the maximum rate of gas transport from the gas side of the porous electrode is much larger than the rate corresponding to i_{l,H_2}. The results in Fig. 92 imply that the current density at a given potential is proportional to the partial hydrogen pressure. If the $i-U$ curves of the H_2 oxidation on smooth platinum in Fig. 23 were replotted as $i/i_{l,H_2}-U$ curves, the limiting current density

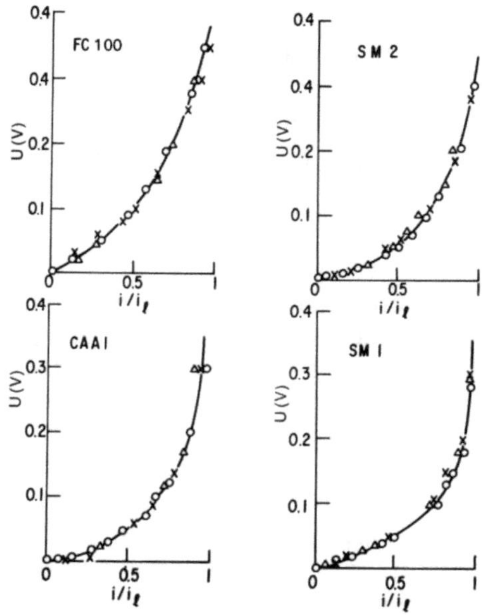

Fig. 92. Potential as a function of i/i_l for the H_2 oxidation on different gas-diffusion electrodes in 0.5 M H_2SO_4 at 25 °C.
○: $p_{H_2}=1$ atm, ×: $p_{H_2}=0.5$ atm, △: $p_{H_2}=0.1$ atm

would practically be reached at 0.1 V. The approaching of the limiting current density with increasing potential occurs in a wider potential range on gas-diffusion electrodes than on completely immersed smooth electrodes of the same electrocatalyst. This is a general property of gas diffusion electrodes. If the value of i/i_l at $U=0.1$ V is taken from the curves in Fig. 92, it increases rapidly from electrode FC 100 to SM2 and more gradually in the order SM2, CAAl, SM1. A similar order was found [12] for the increase of the total porosity and the mean hydraulic diameter. The $i/i_{l,H_2}-U$ curves at 0, 48 °C, and 60 °C look like the curves for 25 °C in Fig. 92.

An extensive study of the effect of platinum deposit and porosity on the performance of flooded porous carbon electrodes for the methanol oxidation in alkaline media was carried out by BLOCH, PRIGENT, and SUGIER [22]. The electrodes of different porosity were discs with a diameter of 3 cm and a thickness of 3 mm. The same amount of platinum

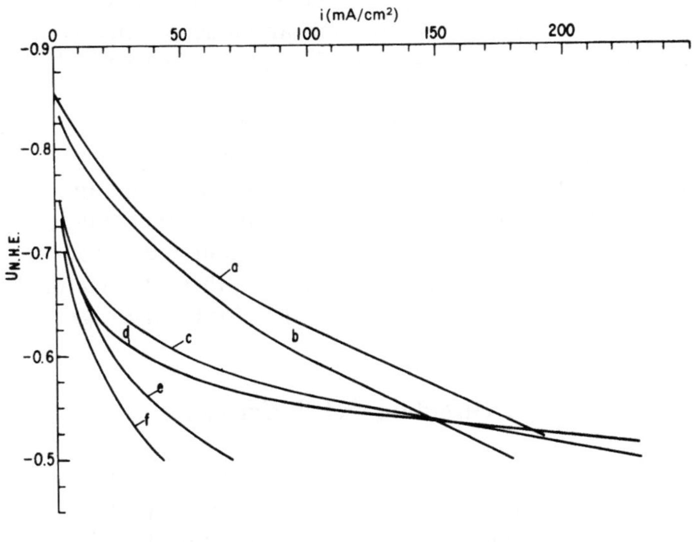

Fig. 93. Steady-state current-potential curves of the methanol oxidation in 6 M KOH + 4 M CH_3OH at 40 °C on different porous carbon electrodes impregnated with platinum (curves a, b, c, e, f) and on platinized platinum (curve d)

(15 mg/cm²) was deposited by chemical reduction of chloroplatinic acid on the electrodes a, b, c and e, f. The electrode d was prepared by depositing electrochemically 15 mg Pt/cm² on platinum foil. The steady-state $i-U$ curves measured in 6 M KOH + 4 M CH_3OH at 40 °C on these electrodes are shown in Fig. 93 in a linear plot which was constructed

from the original plot of $U_{N.H.E.}$ versus log i. The potential is given versus the normal hydrogen electrode ($a_{H^+} = 1$, $p_{H_2} = 1$ atm).

The general shape of the $i - U$ curves in Fig. 93 is similar to that [23] for porous Raney platinum electrodes in alkaline media. The potential increases at first rapidly with the current and afterwards more slowly. There is no indication in the linear plot that a maximum oxidation rate is reached in the potential range under consideration. Current-potential curves [24] of methanol oxidation on smooth and platinized platinum also consist of two parts. While certain curves (a and b, c and d, e and f) are close to each other, the three groups of curves differ greatly. The strong influence of the porosity is reflected by this result. It is not possible to make a curve of one group coincide with a curve of another group by multiplying the current densities of one curve by a constant factor. The latter procedure is equivalent to attempting to attribute the observed changes to different real surface areas.

As discussed in chapter XI, the exchange current densities of one or several heterogeneous steps in the O_2 reduction are very small. Mass transport processes begin to exert an influence only when the rate of O_2 reduction approaches that of i_{d,O_2} in concentrated solutions. Current-potential curves of the O_2 reduction in alkaline solution on porous electrodes of nickel, carbon, and silver are shown in Fig. 79. Since these curves reflect the relative reactivity of the electrocatalyst, it may be assumed that the differences in the porous structure are smaller than the differences in the exchange currents of the rate-determining steps. Both the exchange current densities and the porous structure affect the performance. A distinction between these two factors is more difficult [7, 10] for the oxygen electrode than for the hydrogen electrode.

4. Structure and Performance

Parameters which are characteristic of the porous structure of the electrodes used for the measurement of the current-potential curves in Fig. 92 are given in Table 10 with performance parameters. The electrodes FC 100 and CAAl were supplied by Pure Carbon Company and American Cyanamid Company respectively, while the electrodes SM 1 and SM 2 were prepared in the laboratory [12].

If the parameter i/(Pt content) at 0.1 V is taken as a simple measure of the ratio between performance and cost of production, the electrode CAAl possesses the best properties. The parameter i/(Pt content) is preferred as a measure rather than i_{l,H_2}/(Pt content) since the latter ratio does not take into account the different polarization characteristics of the electrodes. If the cost of fabrication does not matter, as in space

Table 10. *Characteristic parameters of electrodes* FC 100, SM 2, CAAl, *and* SM 1

electrode	material contents mg/cm^2	thick-ness μ	P_t	mean hydrau-lic dia. μ	$_sQ_H$ $mcoul/cm^2$	$\frac{\text{Pt content}}{_sQ_H}$ g/coul	i_{l,H_2} at 1 atm H_2 mA/cm^2	$U=0.1\,V$ $i/_sQ_H$ sec	$\frac{i}{\text{Pt content}}$ a/g	$\frac{i_{l,H_2}}{\text{Pt content}}$ a/g
FC 100	Pt 5.5 carbon + wetproofing 65	254	0.20	0.2	17	0.32	300	8.4	26	55
SM 2	Pt 13.5 Graphite 13 Teflon 3.5 Pt screen 35	254	0.60	1.3	535	0.025	730	0.84	33	55
CAAl	Pt 8.5 Teflon 6	150	0.73	5.6	516	0.016	1050	1.35	82	123
SM 1	Pt 55 Teflon 8 Pt screen47	300	0.72	6.4	2085	0.026	2900	1.0	38	53

applications, the electrode SM 1 with the largest value of i_{l,H_2} is the best performer.

The column with $i/_{S}Q_{H}$ values demonstrates that only a fraction of the maximum available surface participates in the H_2 oxidation. Otherwise the factor of 10 by which $i/_{S}Q_{H}$ differs for FC 100 and SM 2 should have a greater effect on the values of the ratio i/(Pt content). While the electrode CAAl has a similar porosity as the electrodes SM 2 and SM 1 and about the same mean hydraulic diameter as SM 1, its thickness is about half the thickness of the other electrodes and the ratio (Pt content)/$_{S}Q_{H}$ is the smallest. The same limiting current and the same current-potential curve were obtained [12] when several thicknesses of electrode were compressed and used as one electrode. This result rules out any effect of the thickness on performance and suggests [12] in connection with the small ratio (<0.05) between the double layer capacity of the same electrode under diffusion conditions and in the flooded state for elec-

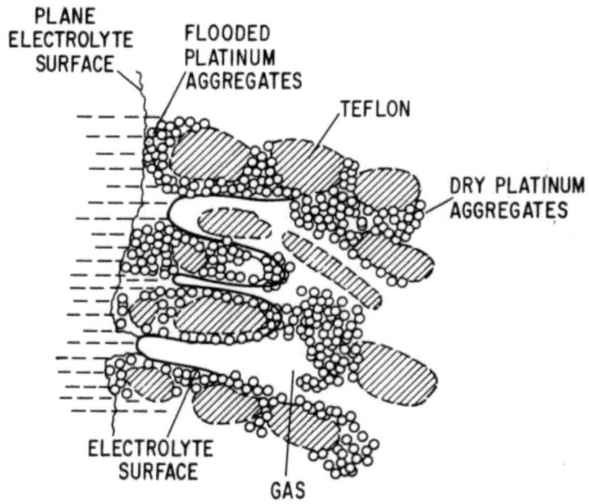

Fig. 94. Schematic drawing of the configuration of Teflon-bonded platinum black electrodes close to the external surface

trodes SM 2, CAAl, and SM 1 that a small fraction (less than 5 %) of the internal area of electrocatalyst is wetted by electrolyte in the gas diffusion condition. It appears that a good utilization of platinum ((Pt content)/$_{S}Q_{H}=0.016$) in CAAl is responsible for the high value of i/(Pt content).

If electron micrographs of the electrodes were available, conclusions on the desirable distribution and the size of the platinum clusters could be drawn.

The electrode FC 100 has a small hydraulic diameter in comparison to the other electrodes. The ratio of the double layer capacities suggests that about 25% of the internal surface of electrocatalyst is wetted. The operation of this electrode seems to differ from that of the other three electrodes in the participation of a large number of pores with relatively small radii.

The ratio $i_{1,H_2}(\vartheta)/i_{1,H_2}(\vartheta_0)$ with $\vartheta_0 = 25\,°C$ was found [12] to agree roughly up to 60 °C for each of the four electrodes with the ratio $D_{H_2}(\vartheta) \cdot {}_bc_{H_2}(\vartheta)/D_{H_2}(\vartheta_0) \cdot {}_bc_{H_2}(\vartheta_0)$. The latter ratio was obtained by interpolation from measurements [25] on smooth platinum in 0.025 M H_2SO_4 and 2.5 M H_2SO_4. The results on the temperature dependence imply that mass transport of H_2 molecules in the electrolyte plays an important role. AUSTIN and ALMAULA [12] suggested that it is the mass transport into those micropores which are filled with electrolyte and

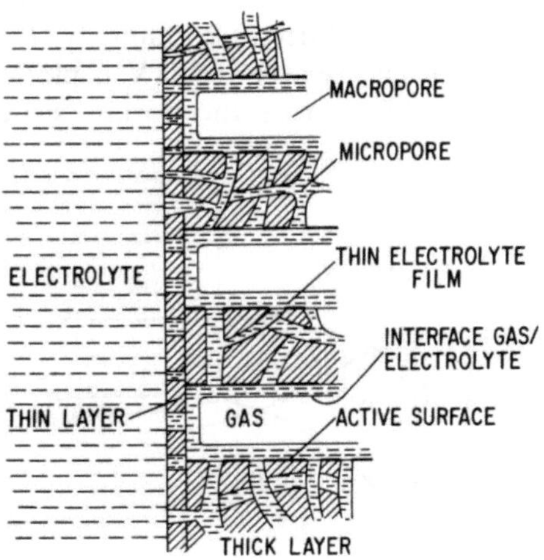

Fig. 95. Schematic drawing of a two-layer electrode

open into gas-filled macropores. A possible configuration of the gas diffusion electrode is shown in Fig. 94 for Teflon-bonded platinum black electrodes. Teflon fibers were omitted in the schematic drawing for simplicity. Electrolyte is present a short distance from the plane electrode surface. The platinum aggregates are wetted there; outside this short distance the platinum clusters are dry.

The H_2 oxidation and the O_2 reduction were studied on porous electrodes of the Bacon type [26] by BURSHTEIN and coworkers [7, 27—29]. These electrodes consist of a thin layer of nickel (about 0.3 mm thick) with pores of radii $r < 10\,\mu$ and a thick layer (about 3 mm thick) of nickel electrocatalyst like Raney nickel and carbonyl nickel with pores of radii $r > 10\,\mu$. A schematic diagram of a gas diffusion electrode of this type according to BURSHTEIN, PSHENICHNIKOV, SABIROV, and ZHURAVLEVA [29] is represented in Fig. 95. The pores of the thin film which is in contact with the electrolyte are shown as going straight through for simplicity. The same performance was achieved [7, 27] whether the thin film was made of nickel or of glass with the same porosity. The electrochemical reactions occur largely inside the pores of the thick film which is exposed with its free surface to the fuel gas. It was suggested [7, 27] that the micropores of the thick film are filled with electrolyte and serve for the transport of current. A thin layer of electrolyte is assumed to exist on the walls of the macropores which are filled with gas.

As reported by JUSTI, PILKUHN, SCHEIBE, and WINSEL [20] for the two-layer electrode, the current of the H_2 oxidation at constant potential or the potential at constant current increase initially with pressure, pass through a maximum and decrease gradually. This is demonstrated in Fig. 96. The current density at $U = 0.18$ V is plotted as a function of the pressure difference Δp across the porous structure for different

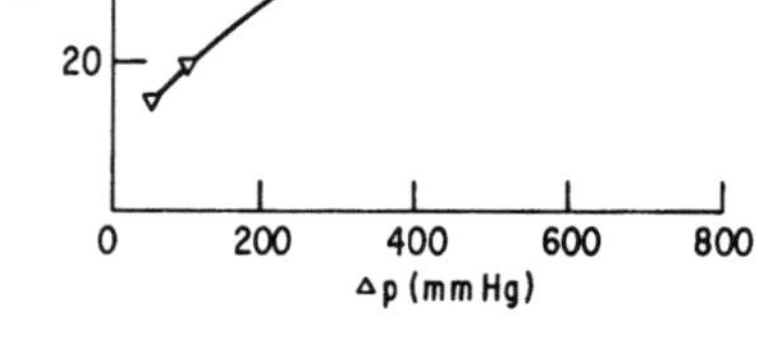

Fig. 96. Current density of the H_2 oxidation on different two-layer electrodes of nickel in 7 M KOH at 95 °C and $U = 0.18$ V as a function of Δp. $\triangle$: $r_{ma} = 6\,\mu$, $\bigcirc$: $r_{ma} = 11\,\mu$, $\square$: $r_{ma} = 14\,\mu$, ∇: $r_{ma} = 30\,\mu$

electrodes [7, 27]. The two-layer electrodes had roughly the same total porosity ($P_t \approx 0.6$) and the same porosity of the thin film ($P_{\text{thin film}} \approx 0.37$), while the size of the macropores differed (see caption to Fig. 96). The maximum of the $i - \Delta p$ curves is shifted towards the right and becomes less pronounced with increasing radius of the macropores. The above $i - \Delta p$ dependence is very different from the linear dependence found

[12] for Teflon-bonded platinum black electrodes. The large difference in the i_0 values of the rate-determining step of the H_2 oxidation (larger than $0.1\,A/cm^2$ for platinum and about $10^{-5}\,A/cm^2$ for nickel) and in the depth of penetration of electrolyte are responsible for the different extent to which macropores and micropores participate at given values of Δp and U in the two porous structures.

The maximum in $U-\Delta p$ curves at constant i was interpreted [20] by a shortening of the three-phase boundary (gas/electrolyte/electrocatalyst) at small and large gas pressures on the basis of an idealized structure consisting of parallel cylindrical pores of different radii and equal length. Such a structure is called "homoporous". The electrochemical reactions were assumed to occur at this boundary. The homoporous structure is nearly drowned at low gas pressures while gas fills too many pores at high pressures. The insufficiency of the above interpretation of the $U-\Delta p$ curves and $i-\Delta p$ curves was pointed out [7]. A linear relation

$$i = K\sqrt{\frac{2\,V_{ma}\,V_{mi}}{r_{ma}}} \tag{8}$$

was found [7, 27] to hold at constant Δp. The parameter $2\,V_{ma}/r_{ma}$ represents the total surface of the macropores which are filled with gas at a given Δp. It is obtained from the pore size distribution curve of the mercury intrusion technique. The volume of micropores which are filled with electrolyte at a given Δp is proportional to the total perimeter of micropores in the idealized structure. The best performance is achieved [27] for the Δp value at which the quantity $\frac{2\,V_{ma}\,V_{mi}}{r_{ma}}$ has a maximum.

An increase of the current density of the O_2 reduction at constant potential with the micropore volume was reported for porous carbon electrodes in alkaline solutions by Schwabe and coworkers [10]. The activation of the electrodes with carbon dioxide or water vapor increased the value of A_B and produced a certain degree of hydrophobicity.

Table 11. *Characteristic parameters of electrodes a to f*

Electrode	a	b	c	e	f
Mean pore radius in μ	9.0	5.0	2.0	1.5	0.7
Pore volume between 0.017 and 50 μ in cm^3/g	0.72	0.61	0.20	0.49	0.12
Micropore volume in cm^3/g	0.62	0.31	0.007	0.072	0.13
A_B in m^2/g	600	500	0.8	10	200
Density in g/cm^3	0.53	0.7	1.42	0.9	1.4

The physical parameters of the impregnated carbon electrodes on which the $i-U$ curves in Fig. 93 were measured [22] are given in Table 11. The performance of the porous structures in the flooded state at $U_{\mathrm{N.H.E.}} < -0.65\,\mathrm{V}$ decreases in the same sequence as the mean pore radius. This result reflects the influence of the macropores because of the magnitude of the mean radius. The shape of the curves a, b, e, and f in Fig. 93 is similar, while electrode c displays a somewhat different current-voltage characteristic resembling that of curve d for platinized platinum. Electrode c differs from the other electrodes in the smallness of the micropore volume. While Eq. 8 with V_{ma}/r_{ma} = (macropore volume)/(mean radius) and V_{mi} = micropore volume is not fulfilled, the performance decreases in the same sequence as $\frac{V_{ma} V_{mi}}{r_{ma}}$ for electrodes a, b, e, and f. These results suggest that the performance is also affected by the micropore volume.

References

1. NIEDRACH, L. W., and H. R. ALFORD: J. Electrochem. Soc. **112,** 117 (1965).
2. DROSSBACH, P., and J. SCHULZ: Electrochim. Acta **9,** 1391 (1964).
3. LASKO, W. R., and G. P. MCCARTHY: Analyt. Chem. **38,** 894 (1966).
4. TRACEY, V. A., and N. J. WILLIAMS: Electrochem. Techn. **3,** 17 (1965).
5. NIEDRACH, L. W.: private communication.
6. BURSHTEIN, R. H., A. G. PSHENICHNIKOV, and N. A. SHUMILOVA: Zhur. Fiz. Khim. **32,** 697 (1958).
7. — — — Doklady Akad. Nauk S.S.S.R. **143,** 1409 (1962).
8. PAXTON, R. R., J. F. DEMENDI, G. J. YOUNG, and R. B. ROZELLE: J. Electrochem. Soc. **110,** 932 (1963).
9. SALKIND, A. J., H. J. CANNING, and M. L. BLOCK: Electrochem. Techn. **2,** 254 (1964).
10. SCHWABE, K., W. WIESNER, E. HOLLAX, A. TSCHÖ, G. ANDERS, J. SCHREIBER, G. MEITZNER, and L. ZOLLFRANK: Electrochem. Techn. **3,** 189 (1965).
11. WEISSMAN, E. Y.: J. Electrochem. Soc. **114,** 659 (1967).
12. AUSTIN, L. G., and S. ALMAULA: J. Electrochem. Soc. **114,** 927 (1967).
13. BONNEMAY, M., G. BRONOËL, E. LEVART, and G. PESLERBE: J. Electrochem. Soc. **111,** 265 (1964).
14. — — — Electrochim. Acta **9,** 727 (1964).
15. APPELT, K., and A. NOWACKI: Electrochim. Acta **11,** 137 (1966).
16. RITTER, H. L., and L. C. DRAKE: Ind. Eng. Chem., Anal. Ed. **17,** 782 (1945).
17. WHEELER, A.: Advan. Catalysis **3,** 249 (1951).
18. JOYNER, L. G., E. P. BARRET, and R. SKOLD: J. Am. Chem. Soc. **73,** 3155 (1951).
19. BRUNAUER, S., R. SH. MIKHAIL, and E. E. BODOR: J. Colloid and Interface Sci. **24,** 451 (1967).
20. JUSTI, E., M. PILKUHN, W. SCHEIBE, and A. WINSEL: Hochbelastbare Wasserstoff-Diffusions-Elektroden für Betrieb bei Umgebungstemperatur und Niederdruck, Abh. Mainzer Akad. No. **8.** Wiesbaden: Steiner 1959.
21. BRUNAUER, S., P. H. EMMET, and E. TELLER: J. Am. Chem. Soc. **60,** 309 (1938).
22. BLOCH, O., M. PRIGENT, and A. SUGIER: Les Piles à Combustile, Chap. II. 8. Paris: Technip 1965.

23. KRUPP, H., H. RABENHORST, G. SANDSTEDE, G. WALTER, and R. MCJONES.: J. Electrochem. Soc. **109,** 553 (1962).
24. BAGOTZKY, V. S., and YU. B. VASILIEV: Electrochim. Acta **12,** 1323 (1967).
25. BREITER, M., and K. HOFFMANN: Z. Elektrochem. **64,** 462 (1960).
26. BACON, F. T.: Times Rev. Quart. **2,** 6 (1956).
27. BURSHTEIN, R. H., V. S. MARKIN, A. G. PSHENICHNIKOV, V. A. CHISMADGEV, and Y. G. CHIRKOV: Electrochim. Acta **9,** 773 (1964).
28. PSHENICHNIKOV, A. G., G. I. SHNAIDER, and R. H. BURSHTEIN: Elektrokhim. **1,** 418 (1965).
29. BURSHTEIN, R. H., A. G. PSHENICHNIKOV, F. S. SABIROV, and V. N. ZHURAVLEVA: Elektrokhim. **4,** 13 (1968).

XVI. Models of Porous Electrodes

As described in chapter XV, the structure of porous electrodes is complex. A rigorous mathematical analysis of the operation of porous electrodes in the flooded state or as gas-diffusion electrodes is not feasible for this reason. The usual approach in such a situation has been taken in the theoretical treatment of porous electrodes. The real structure and also the reaction mechanism are approximated by simpler models amenable to a mathematical analysis. The applicability of the model and of the assumptions in the mathematical derivation is judged by a comparison of the predictions of the theoretical treatment with the experimental results.

Simple models of porous structures involve straight cylindrical pores. An understanding of the processes which determine the operation of a single pore under certain conditions is important. The discussion of the properties of single pores represents the first step. It is followed by the consideration of more elaborate models of porous structures. A comprehensive review of the progress made in the theory of porous electrodes in the last three decades was given by CHISMADZHEV [1] recently. Transient responses of porous electrodes are not discussed in this chapter since steady-state conditions are chiefly of interest for the operation of fuel cells. The reader is referred to the review by DE LEVIE [2].

1. Potential Distribution in the Flooded Single Pore without Influence of Mass Transport Processes

Let us consider a single pore with uniform cross-section filled with electrolyte. The orifice of the pore is immersed in a reservoir of electrolyte. The interior walls of the pore are covered by a layer of electrocatalyst. A wire is connected to this layer at the other end of the pore. A counter electrode is located in the reservoir. When a voltage is applied between the layer and the counter electrode, an electrochemical reaction will occur on the layer. The electrochemical reaction should be chosen in such a way that concentration gradients resulting from mass transport processes are negligible inside the pore. The actual potentials inside the pore are replaced [3] by average potentials in planes perpendicular to the axis. This procedure, in which the curvature of equipotential surfaces is neglected, transforms the tree-dimensional problem into a one-dimensional one.

Following the method of DANIEL'-BEK [4], the pore may be represented by the transmission line in Fig. 97. The subscripts 1 and 2 are used in Fig. 97 to differentiate between the values of potential, current, and resistance per unit pore length (Ωcm^{-1}) in the electrolyte and in the

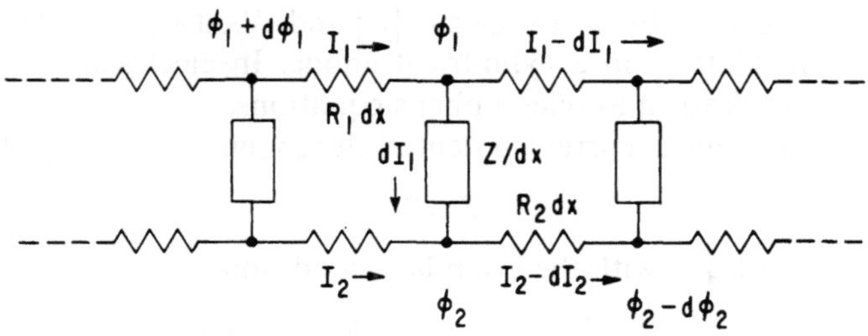

Fig. 97. Network simulating a flooded single pore

layer of electrocatalyst. The impedance of the interface electrocatalyst layer/solution per unit pore length (Ωcm) has been designated by Z and the direction of the pore axis by x, starting with x=0 at the orifice of the pore. Eq. 1 to 4 are derived from Fig. 97:

(1) $$d\phi_1 = -I_1 R_1 dx,$$

(2) $$d\phi_2 = -I_2 R_2 dx,$$

(3) $$\phi_1 - \phi_2 = -Z\frac{dI_1}{dx} = Z\frac{dI_2}{dx},$$

(4) $$I_{x=0} = I_1 + I_2.$$

The above expressions lead to a system of differential equations [2]:

(5) $$\frac{d^2(\phi_1-\phi_2)}{dx^2} - \frac{d\ln R_1}{dx}\frac{d\phi_1}{dx} + \frac{d\ln R_2}{dx}\frac{d\phi_2}{dx} - \frac{R_1+R_2}{Z}(\phi_1-\phi_2) = 0,$$

(6) $$\frac{d^2 I_1}{dx^2} + \frac{d\ln Z}{dx}\frac{dI_1}{dx} - \frac{R_1+R_2}{Z}I_1 + \frac{R_2}{Z}I_{x=0} = 0,$$

(7) $$\frac{d^2 I_2}{dx^2} + \frac{d\ln Z}{dx}\frac{dI_2}{dx} - \frac{R_1+R_2}{Z}I_2 + \frac{R_1}{Z}I_{x=0} = 0.$$

The differential equations simplify in the usual case that $R_2 \ll R_1$, $R_2 \ll Z$, $dR_1/dx = 0$, $dZ/dx = 0$; $I = I_1$:

(8)
$$\frac{d^2(\phi_2-\phi_1)}{dx^2}-\frac{R_1}{Z}(\phi_2-\phi_1)=0\,,$$

(9)
$$\frac{d^2 I}{dx^2}-\frac{R_1}{Z}I=0\,.$$

The preceding model (Eq. 8 and 9) for a one-dimensional pore is mathematically equivalent to those discussed by COLEMAN [5] and FRUMKIN [3] and to that used by WEISSELBERG [6] and WABER [7] in the context of chromium plating on a cylindrical anode. In electrical engineering Eq. 8 and 9 are known as telegrapher's equations.

Assuming a linear current-potential dependence, implying that

(10)
$$Z=R'$$

the solution of Eq. 8 with the boundary conditions

(11)
$$x=0:\ \ \phi_2-\phi_1=(\phi_2-\phi_1)_{x=0}=\text{const}\,,$$

(12)
$$x=L:\ \ \frac{d(\phi_2-\phi_1)}{dx}=0$$

is straightforward for a pore of length L and $\kappa=(R_1/R')^{\frac{1}{2}}$:

(13)
$$\phi_2-\phi_1=(\phi_2-\phi_1)_{x=0}\frac{\cosh\kappa(x-L)}{\cosh\kappa L}\,.$$

The current entering the pore at $x=0$ is:

(14)
$$I_{x=0}=-\frac{1}{R_1}\left(\frac{d(\phi_2-\phi_1)}{dx}\right)_{x=0}=(\phi_2-\phi_1)_{x=0}(R_1R')^{-\frac{1}{2}}\tanh\kappa L\,.$$

The main results of the treatment of the simple model are:

a) While the total current at a flat electrode is proportional to $1/R'$, it is proportional to $(R_1R')^{-\frac{1}{2}}$ for a pore.

b) The contribution of the surface of the electrocatalyst inside the pore decreases in an exponential way with x, the parameter $1/\kappa$ being a characteristic penetration depth.

The following expressions hold for a cylindrical pore of radius r:

(15)
$$R_1=\frac{1}{\sigma\pi r^2}\,,$$

(16)
$$R'=\frac{RT}{zF}\,\frac{1}{i_0 2\pi r}\,.$$

The penetration depth

$$\frac{1}{\kappa}=\left(\frac{\sigma r}{2i_0}\frac{RT}{zF}\right)^{\frac{1}{2}} \tag{17}$$

decreases with decreasing specific conductance and pore radius and with increasing exchange current density of the rate-determining step of the electrode reaction. In agreement with the estimate of FRUMKIN [3], the comparison [2] between the one-dimensional and a two-dimensional model demonstrates that the values of $I_{x=0}$ agree within 5% for $r\kappa<0.5$. As pointed out by EULER [7], the transmission line in Fig. 97 can be simulated by a network of resistors when the linear current-potential dependence is used.

In a cylindrical pore the current density of the electrode reaction depends upon the current inside the pore:

$$-2\pi r i=\frac{dI}{dx}=-\pi r^2\sigma\frac{d^2(\phi_2-\phi_1)}{dx^2}=-\pi r^2\sigma\frac{d^2\eta}{dx^2}. \tag{18}$$

The combination of the current-potential relation

$$i=i_0\left(\exp\frac{\alpha zF\eta}{RT}-\exp\frac{-(1-\alpha)zF\eta}{RT}\right) \tag{19}$$

with Eq. 18 yields under consideration of Eq. 17:

$$\frac{d^2\eta}{dx^2}-\frac{\kappa^2 RT}{zF}\left(\exp\frac{\alpha zF\eta}{RT}-\exp\frac{-(1-\alpha)zF\eta}{RT}\right)=0. \tag{20}$$

If $\eta\gg\frac{RT}{(1-\alpha)zF}$, the first integration of Eq. 20 with the boundary condition 12 leads to Eq. 21 if the second term is neglected (Tafel behavior):

$$\frac{d\eta}{dx}=-\left(\frac{2}{\alpha}\right)^{\frac{1}{2}}\frac{\kappa RT}{zF}\left(\exp\frac{\alpha zF\eta}{RT}-\exp\frac{\alpha zF\eta_{x=L}}{RT}\right)^{\frac{1}{2}}. \tag{21}$$

The current at the orifice of the pore follows from Eq. 21:

$$I_{x=0}=\left(\frac{4\pi^2 r^3\sigma RT}{\alpha zF}\right)^{\frac{1}{2}}\left(\exp\frac{\alpha zF\eta_{x=0}}{RT}-\exp\frac{\alpha zF\eta_{x=L}}{RT}\right)^{\frac{1}{2}}. \tag{22}$$

The slope of the $\eta-\log I$ curve

$$\frac{d\eta_{x=0}}{d\log I_{x=0}}=\frac{4.6RT}{\alpha zF}\left(1-\exp\frac{\alpha zF}{RT}(\eta_{x=L}-\eta_{x=0})\right) \tag{23}$$

becomes twice that for the corresponding flat electrode when the exponential term in Eq. 23 is much smaller than 1.

Analytical solutions of Eq. 20 with $\alpha = \frac{1}{2}$ were given by FRUMKIN [3] for the semiinfinite pore and by KSENZHEK [9] and WINSEL [10] for the finite pore while POSEY [11] reported solutions for several other α values and an approximate solution for any α. Eq. 24 and Eq. 25 are the result of the integration of Eq. 20 for $\alpha = \frac{1}{2}$ and a finite pore with the boundary condition 12:

$$\frac{d\eta}{dx} = -\frac{2^{3/2}\sigma R T}{zF}\left(\cosh\frac{zF\eta}{2RT} - \cosh\frac{zF\eta_{x=L}}{2RT}\right)^{\frac{1}{2}}, \tag{24}$$

$$I_{x=0} = \left(\frac{8\pi^2 r^3 \sigma i_0 R T}{zF}\right)^{\frac{1}{2}}\left(\cosh\frac{zF\eta_{x=0}}{2RT} - \cosh\frac{zF\eta_{x=L}}{2RT}\right)^{\frac{1}{2}}. \tag{25}$$

The current $I_{x=0}$ depends linearly upon $(R_1 R')^{-\frac{1}{2}}$ as in Eq. 14.

The above considerations on the flooded pore suggest for practical applications with negligible influence of mass transport processes that the porous electrodes should not be thicker than $1/\kappa$.

2. Concentration Distribution in the Flooded Single Pore under the Influence of Diffusion

KSENZHEK [13] treated the case where one substance diffuses in the direction of the axis of an infinite pore. AUSTIN and LERNER [14] considered the diffusion of both oxidized and reduced species in a finite pore. EULER [15] pointed out that the longitudinal diffusion can be simulated by a constant resistance per unit pore length in the electrical analog of a pore. Diffusion into the pore from one side is discussed in this section. Supply of fuel from two sides of flooded porous structures and also forced reactant supply were analyzed by GUREVICH and BAGOTZKY [16].

Let us assume that the electrode reaction involves the fuel species B_1 and the oxidized species B_2 in a cylindrical pore of finite size:

$$B_1 = B_2 + ze^- . \tag{26}$$

While the anodic oxidation of molecular hydrogen in acid solutions and without taking account of the water molecules in alkaline solutions can be represented by Eq. 26, the oxidation of liquid fuels and the reduction of molecular oxygen involve the production and further reaction of stable intermediates. As recognized by MICKA [16] for the O_2 reduction, the theoretical treatment should take the role of stable intermediates into account. The subsequent discussion will reveal the main features of the concentration distribution under the influence of diffusion for reactions of type 26.

Denoting the average concentration of B_1 and B_2 in planes perpendicular to the axis by c_1 and c_2 respectively, the following equations hold:

$$I = -D_1 \pi r^2 zF \frac{dc_1}{dx} = D_2 \pi r^2 zF \frac{dc_2}{dx}, \tag{27}$$

$$\frac{dI}{dx} = 2\pi r i_0 \left(\frac{c_1}{{}_bc_1} \exp \frac{\alpha zF\eta}{RT} - \frac{c_2}{{}_bc_2} \exp \frac{-(1-\alpha)zF\eta}{RT} \right). \tag{28}$$

The combination of Eq. 27 and 28 yields:

$$\frac{d^2 I}{dx^2} - \kappa^2 I = 0 \tag{29}$$

with

$$\kappa^2 = \frac{2i_0}{rzF} \left(\frac{\exp \frac{\alpha zF\eta}{RT}}{{}_bc_1 D_1} + \frac{\exp \frac{-(1-\alpha)zF\eta}{RT}}{{}_bc_2 D_2} \right) \tag{30}$$

Neglecting [13] the potential drop inside the pore, the integration of Eq. 29 with the boundary conditions

$$x=0: I = I_{x=0}, \quad c_1 = c_1^*, \quad c_2 = c_2^*, \tag{31}$$

$$x = L: I = 0 \tag{32}$$

leads to:

$$I = I_{x=0} \frac{\sinh \kappa(x-L)}{\sinh(-\kappa L)}. \tag{33}$$

The current at the orifice is given by:

$$I_{x=0} = \frac{2\pi r i_0}{\kappa} \left(\frac{c_1^*}{{}_bc_1} \exp \frac{\alpha zF\eta}{RT} - \frac{c_2^*}{{}_bc_2} \exp \frac{-(1-\alpha)zF\eta}{RT} \right) \tanh \kappa L. \tag{34}$$

The penetration depth $1/\kappa$ depends upon several parameters according to Eq. 30. It has its maximum at:

$$\eta_{\kappa=\max} = \frac{RT}{zF} \ln \frac{(1-\alpha)\,{}_bc_1 D_1}{\alpha\,{}_bc_2 D_2}. \tag{35}$$

The value of $\eta_{\kappa=\max}$ does not differ much from $\eta=0$ (corresponding to U_{rev}).

If $\frac{\eta z F}{R T} \gg 1$, Eq. 34 takes the form:

$$I_{x=0} = \kappa \pi r^2 c_1^* D_1 z F \tanh \kappa L \tag{36}$$

The Tafel slope for the semiinfinite pore ($\kappa L \gg 1$) becomes:

$$\frac{d\eta}{d \log I_{x=0}} = \frac{4.6 R T}{\alpha z F}. \tag{37}$$

A comparison of Eq. 37 with Eq. 27 demonstrates that similar Tafel slopes are predicted if the operation is either limited by diffusion or by the discharge step under consideration of the solution resistance.

The concentration c_1^* depends upon the bulk concentration:

$$I_{x=0} = \frac{\pi r^2 z F D_1}{\delta} ({}_b c_1 - c_1^*). \tag{38}$$

Eq. 36 can be transformed to:

$$I_{x=0} = \frac{\kappa\, {}_b c_1 D_1 \pi r^2 z F \tanh \kappa L}{1 + \kappa \delta \tanh \kappa L}. \tag{39}$$

The current $I_{x=0}$ approaches the limiting diffusion current in the electrolyte reservoir when η becomes large. The limiting diffusion current for a flooded pore is the same as that for a flat electrode of identical external surface. Therefore the region of practical interest for a flooded pore is given by:

$$I_{x=0} < 0.1 \frac{\pi r^2 z F D_1}{\delta}\, {}_b c_1. \tag{40}$$

3. Potential Distribution in the Flooded Single Pore in the Presence of Mass Transport Processes

Although the problem can be formulated [1, 16] for a flooded pore, general solutions of the nonlinear differential equations are not known because of the complicated mathematics. Numerical solutions could be obtained with the aid of electronic computers. However, it appears very questionable whether the simplifying assumptions on the structure and on the mechanism of the electrochemical reaction justify such an attempt.

The special case where the conductivity of the electrolyte inside the pore is constant, possibly due to a large excess of inert electrolyte, was treated by AUSTIN [17]. Let us introduce q as:

(41) $$q = \frac{\alpha z^2 F^2 D_1 c_1^*}{R T \sigma}.$$

Austin demonstrated that the solution of the problem described reduces to Eq. 22 for control by the discharge step and electrolytic conductance at $q > 5$, while diffusion (Eq. 36) is predominant at $q < 0.5$. Both the potential and concentration vary with x for intermediate values of q.

GUREVICH and BAGOTZKY [16] considered the case where the flux of neutral species participating in the electrode reaction is much smaller than the flux of charged species for a pore operating from both sides. Approximate solutions were given in a closed form for a linear current-potential dependence and for a Tafel type dependence. The reader is referred to the original paper for the discussion of the influence of various parameters on performance.

4. Continuum Models of Flooded Porous Electrodes

The porous structure filled with electrolyte is considered as the superposition of two continua, the electrode matrix and the solution in the unoccupied spaces within the matrix. The two phases, which complement one another, are supposed to be homogeneous and isotropic. Effective parameters rather than actual parameters are used for the description of the properties like pore size, conductivity, diffusion coefficient etc. The problem is treated as a one-dimensional one. This is equivalent to the assumption that the penetration depth of the current is larger than the size of the structural units (grains, holes) of the porous electrode. Continuum models were analyzed under different assumptions in references 1, 4, 9, 12 to 14, 16 to 20. Comprehensive treatments with strict derivations were published by TOBIAS and coworkers [21, 22] and by MICKA [23].

Let us discuss the two cases corresponding to the potential distribution without influence of mass transport processes and to the concentration distribution in the single pore under the influence of diffusion. If φ designates the average electric potential inside the continuum and I the average current, it is:

(42) $$I = -\sigma_P \frac{d\varphi}{dx}.$$

The effective conductance σ_P depends upon the electrolytic conductivity, the conductivity of the electrode matrix, and the porous structure. Eq. 43 follows from Eq. 42 under the assumption that the potential does not change abruptly at the branching of any pore.

$$\sigma_P \frac{d^2\varphi}{dx^2} = A_P i(\varphi)\,. \tag{43}$$

Here A_P is the effective surface of the electrocatalyst in $cm^2/cm^3 = cm^{-1}$. The current density of the electrode reaction depends upon φ in the continuum model.

Eq. 43 and Eq. 20 have the same form. If the penetration depth $1/\kappa$ is replaced by an effective penetration depth $1/\kappa_P$

$$\frac{1}{\kappa_P} = \left(\frac{\sigma_P}{A_P}\frac{RT}{zF}\right)^{\frac{1}{2}} \tag{44}$$

the solutions of Eq. 20 are applicable here. Thus the conclusions derived for the potential distribution of a single pore in the absence of mass transport processes are transferable in a simple way to porous electrode structures.

It can be shown in a similar way that the conclusions from section 2 on the concentration distribution inside a single pore under the influence of diffusion are transferable to the continuum model. Eq. 30 reads for the continuum model:

$$\kappa_P^2 = \frac{2i_0}{r_P zF}\left(\frac{\exp\dfrac{zF\eta}{RT}}{{}_bc_1 D_{1,P}} + \frac{\exp\dfrac{-(1-)zF\eta}{RT}}{{}_bc_2 D_{2,P}}\right) \tag{45}$$

Here r_P is the effective pore radius. The effective diffusion coefficients $D_{1,P}$ and $D_{2,P}$ depend upon D_1 and D_2 and the porous structure (tortuosity factor). Since the porous electrode is assumed to be at equal potential, it is not necessary to replace η by φ in Eq. 30.

While the discussion was restricted to the performance of a single pore in sections 1 to 3 of this chapter, the cases referred to there were frequently treated in the one-dimensional continuum model. No distinction was made between a derivation for the model of a single pore and the respective continuum model in sections 1 to 3 since the mathematical equations have the same form in each of the cases.

5. The Thin Film Model of the Gas-Diffusion Electrode

The interface gas/electrolyte in a cylindrical pore consists of a meniscus and a thin film of electrolyte above the meniscus (see Fig. 98a). The formation of the meniscus in a narrow capillary partially filled with a liquid is a well-known phenomenon. Evidence for the presence of a thin film of electrolyte above the meniscus was established by WILL [24] and by KNASTER and TEMKIN [25] for partially immersed electrodes. Good agreement was found between the experimental data on the H_2 oxidation

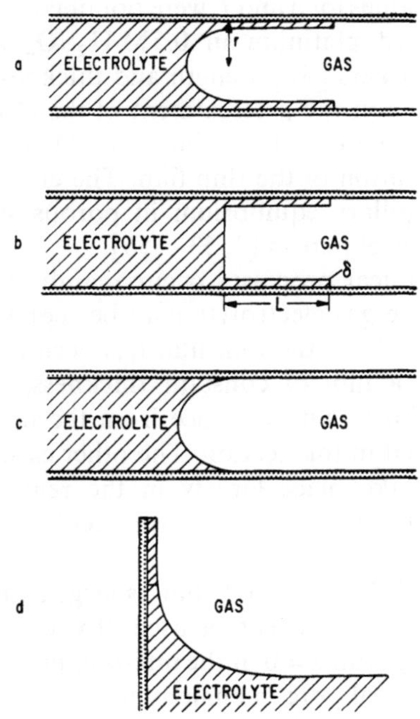

Fig. 98. Interface gas/electrolyte in a partially filled single pore (a, b, c) and on a partially immersed electrode (d).
a: real interface, b: interface in thin-film model, c: interface in the meniscus model

and theoretical results derived under the assumption of the H_2 diffusion through the film as rate-determining step. The similarity between the interface gas/electrolyte on a partially immersed electrode which is schematically drawn in Fig. 98d and that in a pore is evident. Partially immersed electrodes have been used successfully as "inverse pores" to study the H_2 oxidation or O_2 reduction in aqueous solutions [24—41] and carbonate electrolytes [42—46] in dependence of the meniscus and thin film. The experiments of BONNEMAY and coworkers [47, 48] with a partially filled, cylindrical tube (inner radius = 0.5 mm) as a model for a single pore with meniscus and those of KATAN, SZPAK, and GRENS [49] with a partially immersed bed electrode of small silver spheres support the concept of a thin electrolyte film above the meniscus inside the pores of gas-diffusion electrodes. Optical evidence for stable films of electrolyte on polished silver and nickel electrodes which were partially immersed in alkaline solutions was reported by MULLER [50]. The thickness [50] δ of the film (0.5 µ to 1 µ) agrees well with the thicknesses used by WILL [24] in his computations. The length L of the film was found [50] to be about 3 cm. Similar values for δ and L were obtained by optical techniques on partially immersed platinum in 0.5 M H_2SO_4 and 1 M KOH by BURSHTEIN and coworkers [51]. Length and thickness of the electrolyte film depended [51] upon the pretreatment of the electrode. The nature and the type of the reaction (H_2 oxidation or O_2 reduction) seem [52] to influence the formation of the thin film. The closely related problem of the gas-liquid capillary equilibrium in porous structures has been discussed in a number of papers [1, 53—56].

If the electrochemical processes occur mainly in the region of the thin film, the interface gas/electrolyte may be represented by Fig. 98b. The real profile [50, 51] of the thin film has been approximated in the thin film model by a film of constant thickness, ignoring the slight increase of film thickness with distance from the upper edge. The thin film model is discussed in this section. The other case where the electrochemical processes take place largely in the region of the meniscus (see Fig. 98c) is treated in section 6 as the meniscus model of the gas-diffusion electrode.

Since the length of the electrolyte film is large compared to its thickness, a one-dimensional model can be used. If x designates the direction along the wall, starting with x = 0 at the intersection of film and meniscus, it follows for a cylindrical pore in analogy to Eq. 18:

$$\frac{d^2\eta}{d\mathrm{x}^2} - \frac{1}{\sigma\delta\left(1-\dfrac{\delta}{2r}\right)}\,i = 0\,. \tag{46}$$

If the diffusion through the film is supposed [24, 25] rate-determining and if $\delta < 0.1\,r$, it is:

$$i = \frac{\mathrm{z}F\mathrm{D}}{\delta}\left(1 - \exp\frac{-\mathrm{z}F\eta}{RT}\right). \tag{47}$$

In Eq. 47 the concentration of gas at the solution side of the gas/liquid interface is made equal to the equilibrium concentration of gas dissolved at the given partial pressure in the solution, since the rate of gas dissolution is high [57] compared to that of the diffusion through the film. While the assumption $\delta < 0.1\, r$ is justified for partially immersed electrodes, it is unlikely to hold for pores with radii $r < 1\,\mu$. The steady-state solution of the diffusion equation in cylindrical coordinates has to be taken instead of Eq. 47 in the case [58] where $r < 1\,\mu$. Combining Eq. 46 with Eq. 47 leads to:

$$\frac{d^2\eta}{dx^2} - \frac{zFD_b c}{\sigma\delta^2}\left(1 - \exp\frac{-zF\eta}{RT}\right) = 0\,. \tag{48}$$

If σ is constant in the film, the first integration of Eq. 48 with the boundary condition

$$x = L:\ \eta = \eta_{x=L}\,, \qquad \frac{d\eta}{dx} = 0 \tag{49}$$

yields:

(50)

$$\frac{d\eta}{dx} = -\left(\frac{2zFD_b c}{\sigma\delta^2}\right)^{\frac{1}{2}}\left((\eta - \eta_{x=L}) + \frac{RT}{zF}\left(\exp\frac{-zF\eta}{RT} - \exp\frac{-zF\eta_{x=L}}{RT}\right)\right)^{\frac{1}{2}}.$$

The current which enters the film at $x = 0$ is given by:

$$I_{x=0} = 2\pi r\left(\frac{2\sigma RTD_b c}{zF}\right)^{\frac{1}{2}}\left(\frac{zF}{RT}(\eta_{x=0} - \eta_{x=L}) + \exp\frac{-zF\eta_{x=0}}{RT} - \exp\frac{-zF\eta_{x=0}}{RT}\right)^{\frac{1}{2}} \tag{51}$$

The current $I_{x=0}$ is independent of the thickness of the thin film under the initial assumption $\delta < 0.1\, r$. It is proportional to $(\eta_{x=0} - \eta_{x=L})^{\frac{1}{2}}$ for $\eta_{x=0} \gg \frac{RT}{zF}$ and to $(\eta_{x=0} - \eta_{x=L})$ for $\eta_{x=0} \ll \frac{RT}{zF}$. The assumption of the constancy of σ was recently verified by WILL [59] for a thin electrolyte film on top of a large horizontal platinum foil. Convection in the film and water transport above the film were shown to exist. Both phenomena appear to keep the concentration of the ionic species constant in the film. A similar situation may be expected for the electrolyte film above the meniscus in a pore. A preliminary survey of the role of convection in the film was published by LIGHTFOOT [60].

The case of molecular diffusion in the film combined with a hindered electrode reaction was considered for constant σ by KSENZHEK [61] and by LEVICH and coworkers [62]. The expression for the discharge step was relatively simple. In addition to the above conditions, a change of the concentration of the ionic species in the film was taken into account by ICZKOWSKI [58] and BENNION and TOBIAS [34]. Finally ICZKOWSKI [63] computed the total current density of the H_2 oxidation on porous nickel electrodes as the sum of the currents which would be produced at the given overvoltage by each of the individual pores in 1 cm^2 of electrode material

$$i(\eta)=\int_{r_c}^{\infty} I_{x=0}(\eta,r)F(r)\mathrm{d}r \tag{52}$$

under consideration of the pore size distribution function F(r). Pores of radii $r<r_c$ are filled with electrode and therefore assumed to be inoperative in the thin film model. The theoretical $i-\eta$ curves were in moderate agreement with the experimental curves and predict too large overvoltages at small currents and too small overvoltages at large currents.

6. The Meniscus Model of the Gas-Diffusion Electrode

The condition that the electrochemical processes occur largely in the region of the meniscus is only met if the thin film of electrolyte is absent or if the thickness is very small. The simple-pore model [64] is an example of the first case. The meniscus is assumed to form at the intersection of micropores with macropores. While the micropores are filled with electrolyte up to the intersection, the macropores are filled with gas. The meniscus may be treated as flat in a first approximation. The walls of micropores are the seat of the electrode reaction. The simple-pore model was suggested [64] as applying to non-wetted systems like the Teflon-bonded platinum black electrodes. The limiting current due to the diffusion of species into a micropore was derived [64] as the steady-state solution of the two-dimensional diffusion equation in cylindrical coordinates. The summation of the currents of the individual pores leads to:

$$I_{1,j}=\frac{12zFD_jq}{r}\,{}_{b}c_j\,. \tag{53}$$

Here r is the mean radius of the micropores and q the total area of the mouths at the interface gas/electrolyte. The $I_{1,j}$ values computed from Eq. 53 depend strongly upon the estimates of r and q. Thus a verification of the simple-pore model by comparison of the $I_{1,j}$ values from Eq. 53 and the experimental values in Table 10 is not convincing.

SRINIVASAN, HURWITZ, and BOCKRIS [57] derived expressions for the current-distribution and the $I_{x=0}-\eta_{x=0}$ relation in one- and two-dimensional models for a single pore with a flat meniscus. The influence of electrolytic resistance, diffusion, and an inhibited discharge step with a simplified kinetic equation were considered. It was shown that the one-dimensional treatment may be used for current densities lower by a factor of 4 than the limiting current density. The limiting current density was obtained by the two-dimensional treatment. Special cases of predominance of one or two processes were discussed. As expected, the conclusions for a special case are the same as those for the corresponding case of the flooded pore (see sections 1 to 3) in the one-dimensional model.

An alternative mechanism in which the gas is adsorbed on the dry metal and moves towards the interface electrocatalyst/electrolyte by surface diffusion was also treated [66, 67] in the model of a cylindrical pore with flat meniscus. However, such a mechanism is not compatible [1, 24, 58] with the order of magnitude of the currents observed.

7. Simultaneous Consideration of Thin Film and Meniscus

This case was treated in one-dimensional model of the partially immersed electrode by WILL [24] for $\eta > 0.1\,V$ with the H_2 diffusion as rate-determining step. The meniscus was approximated by a parabola. The relative contributions of the meniscus and film depend upon $\eta_{x=0}$, the conductivity, and the film thickness. The maximum contribution of the meniscus increases roughly with $\delta^{-\frac{1}{2}}$. A more rigorous determination of the reaction zone in the same model was given for large and small values of $\eta_{x=0}$ by ALEKSEEV and POPOV [68]. The solution for any value of $\eta_{x=0}$ was reported [41] recently.

The one-dimensional solutions appear adequate for partially immersed electrodes with $\delta > 0.5\,\mu$. However, the influence of the curvature cannot be neglected in pores with radii $r < 1\,\mu$.

8. Model for the Two-Layer Electrode

PSHENICHNIKOV [69] suggested that the electrochemical reactions occur in the film of electrolyte covering the walls of the macropores of a two-layer electrode (see Fig. 95). Since many micropores intersect the macropores, the electrolytic resistance of the solution film is shunted by the much smaller electrolytic resistance of the micropores which are completely filled with electrolyte. It is assumed that mass transport of the neutral species through the film is rapid. Applying the concept of the one-dimensional continuum model to the micropores of the thick layer, Eq. 54 results:

$$\frac{d^2\varphi}{dx^2} = K_t \frac{\Lambda}{q\sigma} i(\varphi)\,. \tag{54}$$

The x axis is parallel to the axis of the macropores, q designates the sum of the areas of the mouths of the micropores, and Λ is the sum of the perimeters of the macropores. The origin of the x axis is at the external electrolyte/electrode boundary. This can be done since the influence of the thin layer of the two-layer electrode is negligible [70]. A linear dependence between i and φ

$$i = \frac{\varphi}{K} \tag{55}$$

leads to:

$$I_{x=0} = \left(\frac{q\Lambda}{K_t K \sigma}\right)^{\frac{1}{2}} \varphi_{x=0}\,. \tag{56}$$

The product $q\Lambda$ is proportional to $\frac{V_{ma} V_{mi}}{r_{ma}}$ (see section 4 in chapter XV) for cylindrical pores. Thus the maximum [70] in the $I_{x=0} - \varphi_{x=0}$, curves and the linear dependence of $I_{x=0}$ upon $(q\Lambda)^{\frac{1}{2}}$ is explained by the theory. The computation of the exchange current density of the rate-determining step of the H_2 oxidation on nickel according to Eq. 56 and the internal surface of the macropores yields $i_0 \approx 4 \cdot 10^{-5}\,A/cm^2$. This value is in agreement with i_0 values for smooth nickel (compare section 2 in chapter VII).

References

1. CHISMADZHEV, YU. A.: Elektrokhim. **2,** 3 (1966).
2. DE LEVIE, R.: Advances in Electrochemistry, Vol. **6,** p. 329. P. Delahay, ed., New York: Interscience Publishers 1967.
3. FRUMKIN, A. N.: Zhur. Fiz. Khim. **23,** 1477 (1949).
4. DANIEL'-BEK, V. S.: Zhur. Fiz. Khim. **22,** 697 (1948).
5. COLEMAN, J. J.: Trans. Electrochem. Soc. **90,** 545 (1946).
6. WEISSELBERG, A., and STAFF: Trans. Electrochem. Soc. **90,** 235 (1946).
7. WABER, J. T.: Trans. Electrochem. Soc. **90,** 243 (1946).
8. EULER, J.: Naturwiss. **45,** 537 (1958).
9. KSENZHEK, O. S.: Zhur. Fiz. Khim. **36,** 633 (1962).
10. WINSEL, A.: Z. Elektrochem. **66,** 287 (1962).

11. Posey, F. A.: J. Electrochem. Soc. **111,** 1173 (1964).
12. Daniel'-Bek, V. S.: Elektrokhim. **1,** 1319 (1965).
13. Ksenzhek, O. S.: Zhur. Fiz. Khim. **36,** 243 (1962).
14. Austin, L. G., and H. Lerner: Electrochim. Acta **9,** 1469 (1964).
15. Euler, J.: Electrochim. Acta **8,** 409 (1963).
16. Gurevich, I. G., and V. S. Bagotzky: Electrochim. Acta **9,** 1151 (1964).
17. Austin, L. G.: Trans. Faraday Soc. **60,** 1319 (1964).
18. Ksenzhek, O. S., and V. V. Stender: Doklady Akad. Nauk S.S.S.R. **107,** 280 (1956).
19. Perskaya, R. M., and I. A. Saidenman: Doklady Akad. Nauk S.S.S.R. **115,** 548 (1957).
20. Euler, J., and W. Nonnemacher: Electrochim. Acta **2,** 268 (1960).
21. Newman, J. S., and C. W. Tobias: J. Electrochem. Soc. **109,** 1183 (1962).
22. Grens II, E. A., and C. W. Tobias: Ber. Bunsenges. Physik. Chem. **68,** 236 (1964).
23. Micka, K.: Collection Czech. Chem. Commun. **29,** 1998 (1964).
24. Will, F. G.: J. Electrochem. Soc. **110,** 152 (1963).
25. Knaster, M. B., and M. I. Temkin: Doklady Akad. Nauk S.S.S.R. **152,** 658 (1963).
26. Hersch, P.: Nature **169,** 792 (1952).
27. Weber, H. C., H. P. Meissner, and D. A. Sama: J. Electrochem. Soc. **109,** 884 (1962).
28. Will, F. G.: J. Electrochem. Soc. **110,** 145 (1963).
29. Mazitov, Yu. A., K. I. Rozental', and V. I. Veselovsky: Zhur. Fiz. Khim. **38,** 448 (1964).
30. Pshenichnikov, A. G., G. I. Shnaider, and R. K. Burshtein: Elektrokhim. **1,** 418 (1965).
31. — — Elektrokhim. **1,** 640 (1965).
32. Maget, H. J. R., and R. Roethlein: J. Electrochem. Soc. **112,** 1034 (1965).
33. Bennion, D. N., and C. W. Tobias: J. Electrochem. Soc. **113,** 589 (1966).
34. — — J. Electrochem. Soc. **113,** 593 (1966).
35. Belokopytov, V. P., and N. A. Aladzhalova: Elektrokhim. **2,** 1255 (1966).
36. Povarov, Yu. M., and P. D. Lukovstev: Elektrokhim. **3,** 220 (1967).
37. Osetrova, N. V., and V. S. Bagotzky: Elektrokhim. **3,** 521 (1967).
38. Pshenichnikov, A. G., and G. I. Shnaider: Elektrokhim. **3,** 842 (1967).
39. Davitt, H. J., and L. F. Albright: J. Electrochem. Soc. **114,** 531 (1967).
40. Shepelin, V. A., Z. I. Salkind, and V. I. Veselovsky: Elektrokhim. **4,** 66 (1968).
41. Vasiliev, A. A., and Yu. A. Popov: Elektrokhim. **4,** 444 (1968).
42. Klevtsov, L. P., G. G. Arkhipov, and G. K. Stepanov: Elektrokhim. **1,** 1304 (1965).
43. Chismadzhev, Yu. A., Yu. G. Chirkov, and L. P. Klevtsov: Elektrokhim. **3,** 645 (1967).
44. Klevtsov, L. P., G. G. Arkhipov, and G. K. Stepanov: Elektrokhim. **3,** 888 (1967).
45. Popov, Yu. A., and Yu. V. Alakseev: Elektrokhim. **3,** 294 (1968).
46. Cobb, Jr., J. T., and L. F. Albright: J. Electrochem. Soc. **115,** 2 (1968).
47. Bonnemay, M., G. Bronoël, and E. Levart: Electrochim. Acta **9,** 727 (1964).
48. — — —, and E. Nakache: Compt. Rend. **257,** 3885 (1963).
49. Katan, T., S. Szpak, and E. A. Grens II: J. Electrochem. Soc. **112,** 1166 (1965).
50. Muller, R. H.: J. Electrochem. Soc. **113,** 943 (1966).
51. Burshtein, R. K., M. R. Tarasevich, S. F. Chernishov, and V. V. Karasev: Elektrokhim. **3,** 907 (1967).
52. Shepelin, V. A.: Elektrokhim. **3,** 1004 (1967).
53. Markin, V. S., Yu. A. Chismadzhev, and Yu. G. Chirkov: Doklady Akad. Nauk S.S.S.R. **150,** 596 (1963).
54. — Doklady Akad. Nauk S.S.S.R. **151,** 620 (1963).
55. — Izv. Akad. Nauk S.S.S.R., Otd. Khim. Nauk 1690 (1963).

56. —, A. A. Chernenko, Yu. A. Chismadzhev, and Yu. G. Chirkov: Fuel Cells, p.21. V. S. Bagotzky and Yu. B. Vasiliev, ed., New York: Consultants Bureau 1966.
57. Srinivasan, S., H. D. Hurwitz, and J. O'M. Bockris: J. Chem. Phys. **46**, 3108 (1967).
58. Iczkowski, R. P.: J. Electrochem. Soc. **111**, 1078 (1964).
59. Will, F. G.: J. Electrochem. Soc. **114**, 138 (1967).
60. Lightfoot, E. N.: J. Electrochem. Soc. **113**, 615 (1966).
61. Ksenzhek, O. S.: Ukr. Khim. Zhur. **30**, 810 (1946).
62. Levich, V. G., Yu. A. Chismadzhev, and Yu. G. Chirkov: Doklady Akad. Nauk S.S.S.R. **157**, 404 (1964).
63. Iczkowski, R. P.: J. Electrochem. Soc. **111**, 605 (1964).
64. Austin, L. G., M. Ariet, R. Walker, G. B. Wood, and R. H. Comyn: I and EC Fundamentals **4**, 321 (1965).
65. —, and S. Almaula: J. Electrochem. Soc. **114**, 927 (1967).
66. Justi, E., M. Pilkuhn, W. Scheibe, and A. Winsel: Hochbelastbare Wasserstoff-Diffusions-Elektroden für Betrieb bei Umgebungstemperatur und Niederdruck. Abh. Mainzer Akad., Nr. **8**. Wiesbaden: Steiner 1959.
67. Justi, E. W., and A. W. Winsel: Kalte Verbrennung — Fuel Cells. Wiesbaden: Steiner 1962.
68. Alekseev, Yu. V., and Yu. A. Popov: Elektrokhim. **1**, 422 (1965).
69. Pshenichnikov, A. G.: Doklady Akad. Nauk S.S.S.R. **148**, 1121 (1963).
70. Burshtein, R. K., V. S. Markin, A. G. Pshenichnikov, V. A. Chismadzhev, and Yu. G. Chirkov: Electrochim. Acta **9**, 773 (1964).

Subject Index

Anorganische und allgemeine Chemie in Einzeldarstellungen

Herausgegeben von Margot Becke-Goehring

II. Band: **Reaktionen in und an festen Stoffen**
Von Dr. **Karl Hauffe,** Professor am Institut für Physikalische Chemie der Universität Göttingen, unter teilweiser Mitarbeit von Dr. **Carl Seyferth,** Institut für Angewandte Mathematik der Technischen Hochschule Karlsruhe
Zweite, erweiterte Auflage. Mit 525 Abbildungen.
XII, 968 Seiten Gr.-8°. 1966. Gebunden DM 148,—; US $ 37.00

III. Band: **Chemische Bindung und Molekülstruktur**
Von **L. E. Sutton,** Dr. phil., F. R. S., Oxford. Übertragen von Dr. Ekkehard Fluck, Heidelberg
Mit 77 Abbildungen. VIII, 122 Seiten Gr.-8°. 1961. Vergriffen

IV. Band: **Inorganic Adduct Molecules with Oxo-Compounds**
By Dr. **Ingvar Lindquist,** Uppsala/Schweden
With 18 figures. VI, 129 pages 8vo. 1963. Cloth DM 20,—; US $ 5.00

V. Band: **Die kernmagnetische Resonanz**
und ihre Anwendung in der anorganischen Chemie
Von Dr. **Ekkehard Fluck,** Privatdozent in der Nat.-Math. Fakultät der Universität Heidelberg
Mit 142 Abbildungen. VIII, 290 Seiten Gr.-8°. 1963
Gebunden DM 48,—; US $ 12.00

VI. Band: **Boron-Nitrogen Compounds**
By Dr. **Kurt Niedenzu** and Dr. **John W. Dawson,** U. S. Army Research Office, Duke University Durham, North Carolina
VIII, 176 pages 8 vo. 1965. Cloth DM 27,—; US $ 6.75

VII. Band: **Anwendungen der Schwingungsspektroskopie in der anorganischen Chemie**
Von Dr. **Hans Siebert,** apl. Prof. an der Fakultät für Natur- und Geisteswissenschaften der Bergakademie Clausthal, Technische Hochschule
Mit 28 Abbildungen. VIII, 209 Seiten Gr.-8°. 1966
Gebunden DM 39,60; US $ 9.90

VIII. Band: **Europium**
By Dr. **Shyama P. Sinha,** Department of Inorganic Chemistry, The University of Leeds, Leeds/England
With 23 figures. VII, 164 pages 8 vo. 1967. Cloth DM 39,—; US $ 9.75

IX. Band: **Electrochemical Processes in Fuel Cells**
By Dr. **Manfred W. Breiter,** General Electric Research and Development Center, Schenectady, N. Y./USA
With 98 figures. XI, 274 pages 8 vo. 1969. Cloth DM 48,—; US $ 12.00

MIX
Papier aus verantwortungsvollen Quellen
Paper from responsible sources
FSC® C105338

If you have any concerns about our products,
you can contact us on
ProductSafety@springernature.com

In case Publisher is established outside the EU,
the EU authorized representative is:
Springer Nature Customer Service Center GmbH
Europaplatz 3, 69115 Heidelberg, Germany

Printed by Libri Plureos GmbH
in Hamburg, Germany